Models

Contributors

Malcolm Baker

Marcel Boumans

Soraya de Chadarevian

Christopher Evans

Eric Francoeur

James Griesemer

Nick Hopwood

Ludmilla Jordanova

Renato G. Mazzolini

Herbert Mehrtens

Christoph Meinel

Mary S. Morgan

Lynn K. Nyhart

Simon Schaffer

Thomas Schnalke

James A. Secord

Jérôme Segal

Models

THE THIRD DIMENSION OF SCIENCE

Edited by

SORAYA DE CHADAREVIAN AND NICK HOPWOOD

STANFORD UNIVERSITY PRESS
STANFORD, CALIFORNIA

© Stanford University Press
Stanford, California

Printed in the United States of America
on acid-free, archival-quality paper

Library of Congress Cataloging-in-Publication Data

Models : the third dimension of science / edited by Soraya de Chadarevian and Nick Hopwood.
p. cm.—(Writing science)
Includes bibliographical references and index.
ISBN 0-8047-3971-4 (alk. paper)—ISBN 0-8047-3972-2 (pbk. : alk. paper)
1. Communication in science. 2. Visual communication. 3. Geometrical models. 4. Polyhedra—Models. I. Chadarevian, Soraya de. II. Hopwood, Nick. III. Series.

Q223 .M58 2004
501′.4—dc22

2003025168

Typeset by TechBooks in 10/13 Sabon

Original Printing 2004

Last figure below indicates year of this printing:
13 12 11 10 09 08 07 06

CONTENTS

FIGURES

EDITORS' ACKNOWLEDGMENTS

This book began in discussions with Simon Schaffer, Jim Secord and Svante Lindqvist in Cambridge, and took shape during a symposium in London at the then Wellcome Institute for the History of Medicine and the Science Museum in November 1998. We are very grateful to the Wellcome Trust for funding the conference; Vivian Nutton, Alan Morton, and their colleagues for hosting it; Frieda Houser for efficient organisation; and the participants, especially Deanna Petherbridge and Dominique Pestre, for their stimulating contributions. We thank the authors for hard work in writing original chapters and commentaries around a common theme, and for their patience; Tim Lenoir for including the volume in the 'Writing Science' series and for his unflagging support; Nathan MacBrien, Kim Lewis Brown, and Norris Pope in the Acquisitions Department at Stanford; and Anna Eberhard Friedlander at the Press, and Soma Chatterjee and Vineet Arora at TechBooks, for overseeing the production of the book.

CONTRIBUTORS

Malcolm Baker is Professor of Art History, University of Southern California and Professorial Research Fellow, Victoria and Albert Museum, London. His publications include *Figured in Marble: The Making and Viewing of Eighteenth-Century Sculpture* (London: V&A Publications; Los Angeles: J. Paul Getty Museum, 2000) and (with David Bindman) *Roubiliac and the Eighteenth-Century Monument: Sculpture as Theatre* (New Haven: Yale University Press, 1995), which was awarded the 1996 Mitchell Prize for the History of Art.

Marcel Boumans is Associate Professor of Philosophy and History of Economics at the University of Amsterdam. Recent publications include "Built-in justification", in *Models as Mediators: Perspectives on Natural and Social Science*, edited by Mary S. Morgan and Margaret Morrison (Cambridge: Cambridge University Press, 1999), and "Fisher's instrumental approach to index numbers", in *The Age of Economic Measurement*, edited by Judy L. Klein and Mary S. Morgan (Durham, N.C.: Duke University Press, 2001).

Soraya de Chadarevian is Senior Research Associate and Affiliated Lecturer in the Department of History and Philosophy of Science at the University of Cambridge. She is the author of *Designs for Life: Molecular Biology after World War II* (Cambridge: Cambridge University Press, 2002) and co-editor of *Molecularizing Biology and Medicine: New Practicies and Alliances, 1910s–1970s* (London: Harwood, 1998).

Christopher Evans is Director of the Cambridge Archaeological Unit, University of Cambridge. He has undertaken several major excavations in Britain, and is currently involved in ethno-archaeological researches in Nepal and

Inner Mongolia. He has published widely on the history of archaeology and its representation.

Eric Francoeur is Lecturer at the École de technologie supérieure, Montreal. His publications include "The forgotten tool: The design and use of molecular models", in *Social Studies of Science* 27 (1997), and "Molecular models and the articulation of structural constraints in chemistry", in *Tools and Modes of Representation in the Laboratory Sciences*, edited by Ursula Klein (Dordrecht: Kluwer, 2001).

James Griesemer is Professor of Philosophy at the University of California, Davis. He has published widely on the history, philosophy, and sociology of biology, especially on conceptual and experimental practices in evolutionary biology, models in natural history museum work, visual representation in biology, and the foundations of evolutionary theory. He has recently co-edited Tibor Gánti's *The Principles of Life* (Oxford: Oxford University Press, 2003).

Nick Hopwood is Lecturer in the Department of History and Philosophy of Science at the University of Cambridge. His publications include *Embryos in Wax: Models from the Ziegler Studio* (Cambridge: Whipple Museum of the History of Science; Bern: Institute of the History of Medicine, 2002).

Ludmilla Jordanova is Director of the Centre for Research in the Arts, Humanities, and Social Sciences at the University of Cambridge. Recent books include *Nature Displayed: Gender, Science, and Medicine, 1760–1820* (London: Longman, 1999), *History in Practice* (London: Arnold, 2000) and *Defining Features: Medical and Scientific Portraits, 1660–2000* (London: National Portrait Gallery and Reaktion, 2000).

Renato G. Mazzolini is Professor of the History of Science at the University of Trento, Italy. He is the author of *The Iris in Eighteenth-Century Physiology* (Bern: Huber, 1980) and (with Shirley A. Roe) *Science against the Unbelievers* (Oxford: Voltaire Foundation, 1986), and editor of *Non-Verbal Communication in Science prior to 1900* (Florence: Olschki, 1993).

Herbert Mehrtens is Professor of Modern History at the Technical University, Braunschweig, Germany. He is the author of *Moderne—Sprache—Mathematik* (Frankfurt am Main: Suhrkamp, 1990), and co-editor of *Naturwissenschaft, Technik und NS-Ideologie* (Frankfurt am Main: Suhrkamp, 1980) and *Social History of Nineteenth-Century Mathematics* (Boston: Birkhauser, 1981).

Christoph Meinel is Professor of History of Science at the University of Regensburg, Germany. He has published widely in the history of chemistry, including editing several volumes, most recently *Experiment—Instrument. Historische Studien* (Berlin: Verlag für Geschichte der Naturwissenschaften und Technik, 2000).

Mary S. Morgan is Professor of History of Economics at the London School of Economics and Professor of History and Philosophy of Economics at the University of Amsterdam. She is co-editor of *Models as Mediators: Perspectives on Natural and Social Science* (Cambridge: Cambridge University Press, 1999), *Empirical Models and Policy-Making: Interaction and Institutions* (London: Routledge, 2000), and *The Age of Economic Measurement* (Durham, N.C.: Duke University Press, 2001).

Lynn K. Nyhart is Associate Professor of the History of Science at the University of Wisconsin–Madison and the author of *Biology Takes Form: Animal Morphology and the German Universities, 1800–1900* (Chicago: University of Chicago Press, 1995). Her current research is on zoology and the rise of an environmentally and visually oriented approach to nature.

Simon Schaffer is Reader in History and Philosophy of Science at the University of Cambridge. He is the co-editor of *The Uses of Experiment: Studies in the Natural Sciences* (Cambridge: Cambridge University Press, 1989) and *The Sciences in Enlightened Europe* (Chicago: University of Chicago Press, 1999), and has published several papers on mechanics, electricity, and experimental philosophy in eighteenth-century Europe.

Thomas Schnalke is Professor of the History of Medicine and Director of the Medical Historical Museum at the Charité, Berlin. He is the author, among many other works, of *Diseases in Wax: The History of the Medical Moulage* (Berlin: Quintessence, 1995).

James A. Secord is Professor of History and Philosophy of Science at the University of Cambridge. He co-edited *Cultures of Natural History* (Cambridge: Cambridge University Press, 1996), and is the author, most recently, of *Victorian Sensation: The Extraordinary Publication, Reception, and Secret Authorship of* Vestiges of the Natural History of Creation (Chicago: University of Chicago Press, 2000).

Jérôme Segal is Assistant Professor of History of Science and Epistemology at the Institut Universitaire de Formation des Maîtres, Paris. Trained in engineering and in history, he has published on the history of information theory and of molecular biology. He is the author of *Le zéro et le un: Histoire de la notion scientifique d'information au 20e siècle* (Paris: Syllepse, 2003).

Models

CHAPTER ONE

Dimensions of Modelling

Nick Hopwood and
Soraya de Chadarevian

This book is about models: wooden ships and plastic molecules, wax embryos and a perspex economy, monuments in cork and mathematics in plaster, casts of diseases and displays of stuffed animals, anatomies to take apart and extinct monsters rebuilt in bricks and mortar. Produced in the sciences of Western Europe and the United States between the mid-eighteenth and the mid-twentieth century, these artefacts extend from rough working models on a bench, or later a computer screen, to elaborate objects of display in major public exhibitions. Some were unique, others mass-produced; they differ in texture, colour, and size. Some mimicked the natural or artificial world, others projected how it might become. They variously purported to bring the tiny, the huge, the past, or the future within reach, to make fruitful analogies, to demonstrate theories, to look good on show. Yet all are three-dimensional (3-D), and as such, their advocates claimed, displayed relations that could not easily be represented on paper. Here, museum-based researchers and historians, philosophers and sociologists of science and medicine have come together to highlight the three-dimensionality of important groups of models and to explore their roles in making knowledge.

Historians and philosophers of science have long been interested in models and other representations, but focusing on the production and uses of 3-D models challenges received views. Philosophers of science have written extensively about models, especially in explicating what theories are and how they work. Models have been seen as mediators between theories and phenomena; others have viewed theories as collections of models (Black 1962;

Hesse 1966; van Fraassen 1980; Cartwright 1983; Giere 1988; Morgan and Morrison 1999). Though a very wide range of models is potentially of philosophical interest, major classes have tended to be sidelined. Acknowoledging models made with "spring washers, magnets, lots of tin foil and such", Ian Hacking suggested that "[m]ost generally . . . a model in physics is something you hold in your head rather than your hands" (Hacking 1983, 216). "The class of scientific models", Ronald Giere has argued, "includes physical scale models and diagrammatic representations, but the models of most interest are *theoretical* models. These are abstract objects, imaginary entities" (Giere 1999, 5, emphasis in original). While agreeing that models, in all their diversity, are important in staking theoretical claims, we concentrate, not on abstract, mental entities, but precisely on objects that people grasped with their hands.

Recent studies of representation in scientific practice have, paradoxically, tended to reinforce neglect of 3-D models. This work has shown how scientists and their collaborators have used tools and instruments, from pencils to scintillation counters, to produce engravings, traces, and print-outs, standardised representations that can be manipulated and displayed (Latour and Woolgar 1979; Latour 1987; Gooding, Pinch, and Schaffer 1989; Lynch and Woolgar 1990; Mazzolini 1993; Rheinberger 1997; Lenoir 1998). Such transformations have reduced complexity, defined and controlled phenomena, and may even constitute sciences around 'visual languages' and 'working objects' (Rudwick 1976; Daston and Galison 1992). The same devices have been mobilised to convince colleagues, train new practitioners, gain funding, and win public support. Yet, though these studies are deeply concerned with the visual and with bodily action in a material world, scientists are reckoned to discipline nature most effectively by reducing three dimensions to two. Three-dimensional representations would, it has widely been assumed, be too expensive and immobile for routine use. As Bruno Latour has written, "There is nothing you can dominate as easily as a flat surface" (Latour 1990, 45). Though this view has much to recommend it, like a history of art without sculpture it is incomplete. We focus here on the many occasions when scientists, physicians, and engineers have not distributed 2-D representations, or the capacity to produce them, but have insisted on making and displaying 3-D models of the world.

Museum curators and historians of science and medicine have already studied some 3-D models in depth, though usually as the holdings of specific collections. There is, for example, a wealth of research on anatomical waxes (*Ceroplastica* 1977; Boschung 1980; Lemire 1990; Schnalke 1995), and 3-D models in chemistry are gaining increasing attention (Francoeur 1997; Schummer 1999–2000; Klein 2001). Yet even for anatomy, we lack work on the status of models in the discipline and accounts of what difference they made, more generally, to images of the body. This book addresses such questions for various kinds of models, and so seeks to promote a more general account of the roles of 3-D models in the sciences.

One reason why models have not loomed larger in our histories—though it is also a result of their neglect—is simply that the routes by which models are collected in museums or linger in laboratory cupboards are more fortuitous than those by which books enter the libraries where most historians still expect to ply their trade. But even abundantly preserved models have typically been slighted as worth studying if one is primarily interested in pedagogy or popularization, but hardly in the mainstream of the history of science. This is a mistake, for not only was teaching the centrally important means of ratifying and conveying knowledge, and addressing wider audiences crucial to establishing scientific authority, the movements of models also exemplify the impossibility of separating these activities from research. As several authors in this volume compellingly show, models were made at all stages in processes of production, and models used for teaching were often the same as those that guided research; models started as research tools and became teaching aids, but also vice versa. Nor did models travel only between the laboratory and the lecture hall. We find them in the wide variety of settings, from studios and museums to dockyards and congresses, where claims to knowledge have been tested and defended.

Three-dimensional models always embodied and displayed knowledge, but how this was done varied enormously and altered over time. Though the meanings of 'model' changed, too, in most of the examples we discuss the actors used the term. (We have included the exception—dioramas, which were not called models but might include models—because they raise interesting issues of definition.) These objects share certain visual and tactile properties simply as representations in 3-D. Producing and using such models has in many ways been different from making and working with flat images; and through common sites of production and of display these differences have been shared across scientific disciplines and beyond the sciences. Yet although, we argue, it is fruitful to think about 3-D models together, we are wary of setting a particular group of objects aside. That historians of science no longer treat printed works as mere containers of information, but investigate the histories of producing and reading these physical objects, makes it much easier to approach models as operating in networks of production and communication along with drawings, maps, books, and instruments (Darnton 1990; Johns 1998; De Renzi 2000; Frasca-Spada and Jardine 2000). Indeed, it is perhaps most readily in the sciences—where benches tend to be cluttered with all of these items—that historians can discover how they act in concert.

MODELS IN THE SCIENCES

We begin in the middle of things (part I). For, as Malcolm Baker shows, in eighteenth-century Europe, and in England in particular, 'model' already meant a diverse range of widely distributed objects. There were models of the human figure, of the heavens and the earth, of buildings and machines; they both imitated some existing feature of art or nature, and promoted the execution of new projects. Models populated spaces of display, many of which were new or newly prominent—above all collections and museums, but also studios, workshops of the luxury trades, academies, meetings of societies, shows, and lectures—everywhere that commissioning, designing, making, exhibiting, awarding prizes, and teaching was going on. As Christopher Evans discusses later in the book, the multifarious designing, collecting, and lecturing activities of the architect Sir John Soane revolved around models of his own buildings and of ancient monuments. Models mediated relations between patron and client, lecturer and audience, maker and purchaser, and represented the first consumer society to itself.

Models were central, Renato Mazzolini and Simon Schaffer show, to projects of enlightenment. A major class of eighteenth-century models were the 'artificial anatomies' deployed to teach medical students, artists, and midwives, to fill noble collectors' cabinets, and to stock museums that entertained and instructed wide audiences. Mazzolini discusses the Enlightenment's most famous collection of anatomical ceroplastics, which the natural philosopher Felice Fontana assembled in Florence at the Royal Museum of Physics and Natural History ('La Specola') with the aim of stimulating research and promoting social reform. Schaffer reminds us of the politics in modellers' moves between making representations and exercising power. By making small-scale models of ships' hulls and of electric fish, natural philosophers staked claims to govern the large-scale systems of trade and industry that the models represented. More generally, Schaffer argues, the great political debates of the age of reason concerned the legitimacy of applying to natural and social worlds the principles that had been demonstrated in mechanical models.

The new disciplines of the age of revolutions were sciences of collections that presented their objects as compounds, ripe for analysis into elements (Elsner and Cardinal 1994; Pickstone 2000). Some of the most important collectables were models. Above all in post-revolutionary Paris, in chemistry, mineralogy, zoology, botany, pathological anatomy, and engineering, professor-curators based in museums, professional schools, and hospitals were employed to lay out previously royal collections for public edification (Lindqvist 1984, 23–33; Morton and Wess 1993). The new technical schools, the École Polytechnique, the École des Ponts et Chaussées, and the Conservatoire des Arts et Métiers, were filled with scale models of machines (Bugge 1969); professors at the clinical and veterinary schools and the natural history museum made strenuous efforts to revive and reform traditions of anatomical modelling in wax (Lemire 1990). In early nineteenth-century Britain, too, the physical sciences were established around models (Schaffer). Collections at the Royal Society and in key sites of military, especially naval, power provided mathematical practitioners, professors, and officers with a strategic means to the precision management of society and nature.

Models gained ever more prominent places in the sciences, technology, and medicine because they could be adapted to new teaching regimes (part II). Sharing the commitment of Swiss educational reformer Johann Heinrich Pestalozzi to instruction by *Anschauung*, that is by visual and hands-on experience, professors introduced models into once largely pictureless lectures, alongside specimens, wall charts, blackboard drawings, and instrumental demonstrations. Institutes and museums maintained cabinets, from which models would be taken for lectures and individual study: molecules in chemistry; surfaces in mathematics; morphologies and mechanisms in anatomy, physiology, zoology, and botany; moulages (casts) of skin diseases in dermatology and venerology; machines in engineering; buildings in architecture; excavations in archaeology (Meinel, Mehrtens, Hopwood, Schnalke, Evans). Models shared features with other teaching aids—they were readily available at a convenient size and in schematic colours—but offered special advantages in demonstrating complex spatial relationships and as objects that could be taken apart and re-assembled. Students learned to see, understand, and make things from models. And models came to represent not just molecules or muscles, but also the disciplines that studied them; mathematical institutes gained rooms by arguing that they needed to house collections of models (Mehrtens).

Though models were generally sold as 'teaching aids', they travelled both ways between teaching and research. Chemical models, initially used as tools for classification or teaching, were gradually interpreted as representing the atoms' true arrangement within the molecule, and employed heuristically to investigate unknown structures (Meinel). For some scientists, modelling was part and parcel of their research work, and they would later arrange reproduction and marketing through a firm.

Modellers join artists, publishers, instrument makers, technicians, photographers, and dealers as key figures in the scientific enterprise. Commissions might remain one-offs, like that of the sculptor, Benjamin Waterhouse Hawkins, to build extinct animals at the Crystal Palace (Secord). But, from casual beginnings in workshops, hospitals, and laboratories, some modelling enterprises endured; the Ziegler studio made wax embryos for eight decades (Hopwood). Increasingly, collections of pieces made by a single hand or school and copied only a few times (like the Florentine waxes) gave way to commercial production of standardised objects for universities and schools. Model animals, vegetables, minerals, buildings, machines, and phrenological heads also joined dolls' houses, tin soldiers, building blocks, train sets, and globes as aids to rational recreation in bourgeois homes; modelling, in clay and in wax, was a common pastime.

Thus models were a key medium of traffic between the sciences and the wider culture. By using models, Christoph Meinel proposes, synthetic chemists not only adopted a new visual language for their science, but also presented themselves as builders of a new world, participating in a wider culture of construction. Models were demonstrated in popular lectures and above all at the international shows of the age of capital and empire that followed the Great Exhibition of 1851. Many of the exhibits representing manufactures, the sciences, and the arts were models, which here came together most prominently as a class.

The gigantic models of extinct animals made for the grounds of the Crystal Palace, James Secord shows, were intended to combine scientific expertise and entrepreneurship. But it proved hard to get the mix right. While popular shows offered entertainment with an educational veneer (Altick 1978), state museums began to wrestle with the problem of attracting wider audiences without unduly compromising academic credentials. Lynn Nyhart investigates competing views among museum workers in early twentieth-century Germany of how, in creating experiences of greater authenticity, this could be done.

Many of the nineteenth-century institutions in which models had been produced, manipulated, and displayed were still important sites of scientific culture in the twentieth century (part III). Aeronautical research was much concerned with experiments on models in wind tunnels and their relevance to full-size aeroplanes in the sky (Hashimoto 2000); and visitors to hygiene exhibitions and world's fairs continued to see the sciences represented by models. Though by the 1960s long-familiar items such as moulages, wax embryos, and physical models of mathematical surfaces were on the way out, new models continued to serve as tools of scientific research and communication. Building the science of molecular biology was, to a significant extent, a matter of making models that were the main products of the research and its public face (de Chadarevian). The models moved from the laboratory to the Brussels World's Fair of 1958 and to the TV studios. They were reckoned to make a fine science show, and for much the same reasons that nineteenth-century exhibition-makers and lecturers had deployed their predecessors.

Even in disciplines in which, after World War II, 'modelling' came to mean something altogether more abstract, physical models might occupy special niches. Mary Morgan and Marcel Boumans discuss a striking example in economics. Built in the 1940s, the Phillips machine represents the macroeconomy as stocks and flows of coloured water in a system of perspex tanks and channels. Users enthusiastically agreed that it not only visualised, but also helped to clarify long-debated aspects of Keynesian theory.

Nevertheless, in the last few decades the traditional form of a model has been challenged by the new technologies of the small screen. Eric Francoeur and Jérôme Segal explore how it began to appear fruitful, in attempting to solve the problem of protein folding, to move from physical models to computer simulations coupled with interactive displays. The new approach did not take off at once, not least because it was very difficult effectively to display virtual models in lectures or exhibitions. But eventually model rooms were emptied to make way for computer work stations. Computers changed modelling not only in disciplines such as chemistry, anatomy, and architecture, in which 3-D models were deeply entrenched, but also in fields where modelling had meant numerical methods. As the production of electronically generated simulations fused the 'image' and 'logic' traditions in microphysics (Galison 1997; Sismondi and Gissis 1999), so distinctions between physical and mathematical modelling are becoming blurred: both now produce images described as '3-D models', but on the flat screen.

If virtual modelling claims to surpass and threatens to efface physical modelling, at the same time it draws on the earlier techniques and lends enchantment to their products. In chemistry, for example, one can accommodate new parameters, and simulate and compare different structures, at a speed inconceivable in physical modelling—but the displays owe much to physical models and their depictions in print. Similarly, we might think of the 'Visible Human Male' and the 'Visible Human Female', through slices of whom one can zoom on the World Wide Web (Waldby 2000), as virtual transformations of the old technologies of plastic reconstruction from serial sections on the one hand, and the transparent 'glass man' and 'glass woman' that were displayed in health exhibitions in interwar Europe and the United States on the other (Hopwood 1999; Beier and Roth 1990). With feedback mechanisms linked to touch and coupling to machines able to produce solid models, virtual modelling continues to borrow from its physical counterpart, which is deployed in new as well as old modes. There is nostalgia for material models, too; today an institute's cupboards are less likely to be emptied into the bin than decanted straight into prestigious exhibitions (Bredekamp, Brüning, and Weber 2000). These rework the models' pasts even as they engage viewers in fresh ways.

DIMENSIONALITY AND BODILY ENGAGEMENT

A general question about dimensionality runs through all of these changes in the making, distribution, and uses of models. On the one hand, we can ask what made 3-D models special. Building them was usually more labour-intensive than producing printed materials, and models were harder to distribute too. What distinctive kinds of engagement did they allow that justified the extra time and trouble? On the other hand, we can consider how, at every stage in the making of knowledge, these 3-D objects were placed in relation to flat items, such as labels, drawings, and printed pages.

Model-making has tended to be organised by media and techniques. Wax, clay, and plaster, and from the mid-twentieth century plastics, were perhaps the most important materials, though archaeological models were made of the relatively expensive but exceptionally light and so easily transported cork, while Hawkins' massive extinct animals were permanently set in bricks and mortar. Modellers' lives, like those of Rudolf and Leopold Blaschka, the celebrated blowers of glass animals and flowers (Schultes and Davis 1992; Reiling 1998), were often devoted to single processes. Most wax modellers stood by a material long favoured for its malleability and capacity to resemble flesh, but Fontana turned to wood because it was strong enough to take apart and put together again many times. Louis Thomas Jérôme Auzoux would in the 1820s move to mass-produce such 'clastic' anatomies in cheap and robust papier mâché (Davis 1977). Here, the quest for the right material drove innovation. More often, the ability to work in a particular medium allowed artists to trade across disciplinary boundaries, or between sciences, arts, and manufactures, for instance to shift from devotional objects or artificial fruit to body parts, from glass eyes to glass animals, from an artist's studio to a hospital or laboratory. Instrument-makers could readily turn out models in metal, wood, and/or glass. Museum dioramas were challenging to make in part because they involved so many techniques, from collecting naturalia through taxidermy, carpentry, making artificial flowers, and painting to working in glass and plaster.

Modelling in these materials more obviously involved sustained physical interaction than representation in 2-D, and was sometimes claimed to produce knowledge that could not be acquired in other ways. Championing the virtues of knowing by making, Fontana argued that one could know the human body only by assembling it from parts, and this was a reason he preferred robust wood to delicate wax. James Griesemer's commentary discusses the philosophical implications of this closer bodily engagement and the claims that it licensed to deeper knowledge. It had other effects too. As Thomas Schnalke shows, a moulage required far greater intimacy with a sick person than other kinds of medical portrait. In mid-twentieth-century Zürich the mouleuse Elsbeth Stoiber spent many hours alone with a patient over the course of a week, making the mould by applying plaster to the skin, often of the face or genitals, and then returning to paint the wax positive. All the time she explained the procedure and listened to the sick person talk.

Special media notwithstanding, paper was used in making nearly all models, and modelling was deeply entwined with drawing and printing. Transitions between paper and models were also moves between two and three dimensions. By highlighting the variety of these transformations, studies of models expand our understanding of representation in scientific practice. The dominant view has models as optional end-products, but they were in fact built at various steps in sequences of representational operations. Models might come last, as a way of filling out an engraving or synthesising several partial views; many of the anatomical waxes at 'La Specola' began like this, though a large number of corpses was dissected to make them show more. But models could also come first. One example of models' manifold uses as research tools is when people modelled in order better to depict light and shade on the page or on canvas. A watercolour showing a bird's-eye view of a rustically gnarled Stonehenge betrays its likely origin in looking down on a model in cork. A great many illustrations in science books are of models: moulages were displayed in colour atlases; embryological articles depicted and described wax models; and from the 1960s textbooks of biochemistry and molecular biology carried architectural drawings of protein models. More generally, just as drawing involved models—in the sciences no less than in the drawing schools—so a great deal of modelling relied on representational work in 2-D. We can think of the family of techniques that built up models from slices, whether an anatomist's sections of an embryo displayed on wax plates, or a crystallographer's electron-density maps of a protein on perspex sheets, as keeping the labour of each step within two dimensions.

Models were viewed both directly and in pictures. That drawings, photographs, and texts so often show or are about models means that to appreciate how these 2-D representations were made and used we may need to take models into account. The printed 'communications' of late nineteenth-century archaeology are hard to follow unless we realise that they began as lectures commenting on a model in the room. No wonder that modellers saw pictures and texts as poor proxies for models that in many fields were the primary objects of knowledge, and put much effort into multiplying experiences of the models themselves. Some models attracted people to see them and others were carried along and shown around. Unique models could be viewed in the original medium only by limited, if, as most strikingly in the case of the extinct animals at the Crystal Palace, sometimes very large, publics. But by the mid-nineteenth century modellers generally copied and distributed their work, and so gave wider audiences some of the advantages of engagement in three dimensions.

How, then, did viewers engage? Ludmilla Jordanova's commentary points out the complexities of viewing, and how approaches from studies of visual culture can help us recover these. Experiences of looking at models diverged most from those of viewing flat items when models were variously monumentally large, deeply sculpted, designed to be seen from a wide range of points of view, or in motion (Hall 1999). Special to models was also the degree of tactile engagement they encouraged. This allowed the principle of knowing by making to be extended to end users. Fontana's demountable wooden models, designed to be taken apart and, crucially, to be recomposed, were made for teaching that would connect sight and touch. Chemical models were generally sold as kits for students and researchers to assemble; it was only through these actions that they would literally feel the importance of structural constraints (Francoeur 1997). Morgan and Boumans employ the notion of the 'mind's eye' (Ferguson 1977), the 'organ' in which sensory experiences are stored and combined, to stress that far more than by seeing pictures of the Phillips machine, or even the machine at rest, economists learned about the implications of Keynesian theory by using it, or at least seeing it at work.

Those who would communicate with models regarded pictures and texts as inferior substitutes, but as supplements paper representations were essential. They made potentially perplexing objects intelligible, and at the same time helped to limit uses considered inappropriate. Managing the conditions of viewing was supposed to prevent anatomical models becoming pornography, or chemical models mere toys. Some of the most important aids to viewing were catalogues, price lists, labels, descriptions, and keys. Today these may represent historians' only records of models that have been dismantled or lost. They have even been used to reconstruct the models, and so have come to define what the originals were.

What counted as an appropriate model remained contested. Spectacular displays were intended to impress audiences and silence opposition, but some refused to accept the modellers' analogies. Anatomist Richard Owen cast doubt on the tenuous extrapolation from a handful of fossil bones that Hawkins' muscular iguanodon represented, but the gigantic model stayed. Critics warned of the dangers of taking visual representations too literally: would August Wilhelm Hofmann's demonstrations lead students to imagine that atoms were really miniature croquet balls, oxygen in red and nitrogen in blue? Emotion-charged animal groups attracted visitors to natural history museums, but were accused of putting drama before science. Models' radically different uses went beyond clashing interpretations. Mathematics professors took satisfaction in giving their equations solid form; Man Ray's photographs appropriated the models for surrealism. A moulage that for a dermatologist represented diseases he had understood and ideally cured, and for his students meant so many disease pictures to learn for examinations, provided some patients with a votive offering to take to church.

The commentaries that close this book open new perspectives on models. Griesemer finds that considering precisely those models that philosophical discussion has traditionally given shortest shrift challenges current approaches to the roles of models in making knowledge. He insists that the materiality and dimensionality of modelling matters to the ways that claims can be made with models, and explores further the historically changing meanings of 'model'. Reflecting on the visual and material properties of models, Jordanova discusses several concerns that historians of science, medicine, and technology increasingly share with art historians and researchers into visual culture. She stresses the diverse pleasures of models and the range of bodily reactions that they invite.

Three-dimensional models offer the possibility of unusually rich engagement and move routinely between the most private sites of discovery and the most public arenas of display. The variety of models' relations with flat items means that to understand other media we have to take them into account. So models are strategic objects of knowledge from which to investigate the cultures of the sciences, technology, and medicine more generally. By studying their making, distribution, and display, we learn more about models, about representation and dimensionality, and about producing knowledge.

ACKNOWLEDGMENTS

We are very grateful to Anjan Chakravartty, Martin Kusch, Simon Schaffer, Jutta Schickore, and Jim Secord for helpful comments on drafts.

REFERENCES

Altick, Richard D. 1978. *The Shows of London*. Cambridge, Mass.: Harvard University Press, Belknap Press.

Beier, Rosmarie, and Martin Roth, eds. 1990. *Der gläserne Mensch—Eine Sensation. Zur Kulturgeschichte eines Ausstellungsobjekts*. Stuttgart: Hatje.

Black, Max. 1962. *Models and Metaphors: Studies in Language and Philosophy.* Ithaca, N.Y.: Cornell University Press.

Boschung, Urs. 1980. "Medizinische Lehrmodelle. Geschichte, Techniken, Modelleure." *Medita* 10, no. 5: ii–xv.

Bredekamp, Horst, Jochen Brüning, and Cornelia Weber, eds. 2000. *Theater der Natur und Kunst / Theatrum naturae et artis. Wunderkammern des Wissens. Eine Ausstellung der Humboldt-Universität zu Berlin vom 10. Dezember 2000 bis 4. März 2001, Martin-Gropius-Bau, Berlin*, 2 vols. Berlin: Henschel.

Bugge, Thomas. 1969. *Science in France in the Revolutionary Era*, ed. Maurice P. Crosland. Cambridge, Mass.: MIT Press.

Cartwright, Nancy. 1983. *How the Laws of Physics Lie.* Oxford: Clarendon.

La ceroplastica nella scienza e nell'arte. Atti del I Congresso Internazionale, Firenze, 3–7 giugno 1975. 1977. 2 vols. Florence: Olschki.

Darnton, Robert. 1990. "What is the history of books?" In *The Kiss of Lamourette: Reflections in Cultural History*, 107–35. New York: Norton.

Daston, Lorraine, and Peter Galison. 1992. "The image of objectivity." *Representations* 40: 81–128.

Davis, Audrey B. 1977. "Louis Thomas Jerôme Auzoux and the papier mâché anatomical model." In *La ceroplastica nella scienza e nell'arte. Atti del I Congresso Internazionale, Firenze, 3–7 giugno 1975*, vol. 1: 257–79. Florence: Olschki.

De Renzi, Silvia. 2000. *Instruments in Print: Books from the Whipple Collection.* Cambridge: Whipple Museum of the History of Science.

Elsner, John, and Roger Cardinal, eds. 1994. *The Cultures of Collecting.* London: Reaktion.

Ferguson, Eugene S. 1977. "The mind's eye: Non-verbal thought in technology." *Science* 197: 827–36.

Francoeur, Eric. 1997. "The forgotten tool: The design and use of molecular models." *Social Studies of Science* 27: 7–40.

Frasca-Spada, Marina, and Nick Jardine, eds. 2000. *Books and the Sciences in History*. Cambridge: Cambridge University Press.

Galison, Peter. 1997. *Image and Logic: A Material Culture of Microphysics.* Chicago: University of Chicago Press.

Giere, Ronald N. 1988. *Explaining Science: A Cognitive Approach.* Chicago: University of Chicago Press.

———. 1999. *Science without Laws.* Chicago: University of Chicago Press.

Gooding, David, Trevor Pinch, and Simon Schaffer, eds. 1989. *The Uses of Experiment: Studies in the Natural Sciences.* Cambridge: Cambridge University Press.

Hacking, Ian. 1983. *Representing and Intervening: Introductory Topics in the Philosophy of Natural Science.* Cambridge: Cambridge University Press.

Hall, James. 2000. *The World as Sculpture: The Changing Status of Sculpture from the Renaissance to the Present Day*. London: Pimlico.

Hashimoto, Takehiko. 2000. "The wind tunnel and the emergence of aeronautical research in Britain." In *Atmospheric Flight in the Twentieth Century*, ed. Peter Galison and Alex Roland, 223–39. Dordrecht: Kluwer.

Hesse, Mary B. 1966. *Models and Analogies in Science*. Notre Dame, Ind.: University of Notre Dame Press.

Hopwood, Nick. 1999. "'Giving body' to embryos: Modelling, mechanism, and the microtome in late nineteenth-century anatomy." *Isis* 90: 462–96.

Johns, Adrian. 1998. *The Nature of the Book: Print and Knowledge in the Making*. Chicago: University of Chicago Press.

Klein, Ursula, ed. 2001. *Tools and Modes of Representation in the Laboratory Sciences*. Dordrecht: Kluwer.

Latour, Bruno. 1987. *Science in Action: How to Follow Scientists and Engineers through Society*. Cambridge, Mass.: Harvard University Press.

———. 1990. "Drawing things together." In *Representation in Scientific Practice*, ed. Michael Lynch and Steve Woolgar, 19–68. Cambridge, Mass.: MIT Press.

Latour, Bruno, and Steve Woolgar. 1979. *Laboratory Life: The Social Construction of Scientific Facts*. Beverly Hills, Calif.: Sage.

Lemire, Michel. 1990. Artistes et mortels. Paris: Chabaud.

Lenoir, Timothy, ed. 1998. *Inscribing Science: Scientific Texts and the Materiality of Communication*. Stanford: Stanford University Press.

Lindqvist, Svante. 1984. *Technology on Trial: The Introduction of Steam Power Technology into Sweden, 1715–1736*. Uppsala: Almqvist & Wiksell.

Lynch, Michael, and Steve Woolgar, eds. 1990. *Representation in Scientific Practice*. Cambridge, Mass.: MIT Press.

Mazzolini, Renato G., ed. 1993. *Non-Verbal Communication in Science prior to 1900*. Florence: Olschki.

Morgan, Mary S., and Margaret Morrison, eds. 1999. *Models as Mediators: Perspectives on Natural and Social Sciences*. Cambridge: Cambridge University Press.

Morton, Alan Q., and Jane A. Wess. 1993. *Public and Private Science: The King George III Collection*. Oxford: Oxford University Press in association with the Science Museum.

Pickstone, John V. 2000. *Ways of Knowing: A New History of Science, Technology, and Medicine*. Manchester: Manchester University Press.

Reiling, Henri. 1998. "The Blaschkas' glass animal models: Origins of design." *Journal of Glass Studies* 40: 105–26.

Rheinberger, Hans-Jörg. 1997. *Toward a History of Epistemic Things: Synthesizing Proteins in the Test Tube*. Stanford: Stanford University Press.

Rudwick, Martin J. S. 1976. "The emergence of a visual language for geological science, 1760–1840." *History of Science* 14: 149–95.

Schnalke, Thomas. 1995. *Diseases in Wax: The History of the Medical Moulage*, translated by Kathy Spatschek. Berlin: Quintessence.

Schultes, Richard Evans, and William A. Davis, with Hillel Burger. 1992. *The Glass Flowers at Harvard*. Cambridge, Mass.: Botanical Museum of Harvard University.

Schummer, Joachim, ed. 1999–2000. "Models in Chemistry." *Hyle: International Journal for Philosophy of Chemistry*, special issues, 5: 77–160; 6: 1–173.

Sismondi, Sergio, and Snait Gissis, eds. 1999. "Modelling and simulation." *Science in Context*, special issue, 12, no. 2.

Waldby, Catherine. 2000. *The Visible Human Project: Informatic Bodies and Posthuman Medicine*. London: Routledge.

van Fraassen, Bas C. 1980. *The Scientific Image*. Oxford: Clarendon.

PART ONE

Modelling and Enlightenment

CHAPTER TWO

Representing Invention, Viewing Models

Malcolm Baker

In both the history of science and the history of art and visual culture, three-dimensional models have received less than their due. If historians of science have marginalised visual representations in favour of texts, historians of art and design have tended to privilege the 2-D, in the form of the drawing, over the 3-D, in the form of the model. Both approaches, I suggest, would have seemed strange to an Enlightenment viewer, particularly one in eighteenth-century England. This essay is an attempt to retrieve the eighteenth-century model in its various manifestations and to map the shared cultural context in which different forms of model were produced, used, and viewed. It also argues that the use of models in natural philosophy was not a discrete phenomenon, separate from other intellectual endeavours and processes of making but was, rather, closely linked with practices and attitudes that extended well beyond the sciences.

As a starting point two models connected with eighteenth-century transport serve to illustrate the range of forms that models took, the variety of functions they served, and the different settings in which they were viewed. By contrast, their later separate histories show why they have not usually been discussed together. My first example is the model of a four-wheeled wagon (Fig. 2.1) that was offered by Stephen Demainbray, a lecturer in natural philosophy, to the Dublin Society for Promoting Husbandry (Morton and Wess 1993, 152). Although it declined this particular offer, the Dublin Society had been set up in 1731 expressly to encourage innovation in technology and the 'useful arts' and as such was a precursor of the Society for the Encouragement of Arts, Manufactures and Commerce founded in London

Figure 2.1 Dr Demainbray's model of a four-wheeled wagon. Science Museum, London. Photograph: Science and Society Picture Library.

23 years later. The cart model formed part of Demainbray's apparatus, which the Dublin wine-merchant and antiquarian James Simon described in 1751 as "the best & largest I ever saw & I now believe extant", reporting that "Dr Demainbray goes on very successfully here in his lectures on Natural and Experimental Philosophy" (Morton and Wess 1993, 94–95). Made of holly wood, this model was said to represent an Irish improved carriage (though based on a type of cart already known in France) and was intended, according to Demainbray, to show how a cart of this design had all the advantages of a four-wheeled vehicle but was easy to turn. Although it therefore incorporated innovative features and so in one way served as a prototype of a new form of cart, its primary purpose was not as one stage in the development of such a vehicle but as a model to be displayed before an audience in a public place.

Nine years after Simon's comments about Demainbray, Thomas Hollis, who was playing a key role in the Society of Arts and Manufactures and had earlier been involved in the award of a premium for "the best Model in wood of a Ship", made the first of several visits to the studio of the sculptor, Joseph Wilton, to see a model of another form of transport.[1] This is my second example, the state coach of George III (Fig. 2.2), designed by the architect William Chambers, along with Wilton and the decorative painter

Figure 2.2 Model for the state coach of George III, designed by William Chambers. Worshipful Company of Coach and Harness Makers. Photograph: Victoria and Albert Museum, London.

Giovanni Battista Cipriani (Snodin 1984, 187). Made of wood and wax, painted and gilded—only one side is in fact decorated—this model might seem at first sight far removed from Demainbray's cart. Manifestly a costly, luxury object, designed collaboratively by an architect, a sculptor, and a painter who together would be founder members of the Royal Academy of Arts, the model itself exemplified sumptuous display. But as a model of the projected coach its function was to give an impression of the effect that the full-size completed vehicle would have on the people, as well as to persuade the royal patron that this team of artists should receive the commission. So though the relationship between the model and what it represented was somewhat different in the two cases, both Demainbray's and Chambers' models were intended for display before particular audiences. Despite this similarity, historians have not usually considered such models together.

Perhaps it is not surprising that the literatures on these different types of model have made no reference to each other. The cart model, along with much of Demainbray's apparatus, had by the 1780s joined the silver microscopes and scientific instruments that formed George III's collection, which temporarily brought it closer to the state coach model. But, while the objects in the king's cabinet combined the qualities of scientific curiosity with the precious and finely wrought characteristics of the *objet d'art*, their location since 1927 in the Science Museum has situated them within a narrative of science and technology. By contrast, the coach model was for many years exhibited in the Museum of London, where it formed part of a display about the City and its civic ceremonies, and is now shown, alongside other works designed by Chambers and Wilton, in the Victoria & Albert Museum's British Galleries. Yet, despite these very separate contexts—both textual and museological—more recent discussions of both models have been concerned with similar questions, many of which involve (at least implicitly) the issue of the model as a mode of representation.

This chapter is about the relationship between these different types of model, both their common features and their differences. While being historically specific in its use of examples from eighteenth-century Britain, I shall be raising more general questions about how such models were viewed and how models might be thought about. What was meant by the term 'model' in the eighteenth century? What different forms did models take and what uses did they have? What can models represent and what are the conventions of representation involved? How do models that are *for* projects intended for execution differ from those that are *after* existing works or artefacts? And what is the relationship of models to other forms of representation, such as drawings? What is involved in representing something in 3-D and on a smaller scale, and what does it assume about the spectator's ability to understand this type of representation? That prompts further questions about the contexts in which models were viewed and the assumptions brought to their viewing. What social practices surround this? Who constituted their different intended audiences? And, with what further meanings have they been invested since they were made?

My line of approach to these issues is to begin with the notion of the model as it was understood in the eighteenth century, and from there to survey the range of models for which we have either documentary evidence or surviving examples. Then I shall attempt to categorise these various forms of model with an eye to their functions as well as their roles as representations of the design process or, sometimes, as stages in that process. But consideration of a model's function and nature also demands that we pay attention to the conditions in which it was viewed; so my final section deals with the viewing of models, suggesting that while Demainbray's cart and George III's state coach came to be treated as separate, their original viewers shared a common culture allowing models of different sorts to be seen and understood in a similar way.

THE UBIQUITY AND VARIETY OF MODELS

To consider these questions together is in itself to register both the variety and the ubiquity of the model. This combination of ubiquity and variety is especially marked in the eighteenth century. In other words, the ways in which invention and representation were then being reformulated are signalled by the numbers of models produced, the diverse and shifting uses to which they were put, and the prominence they were given. One of the most familiar classes of model represented the human figure. Sometimes these were used for the teaching of anatomy and the explication of medical conditions (Mazzolini, this volume), others were displayed as waxworks to attract the growing public for such 'shows' (Altick 1978), while still others were used both in the studio where they served as the basis for elements in large-scale figurative compositions and for the training of artists in the academy.[2] None of these categories or viewing spaces were exclusive or entirely separate, allowing us to see common features in the use made of models by William Hunter in his lectures about anatomy at the Royal Academy of Arts and the way they were employed in the lectures and exhibitions of the modeller, lecturer in natural philosophy, and showman, Benjamin Rackstrow. But my primary concern here is not with models of individual figures but rather with models in which figurative components are at most subsidiary, for it is here that the interconnections between different types of models—between Chambers' coach and Demainbray's cart—become most apparent as well as most telling.

Models were produced to represent all manner of artefacts, from churches to watchcases, from coaches to tomb monuments, as well as to show the movement of the planets or the operation of a water-pump. Some hint of the diversity of models (as well as the mobility of the word's meaning) is apparent from the entry for "Model" from Ephraim Chambers' *Cyclopaedia* of 1727 (Russell 1997, 269–70). Beginning with its definition as "an Original, or pattern proposed for any one to copy or imitate", using as an example that

St Paul's had been "built on the *Model* of St *Peter's* at *Rome*", Chambers goes on to mention the use of models in building as an "Artificial Pattern ... for the better Conducting and Executing of some great Work, and to give an Idea of the Effect it will have in Large". Having referred to "*Models* for the Building of Ships, *etc.* for extraordinary Stair-cases, *etc.*", he then introduces a further definition relating to "Painting and Sculpture", where it signifies "any thing proposed to be imitated". This can in its turn mean either "a naked Man, disposed in several Postures, to give an opportunity for the Scholars to design him various Views and Attitudes" or the models in "Clay and Wax" employed by sculptors and "Statuaries". Models of this sort can either be after existing works—"Designs of others that are larger in Marble"—or be related to projected works—"Figures in Clay or Wax which are but just fashion'd, to serve by way of Guide for the making of larger, whether of Marble or other Matter". The ambiguity of the term 'model' is very evident here. So too are the range of forms a model might take and the diversity of uses to which models might be put.

The impression given by these texts is reinforced by the evidence provided by contemporary institutions, most notably the Society for the Encouragement of Arts, Manufactures and Commerce. Established in 1754, partly in emulation of the Dublin Society, the Society of Arts and Manufactures (now the Royal Society of Arts) set out to further the development of "those Arts and Sciences which are at a low ebb amongst us" and to benefit the national economy by promoting "ingenuity in the ... Polite and Liberal Arts, useful Discoveries and Improvements in Agriculture, Manufactures, Mechanicks and Chemistry" (Allen and Abbott 1992). In fulfilling its aims of fostering technological innovation and raising standards of design practice, both for the enhancement of commerce, the Society seems to have given the making and display of models a central role. Models of machines, particularly those considered to further agricultural improvement, were regularly presented to the Society and some of these passed eventually to the Science Museum, via the Patent Office. A number of them were described and illustrated in the 1776 publication of the Society's Registrar, Alexander Mabyn Bailey, *The Advancement of Arts, Manufactures and Commerce; or, Descriptions of the Useful Machines and Models Contained in the Repository of the Society* (Bailey 1776). The submission of models formed the basis for one important, perhaps the single most important, aspect of the Society's activities:

the award of premiums for works, usually by the young, in a wide range of classes. Among these various groups were those areas of technical innovation represented in Bailey's book. The improvement and innovation fostered by the Society meant raising standards of design and training in the skills involved in the various design procedures required in the luxury trades. And, as the categories of work in which premiums were awarded indicate, models played a significant, if somewhat uncertain, part. The models submitted to the Society of Arts and Manufactures ranged from Mr Nutz's hydraulic machine to a terracotta relief showing a scene from the life of Queen Eleanor. Here then we see the model's centrality within eighteenth-century culture being registered institutionally.[3]

Using evidence such as this, along with surviving examples, we can turn again to the question, how diverse was the range of models produced and what forms did such models take?[4] One class of model of which we have a number of surviving examples was the terracotta made in connection with the commissioning or execution of a large-scale work of commemorative sculpture, of a sort that formed a distinctive and costly part of eighteenth-century visual culture (Bindman and Baker 1995). Such works included the monument to the 2nd Duke of Argyll (by Louis François Roubiliac) that dominates the south transept of Westminster Abbey—a significant site for polite and public culture in the eighteenth-century metropolis—and the monument (by Michael Rysbrack) to members of the Foley family erected in the newly-built church attached to the family's house at Great Witley in Worcestershire. Models of rather different formats, the complete composition in the case of the Argyll monument and a single figure group for the Foley tomb, survive for both works. It was models such as these that Ephraim Chambers must have had in mind when he referred to "Figures in Clay or Wax which are but just fashion'd, to serve by way of Guide for the making of larger, whether of Marble or other Matter" (Russell 1997, 269–70). But figurative sculpture took other forms too in eighteenth-century London so that our notion of the model also needs to accommodate works such as the wax anatomical figures shown at Rackstrow's Museum in Fleet Street. At the same time as exhibiting these, Rackstrow was producing plaster multiples of more conventional types of figure sculpture, and claiming on his trade card to "[take] off Faces from the Life and [form] them into Busts to an exact Likeness", as well making "Leaden Figures, Vases &c for Gardens &

Fountains". The wide-ranging and uncertainly defined range of such figure production in its turn alerts us to the variety of ways in which 'model' might be understood (Morton and Wess 1993, 58).

Another important category, and one which figured prominently in Ephraim Chambers' definition, was the architectural model. Such models were to be deployed from 1809 onwards by Sir John Soane in his lectures on architecture. However, their use as educational aids, akin to the models of machines employed by Demainbray and earlier scientific lecturers, had been anticipated by Sir Joseph Banks in 1789 when he presented at a meeting of the Society of Antiquaries a far more unusual model—the bronze reproduction (Fig. 2.3) of part of the temple attached to the pillared hall (Tirumala Nayak's Choultry) at Madurai in southern India (Guy 1990). Banks showed this in an antiquarian context to illustrate the description of the building provided in a letter from Adam Blackader, a surgeon resident at Madurai. Unlike the models presented to the Society of Arts and Manufactures, the

Figure 2.3 Bronze model of Tirumala Nayak's Choultry. Victoria and Albert Museum, London.

bronze model of the choultry was thus made to complement a verbal account of an existing structure. But it was itself complemented by a series of drawings, which, according to the inscriptions in both English and Tamil, were originally used by the model-maker.

Architectural models were, however, not primarily for teaching, as Soane made clear when he wrote about their use before his own time: "No building, at least none of considerable size or consequence, should be begun until a correct and detailed Model of all its parts has been made. Such Models would be of great use not only to the Workmen, but to the Architect likewise. Formerly Models were considered as essentially requisite in Civil, as in Military Architecture, and no building was begun without a Model having been previously made thereof" (quoted in Wilton-Ely 1969, 15). As John Wilton-Ely has shown, the architectural model had long been used as an explanatory device to help patrons to understand a proposed design as well as to persuade them to proceed with a scheme (Wilton-Ely 1965; 1967; 1968; 1996). But, first with Brunelleschi in fifteenth-century Florence (Millon 1995) and then in seventeenth-century England, it took on a more central role in the processes of design and execution. Early in the planning of St Paul's Cathedral, Sir Christopher Wren recommended that a "good and careful model" was made "for the encouragement and satisfaction of benefactors that comprehend not designs and drafts on paper as well as for the inferior artificers' clearer intelligence of their business" (quoted in Wilton-Ely 1968, 253); some four years later the even larger "Great Model" was produced, following the precedent of Sangallo's model for St Peter's in Rome. Nicholas Hawksmoor's model for Easton Neston in Northamptonshire, by contrast, with its relative lack of detail and separately made internal room divisions, seems to have been used by the architect to work out the relationship between the exterior design and the interior spaces. Others, such as Wren's model for the chapel of Pembroke College, Cambridge (Fig. 2.4), show the structure of the roof, rather than what would be seen when the building was erected.

Another very different type of model familiar to eighteenth-century spectators and consumers consisted of the plaster versions of embossed gold watchcases. Executed on a small scale, these were very expensive luxury items decorated in a sophisticated and carefully conceived manner (Edgcumbe 2000). Although such plaster reliefs, of which only a few survive, were evidently kept in considerable numbers by the goldsmiths and chasers involved, they would seem to be not preliminary 3-D designs for watchcases but casts

Figure 2.4 Wood model for Pembroke College Chapel, designed by Sir Christopher Wren. Pembroke College, Cambridge.

taken from finished pieces. Retained within the workshop, the plasters could be both shown to potential patrons and used as the basis for plaques to be set in other watchcases. Sometimes, they were framed in ebonised wood, like miniatures or waxes, and displayed in exhibitions, such as those mounted by the Society of Artists in the 1760s and 1770s. At the 1769 exhibition the goldchaser Henry Manly showed a cast of "The story of Coriolanus" (Fig. 2.5) and two years later exhibited two further "Cast[s] of a chasing" (Edgcumbe 2000, 82–83). Plasters of this sort were then models that reproduced already existing works, though unlike that category mentioned by Ephraim Chambers—"Designs by others that are larger in Marble"—they are by the original designer and on the same scale as the finished object.

In the production and marketing of ceramics, models played an even more significant but different role (Young 1999). The starting points for porcelain figures or groups were models in wax or clay; moulds were taken from them and the china clay cast or moulded from these. In the process of producing the moulds, however, the model was usually cut up into pieces which were at that point discarded, so that few survive. Here, then, the model was very much part of the production process, rather than a 3-D design to be

Figure 2.5 Plaster model after a watchcase by Henry Manly. Victoria and Albert Museum, London.

shown to a potential patron or customer. Occasionally a model for a vessel might be in wood, such as the tureen designed by Sir William Chambers for Wedgwood. But, although details were sometimes cast from moulds made of wood, Wedgwood seems to have been dissatisfied with wood models, describing them in 1770 as being "of very little service to us, or none at all farther than patterns to look at for in most cases it is less trouble to make new models than to repair the presses out of moulds taken from Wood models" (Young 1995, 67).

As these examples show, models of various types formed a familiar part of commissioning, designing, making, teaching, and exhibiting in eighteenth-century England. But in all the cases I have mentioned, these 3-D models were produced and used in relation to 2-D representations, even if the relationship was an uncertain and shifting one. One view of this relationship—that between architectural drawings and models—was outlined by the architect Sir Roger Pratt in the mid-seventeenth century:

> Whereas all the other drafts do only superficially and disjointedly represent unto us the several parts of a building, a model does it jointly, and according to all its dimensions. So that in one rightly framed all things both external and internal with all their divisions, connexions, vanes, ornaments etc., are there to be seen as exactly and in their due proportions (quoted in Wilton-Ely 1967, 30).

Likewise, the models for Wedgwood's tureen and Rysbrack's monument to the Foley family need to be considered in relation to various drawings. And this relationship between drawing and model can differ, just as the models do themselves. In the case of the tureen, the drawing by Chambers' draughtsman, John Yen, was originally executed as a design for a silver vessel to be executed by someone else and formed part of the process of design and manufacture. Rysbrack's drawings for the Foley monument, on the other hand, not only represent the complete composition, rather than a part as the model does, but show the complete composition (rather than individual figures represented by the same sculptor's models) in a number of variant versions, to be presented to the patron rather than used by the sculptor. Just as models can take various forms and serve different and multiple purposes, so their relationship to drawings should not be thought of simply in terms of a straightforward development from 2-D to 3-D model. Furthermore, even in a single area of practice the relationship between drawings and models can shift, as it did with architects' use of both. The advantages of the model described by Pratt remained important for Wren and Hawksmoor but became less significant for their successors such as Robert Adam, whose neo-classical schemes could be developed and presented equally effectively on paper; only with Soane's return to complex spatial and lighting effects were models once again brought to the fore.

CATEGORISING THE MODEL

This survey of models in different areas of eighteenth-century design, though based on a somewhat arbitrary selection of examples, provides us with a context in which to place Demainbray's Irish cart and William Chambers' state coach. But what may be inferred about models as a generic type? In one sense this is hardly a legitimate question in that the different types and uses of model glimpsed here were to some degree the outcome of specialised craft and design practices, developed and maintained through distinct traditions of training. Even within the sculpture trade, for example, the different practices used by the French-trained Roubiliac and Flemish-trained Rysbrack may be convincingly linked to their familiarity with the procedures employed in the workshops where they learned their trade. Yet these differences of practice have perhaps been exaggerated by the way in which models and their uses have been considered only in the histories of individual trades or professions—ceramics or architecture, for example—and have gone almost unremarked even in those more wide-ranging discussions of visual culture or economic and social history that have ventured to address questions of design. Surprisingly, there is, as far as I know, no discussion of models *per se* and their uses in eighteenth-century Britain. This amounts to a major gap in our understanding of eighteenth-century culture. It is one that appears all the more striking if we link the models discussed above with the models of art and nature that played such an important role for those establishing the new physical sciences in the late eighteenth and early nineteenth centuries (Schaffer, this volume).

How then might we make sense of models and their roles? The most striking feature evident from the cases I have mentioned is their shifting, mobile nature—the forms they took, the ways in which they were used, and what was meant by the term. The examples I have cited sometimes involve models being employed as stages in a design process. But they were as often used to communicate design information from the person responsible for the invention of the design to those given the task of making the finished work, often on a larger scale and in a different material. Others were not linked with something to be made in the future but reproduced (often in little) something that already existed. Some models were used primarily for presentation to a patron or a potential customer and then employed, often annexed to a contract, as a visual record of what had been legally agreed between the maker and the patron. Here what had been the primary purpose was displaced by another function at a later stage. This notion of the after-life of models comes still more evidently into play in the way that they could be used by the artist or maker as a record of a design to be re-employed or adapted years later. When the contents of artists' and craftsmen's workshops were sold along with their stock, models might well have been appropriated and re-used by their rivals or successors. When freed from the workshop they might have been re-used not only for purposes of manufacture but also for teaching—'models' (in a different sense) for students—or for collection by connoisseurs valuing them as tangible traces of invention.

Many eighteenth-century models—terracottas for monuments or Mr Evers' model for a threshing mill (Bailey 1776, pl. III)—look forward to, or are proposals for, a large-scale work or a machine still to be made. The viewer looking at such models was invited to assume that they anticipate on a small scale something still to be executed in a much enlarged form. But while models such as these represented in 3-D something that was in the future to be executed on a larger scale and often in a different material, others belong to (or move between) several other categories rather different from those in which they are customarily placed or those I have just outlined. One such category would comprise models—the architect John Nash's cork models of Greek temples, for instance—which reproduce on a small scale something that already exists. A third—where we would find some of the models used by Demainbray in his lectures as well as Wren's model for the roof of Pembroke College chapel—might make visible a structural or mechanical principle not evident in the completed large-scale work. And a fourth category might consist of models that represent what is only described in texts, whether this be the Temple of Solomon or one of Newton's principles.

These categories involve models being used to facilitate communication along various lines and between different groups. One line—the commissioning or marketing of a work—connects artist or designer, on the one hand, and patron or consumer, on the other. This line perhaps lengthened or became less direct during the course of the eighteenth century as the patron, in face-to-face contact with the artist, was gradually supplanted by the consumer who saw the model at a public exhibition. Another line—the use of models in the formulation and development of a design—involves the artist and other artists who offer comments, or even a single artist in dialogue with him or herself. A third is between the inventor, designer, or architect, and the artisan or craftsmen who will translate what is represented in a small-scale model into a large-scale finished work. Yet others involve relationships between teacher and taught, or between a collector and those who view the collection in which a model has been re-contextualised.

VIEWING THE MODEL

As well as suggesting the diversity of forms of models and the various and shifting roles they had in the eighteenth century, here I am placing the models I have been discussing in a social setting that involved various modes of viewing. While recent work on scientific models, such as Demainbray's, has very much brought this into play, most of the models I have illustrated have been considered primarily in terms of design practice. All too often they have been regarded rather unproblematically as straightforward elements within the development of a design and so made part of a linear narrative of ideation and making. But other possibilities are suggested if we look at them, firstly, as models that share with other models a diverse set of shifting roles and, secondly, as 3-D representations employed in a variety of settings but often involving similar modes of viewing. Simply asking how a model was viewed tellingly shifts the agency from the maker, artist, or inventor to the spectator, patron, or consumer. And at this point we can recognise how many models were made not directly as part of a design process but, as it were, for use at one remove in negotiations with a patron.

Take, for example, Roubiliac's model and drawings for the monument to the Duke of Argyll (Baker 1998). The two drawings, terracotta model, and finished work, are usually read as a linear sequence, through which the development of the composition might be followed, starting with the rather rudimentary drawing of the figures on the more carefully delineated architectural structure, and progressing via the more highly finished drawing into the 3-D terracotta version, and finally to the finished work. But, as well as ignoring variations in detail which cannot be accounted for in such a sequence, this linear developmental reading fails to take account of evidence about the contexts in which these various representations were used. Both drawings were originally associated with the contract between sculptor and patron, and so both amplify and complement a textual description; they also complement each other, one giving an impression of the figure composition and effects of the different marbles and the other providing more precise information about the dimensions and architectural elements. This relationship seems to have been obscured soon after the completion of the work by the framing of the more finished design and its re-use as an independent image. While the model may be rather later than the drawings, it should be thought of less as a working model and more as a representation to be presented to the patron, probably at the point when one of the staged payments mentioned in the contract was to be made. Even when the development of a design is clearly discernible in the relationship between two models, we need to consider the viewing of these models by patron and sculptor together. The changes made to the dimensions and form of Roubiliac's proposed monument to the Duke of Montagu when it was presented as a wood and plaster model are already suggested by incisions made in the wet clay of the earlier terracotta model (Fig. 2.6). The sculptor may well have made these as he and the patron discussed the still unfired clay model in the studio.

The viewing of models in the studio by the patron or his or her agent was a familiar and well-documented practice. As I mentioned earlier, the model for George III's state coach was inspected on a number of occasions in the studios of both the architect William Chambers and the sculptor Joseph Wilton by Thomas Hollis, who recorded his visits and discussions in his diary. In this case, as with the Montagu terracotta, the model served as a focus for negotiation between sculptor, patron, and others involved. I am arguing here that viewing conditions are important for our understanding of 3-D models and their roles. Furthermore, those who viewed sculptors' models, for example, were often the same people who viewed other types of model, bringing to that act of viewing certain assumptions about reading such 3-D representations. Apart from inspecting Chambers' coach model, Hollis also examined ship models in wood at the Society of Arts while Martin Folkes, who was given responsibility for overseeing the progress of the Montagu monument commission and no doubt inspected Roubiliac's models, would also, in his role as President of the Royal Society, have regularly viewed scientific models.

If one viewing space for scientific and technological models would have been the lecture room in which figures such as Demainbray demonstrated natural philosophy, another was the Society of Arts where such models were seen in the same institutional context as models for sculpture or metalwork, if not physically juxtaposed with them. There was indeed a continuum of viewing. As in Demainbray's lectures, models were here being presented to a wider polite public, but by this date there were also various other public spaces in which models were being displayed and viewed. By the 1730s advertisements were stating that sculpture—usually completed marbles but sometimes models—could be seen in sculptors' studios, and these viewing spaces were shortly supplemented by a rapidly growing number of exhibitions. By 1770 the writer Horace Walpole could remark: "The rage for

Figure 2.6 Terracotta model for the monument to the Duke of Montagu, by Louis François Roubiliac. Victoria and Albert Museum, London.

exhibitions is so great that sometimes one cannot pass through the streets where they are" (Baker 1995, 118). This "rage for exhibitions" was encouraged not only by shows such as Sir Ashton Lever's Museum, the display of wax effigies of the royal family at Westminster Abbey, Cox's mechanical clocks, and the display of the 'mechanical tarantula' at Week's Museum but also by the more elevated exhibitions of the Royal Academy and the Society of Artists. It was at these last venues that the sculptors' models or the impressions of watchcases were shown, as well as at the Society of Arts. Together they formed a range of new exhibiting and viewing spaces, the emergence of which, along with the related development of shops, constituted one of the most significant cultural shifts of the late eighteenth century.

The inspection of sculptors' models and models of machines, I suggest, took place in similar contexts. The examination of both also involved similar modes of viewing, assuming an ability to look at something on a small scale and imagine it executed on a larger scale and often in a different material. But at this point we need also to take note of the differences between the two. New practices of exhibiting and viewing in late-eighteenth-century London may have been prompted by both commercial concerns and a shift towards commodification that are registered in attitudes to artists' models and models of mechanical inventions alike. But the distinctive and more exclusive nature of some of these exhibitions may be linked to the contemporary reformulation of the notion of art. Along with the emergence of this notion in terms of recently developed aesthetic theory went the separation of certain categories of model that could be seen as autonomous aesthetic objects, carrying new meanings as the products of the individual imagination. This is most clearly articulated in the way that models were framed—both literally and metaphorically—within collections. And it is here that the shifting meanings with which models could be invested become most apparent.

Although models of various sorts, like raw materials juxtaposed with artefacts, had earlier been placed within the Wunderkammer where they served as representations of process, artists' models were being rather differently understood when seen in the collection of an eighteenth-century connoisseur. Re-contextualised in a collection of autonomous aesthetic objects they were increasingly read within a narrative of artistic creativity. Like drawings at an earlier date, sculptors' models now became valued in their own right, almost as independent works, rather than as designs that might be reproduced by other sculptors. At the same time, they began to figure more prominently in accounts of sculptors' Lives. When placed in collections they were therefore seen as evidence of the sculptor's own hand, as direct and tangible traces of the artist's invention (Baker 2000a, 34–49). The framing of the model in this way, as well as its rich after-life, is well illustrated by the case of Roubiliac's model for the monument to General Wolfe and its later

Figure 2.7 Drawing by Nathaniel Smith showing Roubiliac's model for the monument to General Wolfe within a frame. National Gallery of Canada, Ottawa.

representation (Bindman and Baker 1995, 336–39). Not long after its execution in 1760 the model was shown in a drawing where it seems to have been enclosed within an elaborate frame that not only emphasised the theatricality of the composition but also enshrined the model itself (Fig. 2.7). About twenty years later (in 1789) the same model was represented (now rather differently framed) in an engraving in the *Gentleman's Magazine*, associated with an article (published in 1783) that both celebrated the model as a work

Figure 2.8 The model of William Evers' threshing mill, engraved by Fougeron, from William Mabyn Bailey's *The Advancement of Arts, Manufactures and Commerce* (London, 1776). British Library, London.

of Roubiliac's own hand and described it (erroneously) as his last work, so linking it with the sculptor's biography. Here then the model is increasingly seen as the 'authentic' trace of the individual hand and invested with the notion of authorship.

Around the same time the representation of mechanical and scientific models took a different turn. While Bailey may have reproduced some of the models of agricultural machines presented to the Society of Arts and Manufactures in the form of engravings in his book of 1776 (Fig. 2.8), neither the manner of their representation nor their textual setting draws the reader's attention to their interest as models. On the contrary, despite the reference to models in the book's title, the text seems to make no distinction between actual machines and the models for or after them.[5] Mr Evers' Mill, for example, was erected "in its full magnitude, for Mr. John Turton, farmer, at Wragby" but the account given here is titled "A Description of the Model... made to a Scale of an Inch and a half to a foot" (Bailey 1776, 54–59). Any distinction between model and machine is further effaced through the way they are engraved. Far from celebrating the model, Bailey seems to deny it, or at least take it for granted, rather as I am doing with the images of models that I have used to illustrate this chapter. While sculptors' models were increasingly being seen as valuable traces of artistic invention, scientific and mechanical models were regarded not as significant representations in their own right but instead as evidence of those scientific and mechanical innovations for which they served merely as surrogates. It is this distinction that has perhaps led to our seeing these various categories of 3-D models so differently. It has also resulted in our neglect of what was common to all of them as representations and so of what they shared in their making, use, and viewing earlier in the eighteenth century. Perhaps it was at this point that Demainbray's cart and Chambers' coach began to take divergent paths. To consider them together once more raises important questions about 3-D models as a distinctive mode of representation.

NOTES

Many colleagues have given me the benefit of their specialist knowledge of particular areas of design history. Among them are Helen Clifford, Richard Edgcumbe, Celina Fox, John Guy, Carolyn Sargentson, Michael Snodin, John Styles, and Hilary Young. I am also grateful to Alan Morton for discussion of these issues common to the histories of science and art over many years and to Nick Hopwood and Soraya de Chadarevian for their insightful and tactful editing.

1. Hollis's unpublished six-volume Diary (Houghton Library, Harvard University, MS Eng 1191) records a number of visits to see the model between 29 March and 28 October. On 2 April Chambers came to breakfast and Hollis "Shewed him much Virtu, & recommended him to apply the designs of some of it, nobly to the Coronation Coach of our ingenuous Monarch, which he seem'd inclin'd to". As well as being involved in the Society of Arts (Abbott 1992), Hollis also had wide-ranging antiquarian interests that included Renaissance bronzes (Baker 2000b). His role in the commissioning of the coach has most recently been discussed in a full account of the work (Marsden and Hardy 2001).

2. The term 'model' is still more ambiguous here, for within the academy it more often than not meant the living model drawn by artists who had earlier learned how to copy using 2-D images of the figure and then casts (Bignamini and

Postle 1991). But anatomical figures and sculptors' models in terracotta also played a role in this training as well as in subsequent studio practice.

3. As far as I am aware there has been no systematic study of the key role that models of various sorts played in the Society's activities. In other ways, however, the Society, especially its founder William Shipley, has been studied in considerable detail (Allen 1979).

4. Most of my examples are drawn from eighteenth-century England but an equally diverse selection could be assembled from continental design practices. These might include wax models for furniture (Pallot 1987), Hans Peter Oeri's late-seventeenth-century copper models for sword fittings (Lanz 1995), the wood and terracotta models for central European altarpieces (Volk 1986), and the remarkable array of architectural models from Madrid to St Petersburg recently assembled in Turin (Millon 1999).

5. In some cases this is left ambiguous, as with the hydraulic machine (Bailey, pl. II), the views of which are first described in the title to the plate as "Models" and then as "taken from a drawing which Dr Ziegler made of it, in its original state at Zurich". Comparable German and Dutch models of a grinding mill and windmill were displayed recently in Turin (Millon 1999, 553–54).

REFERENCES

Abbott, John L. 1992. "Thomas Hollis and the Society, 1756–1774." In *The Virtuoso Tribe of Arts and Sciences*, ed. D. G. C. Allen and John L. Abbott, 38–55. Athens, Ga.: University of Georgia Press.

Allen, D. G. C. 1979. *William Shipley: Founder of the Royal Society of Arts*. London: Scolar.

Allen, D. G. C., and John L. Abbott, eds. 1992. *The Virtuoso Tribe of Arts and Sciences*. Athens, Ga.: University of Georgia Press.

Altick, Richard D. 1978. *The Shows of London: A Panoramic History of Exhibitions, 1600–1862*. Cambridge, Mass.: Harvard University Press.

Bailey, Alexander Mabyn. 1776. *The Advancement of Arts, Manufactures and Commerce; or, Descriptions of the Useful Machines and Models Contained in the Repository of the Society for the Encouragement of Arts, Manufactures and Commerce* .... London: For the Author.

Baker, Malcolm. 1995. "A rage for exhibitions: The display and viewing of Wedgwood's frog service." In *The Genius of Wedgwood*, ed. Hilary Young, 118–27. London: Victoria and Albert Museum.

———. 1998. "Limewood, chiromancy, and narratives of making: Writing about the materials and processes of sculpture." *Art History* 21: 498–530.

———. 2000a. "Narratives of making: The interpretation of sculptors' drawings and models." In *Figured in Marble: The Making and Viewing of*

Eighteenth-Century Sculpture, 34–49. London: V&A Publications; Los Angeles: J. Paul Getty Museum.

———. 2000b. "Some eighteenth-century frameworks for the Renaissance bronze: Historiography, authorship, and production." *Studies in the History of Art* 41, 211–21.

Bindman, David, and Malcolm Baker. 1995. *Roubiliac and the Eighteenth-Century Monument: Sculpture as Theatre*. New Haven: Yale University Press.

Bignamini, Ilaria, and Martin Postle. 1991. *The Artist's Model: Its Role in British Art from Lely to Etty*. Nottingham: University of Nottingham; London: English Heritage.

Edgcumbe, Richard. 2000. *The Art of the Goldchaser in Eighteenth-Century London*. Oxford: Oxford University Press; London: V&A Publications.

Guy, John. 1990. "Tirumala Nayak's Choultry, an eighteenth-century model." In *Makaranda: Essays in Honour of Dr James C. Harle*, ed. C. Bautze-Picron, 207–13. Dehli: Siri Satguru.

Lanz, Hanspeter. 1995. "Training and workshop practice in Zürich." In *Goldsmiths, Silversmiths, Bankers: Innovation and the Transfer of Skill, 1550 to 1750*, ed. David Mitchell, 32–42. London: Alan Sutton and Centre for Metropolitan History.

Marsden, Jonathan, and John Hardy. 2001. "'O fair Britannia hail': The 'most superb' state coach." *Apollo* 152: 3–12.

Millon, Henry, ed. 1999. *The Triumph of the Baroque: Architecture in Europe, 1600–1750*. Turin: Bompiani.

Millon, Henry, and Vittorio Magnago Lampugnani, eds. 1995. *The Renaissance from Brunelleschi to Michelangelo: The Representation of Architecture*. Washington, DC: National Gallery of Art.

Morton, Alan, and Jane Wess. 1993. *Public and Private Science*. London: Science Museum.

Pallot, Bill G. B. 1987. *L'art du siège au XVIIIe siècle en France*. Paris: Grismondi.

Russell, Terence M. 1997. *The Encyclopaedic Dictionary in the Eighteenth Century: Architecture, Arts, and Crafts*, vol. 2: *Ephraim Chambers Cyclopaedia*. Aldershot: Ashgate.

Snodin, Michael, ed. 1984. *Rococo: Art and Design in Hogarth's England*. London: Victoria and Albert Museum.

Volk, Peter, ed. 1986. *Entwurf und Ausführung in der europäischen Barockplastik*. Munich: Bayerisches Nationalmuseum.

Wilton-Ely, John. 1965. *The Architect's Vision: Historic and Modern Architectural Models*. Nottingham: University of Nottingham.

———. 1967. "The architectural model." *Architectural Review* 142: 26–32.

———. 1968. "The architectural model: English Baroque."*Apollo* 88: 250–59.

———. 1969. "The architectural models of Sir John Soane: A catalogue." *Architectural History* 12: 5–38.

———. 1996. "Architectual models." In *The Dictionary of Art*, ed. Jane Turner, vol. 2: 335–38. London: Macmillan.

Young, Hilary, ed. 1995. *The Genius of Wedgwood*. London: Victoria and Albert Museum.

———. 1999. *English Porcelain, 1745–1795: Its Makes, Design, Marketing, and Consumption*. London: V&A Publications.

CHAPTER THREE

Plastic Anatomies and Artificial Dissections

Renato G. Mazzolini

"We are dealing with plastic anatomy. It has been practised for many years in Florence at a high level. However, it can neither be undertaken nor prosper anywhere but where sciences, arts, taste, and technique are perfectly at home and engaged in living activity" (Goethe 1964, 366). Thus the aged Johann Wolfgang Goethe introduced the main topic of his *Promemoria* of 4 February 1832 entitled *Plastic Anatomy*. This was addressed to the Prussian state counsellor Peter Christian Wilhelm Beuth to gain his support for the creation in Berlin of a collection of wax anatomical models of the human body similar to those in Florence. The reason Goethe gave for establishing such a collection was the increasing lack of corpses for teaching purposes, and the horror produced by the news of body-snatching in England and Scotland to meet the demands of the medical market. Given the high cost of copies of the Florentine models and their fragility in transportation, Goethe suggested that an anatomist, a sculptor, and a modeller be sent to Florence to learn the necessary techniques, so that on their return they could devote themselves to building up the collection in Berlin itself. In his view, plastic anatomy was "a worthy surrogate that, ideally, substitutes reality by giving it a hand" (Goethe 1964, 367). A few weeks after writing this memoir Goethe died. Little was done in Berlin to implement his suggestions, but the expression 'plastic anatomy' has survived to denote the art of modelling various materials into the forms of dissected human and animal bodies.

'Artificial anatomy' was another expression employed to indicate three-dimensional anatomical models in wax, wood, papier mâché, and other materials. It had been used about forty years earlier, in 1793, by the French army surgeon Nicolas-René Desgenettes, then active in Italy. In a detailed description of the Florentine collection Desgenettes pointed out that it had been created and directed by the "indefatigable" Felice Fontana (1730–1805) and put forward a number of arguments to convince his own compatriots of its usefulness for teaching and research (Desgenettes 1793, 251). He insisted that copies of the Florentine waxes be made for French institutions just as they had been made for the military medico-surgical academy (Josephinum) in Vienna, given "the difficulty of training artists and of achieving rapidly the same perfection" of execution (Desgenettes 1793, 251).

While 'plastic' stresses the form by which the anatomy is manifested—not an illustration on paper, but a statue—'artificial' makes it clear that one is not speaking of a real anatomy or dissection. In the title of this chapter both terms are used in conjunction to describe two persisting elements in Fontana's prolonged obsession with anatomical models: that of giving volume to the human body and that of substituting an artificial anatomical practice for the real one.

These quotations from Goethe and Desgenette are important not only for their terminology, but also because they document well the fame of Florentine plastic anatomy in the late eighteenth and early nineteenth centuries. And indeed, for the historian the Florentine models of the human body in wax manufactured in the last thirty years of the eighteenth century and preserved at the Zoological Museum 'La Specola' in Florence, the Josephinum in Vienna, the Anatomical Museum in Cagliari, and the University Museum in Pavia constitute the largest set of non-printed sources displaying the anatomical knowledge of the age.[1] However, although these collections have been the subject of considerable historical research and interpretation on the part of medical historians, art historians, and local historians, there are still several matters that warrant investigation. In particular two sets of questions have not yet been studied. The first is how Fontana organised the production of these huge collections, the financial aspects of his endeavour, and, especially, his relations with the artists he employed. The second relates to the reasons why, in old age, Fontana turned to the production of wooden 'demountable' (*decomponibile*) models of the human body, with all internal organs also demountable, which were apparently less artistic and sophisticated than those in wax. These later models are very little known to scholars (but see Lemire 1990, 59–61). They reflect, on the part of Fontana, a search for models that would either enable the viewer to enter the body and, as it were, gain an internal view of it, or manually to take it apart and carry out an artificial dissection. Since Fontana published so little on either wax models or on what he called 'demountable anatomy', and since published discussion of models and their use was also limited during the eighteenth century, the emphasis of the present essay is on a tentative reconstruction—based on archival material and published correspondence—of Fontana's relations with his collaborators and of his search for ideal models for anatomical instruction and public exhibition.

FONTANA AND THE FLORENTINE MUSEUM

A native of the Trentino (i.e., Northern Italy), Felice Fontana is considered, together with Lazzaro Spallanzani, Luigi Galvani, and Alessandro Volta, one of the most distinguished Italian natural philosophers of the second half of the eighteenth century. His main contributions were to physiology, anatomy, microscopy, plant pathology, toxicology, and chemistry.[2] Appointed physicist to the Grand Duke of Tuscany, Peter Leopold, in 1766, he was charged with the realisation of the project he himself had proposed: that of establishing a Royal Museum of Physics and Natural History in Florence next to the Pitti Palace, the Grand Duke's residence. This resulted in what is probably Fontana's most lasting contribution to scientific culture: the creation of the main centre for scientific research in Florence at least until 1859, when it was incorporated into the newly founded Institute of Advanced Studies.[3]

Fontana's activity as director of the museum was amazingly prolific. Not only did he acquire a great number of scientific instruments, but many of them—such as recording barometers and eudiometers—he had made according to his own designs ([Ceruti] 1775; Fontana 1775; Ingen-Housz 1779). He organised a chemical laboratory and one for meteorology, and spent much time building an astronomical observatory. He knew that the chosen site was poor (Contardi 1996, 293), but since he was pursuing the ideal of bringing together all the sciences under one roof, astronomy had to be included too. Furthermore, because the purpose of the museum was also to illustrate the Tuscan tradition in scientific research, Fontana collected historical scientific instruments such as those produced for the experiments of the Accademia del Cimento in the seventeenth century. He assembled what was left of the old Florentine natural collections and purchased new ones. Finally, he set up a workshop for the production of wax anatomical models of the human body which were then exhibited. When the museum opened to the public in

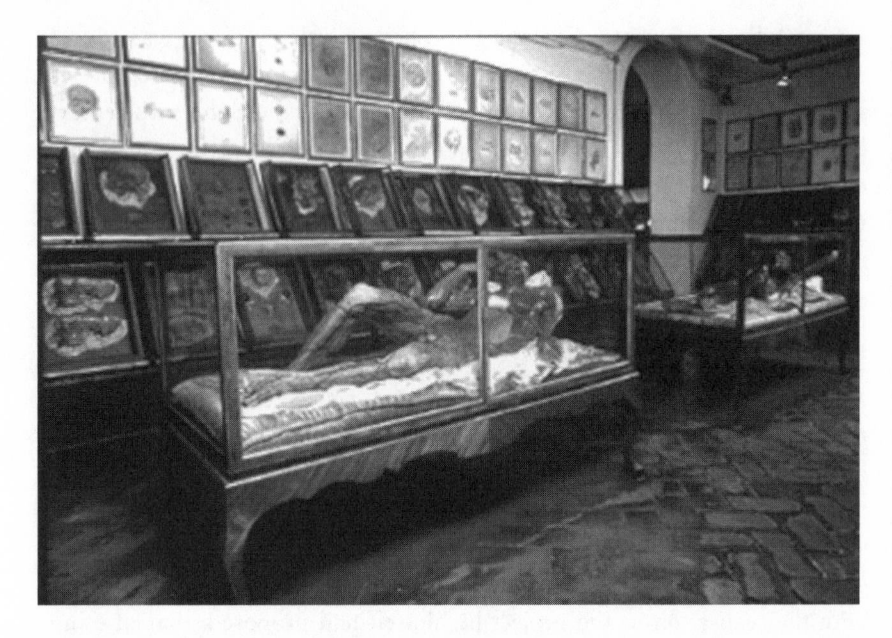

Figure 3.1 Room at the Zoological Museum 'La Specola' containing wax models in their eighteenth-century cases, with watercolours of the models on the walls. Courtesy of Museo Zoologico 'La Specola', Florence.

1775 the collection of waxes consisted of three full-size statues and 137 cases containing 486 models. It occupied six rooms, or according to Johann Bernoulli III, eight rooms ([Ceruti] 1775, 131–38; Bernoulli 1777, 251–52; Martelli 1977, 113; Knoefel 1984, 252). Today the entire collection covers about 700 square metres and contains nineteen full-size statues plus 543 cases containing about 1,400 models (Fig. 3.1).[4]

Fontana intended the museum to be both an institution for the promotion of science to the public and a centre of research. It was, after all, his aim to hold public lectures in the museum and to incorporate a scientific academy that would revive the Accademia del Cimento, a goal he never fully achieved.[5] For this reason an object on display was both a research tool for scholars and an exhibit which was meant, on the one hand, to provide visual evidence of the laws of nature and, on the other, to stimulate public appreciation of the useful sciences. The latter was also the Grand Duke's aim when he decided to fund the museum, since he believed in cultural renewal on the basis of modern science and for the public good. Building the museum was one cornerstone of policies to reform institutions (universities, hospitals, and learned societies) and to enhance the professionalisation of surgeons, physicians, and engineers.[6] As time passed, however, the museum became increasingly a symbolic showcase for the Grand Duke's reform politics and less and less a centre of research.

ORIGIN OF THE FLORENTINE ANATOMICAL MODELS

When dealing with the origins of the Florentine anatomical models in wax it is customary to recall a Tuscan tradition of wax modelling dating back to Lodovico Cardi (called Cigoli, 1559–1613) in the late sixteenth century and Gaetano Giulio Zummo (or Zumbo, 1656–1701) at the end of the seventeenth century (e.g., Belloni 1959–60; Knoefel 1978, 330–31; Lanza et al. 1979, 22–23). It should be borne in mind, however, that when Fontana moved to Florence in 1765 this tradition no longer existed, and that there was no continuity between his own enterprise and those of previous artists who had worked in Tuscany. Rather, it should be emphasised that Fontana was inspired by what he saw in Bologna during his sojourn in that city in the mid-1750s (Mazzolini 1997, 663). At that time a number of sculptors and wax modellers, including Ercole Lelli, Giovanni Manzolini, and his wife Anna Morandi Manzolini, were producing models in wax for the anatomical cabinet at the Institute of the Sciences of Bologna (Armaroli 1981). There is no doubt that during the 1750s and 1760s Bologna was the main centre in Italy for the production of wax models of the human body and that Fontana should be considered a follower and innovator of the Bolognese tradition. By the 1780s, Florence had replaced Bologna thanks to the workshop (*Officina di ceroplastiche*) that Fontana set up in the museum. It gained such expertise that, after his death, it persisted well into the second half of the nineteenth century, producing models in wax not only of the human body, but also of plants, of animal microstructures, and of human pathological conditions for the Hospital of Santa Maria Nuova.[7]

THE WORKSHOPS

Who produced the Florentine wax models? Given their number and the different skills required to produce them, it is clear that they cannot be conceived as the work of one person alone. And indeed, many individuals with different qualifications, such as anatomists, dissectors, modellers, and painters, took part in an enterprise skilfully orchestrated by Fontana.

As Desgenettes pointed out in 1793, the models had to represent the living and healthy body with working organs, not the cadaver. In his description of the Florentine collection he noted that wax was the best medium with which to represent the body, not only because of its transparency and the possibility of giving it every possible colour, but also because it lent itself to producing "that moist appearance which perfectly imitates the state of life" (Desgenettes 1793, 171). Furthermore, it could be preserved effectively and cleaned without being damaged.

Representation of the parts of the body followed two visualisation schemes closely corresponding to two different anatomical preparation techniques: the systematic and the topographic. The aim of the former was to display all the parts belonging to the same system, such as bones, muscles, and nerves, modelling each structure separately from adjacent organs and cleaned of its attachments, as well as of vascular and nervous supplies. Celebrated examples of representations in accordance with this scheme were William Cheselden's *Anatomy of Bones* (Cheselden 1733) and Bernard Siegfried Albinus' *Tables of the Human Skeleton and Muscles* (Albinus 1747). The latter scheme responded to the need to view a given organ *in situ* with all its supplies of nerves, vessels, and lymphatics and its attachments to neighbouring organs. Examples of this scheme are provided by the plates in Albrecht von Haller's *Anatomical Figures of the Parts of the Human Body* (Haller 1743–54) and William Hunter's *Anatomy of the Human Gravid Uterus* (Hunter 1774).

On viewing the Florentine or Viennese collections, one recognises that many of the models are extensive quotations of famous anatomical illustrations by such authors as those mentioned above. At times these quotations are so evident that they appear to be intentional volumetric or 3-D translations of those very illustrations. The surviving records, however, show very clearly that for each model of an organ or life-size statue, anatomical preparations were also made from corpses regularly supplied by the local hospitals. For instance, in one year only—from 2 January to 31 December 1793, by which time most of the models had already been produced—177 "parts" of cadavers were brought to the museum (Martelli 1977, 116).

The main phases in the production of wax models included: 1. the choice of the topic to be represented, often on the basis of one or more authoritative

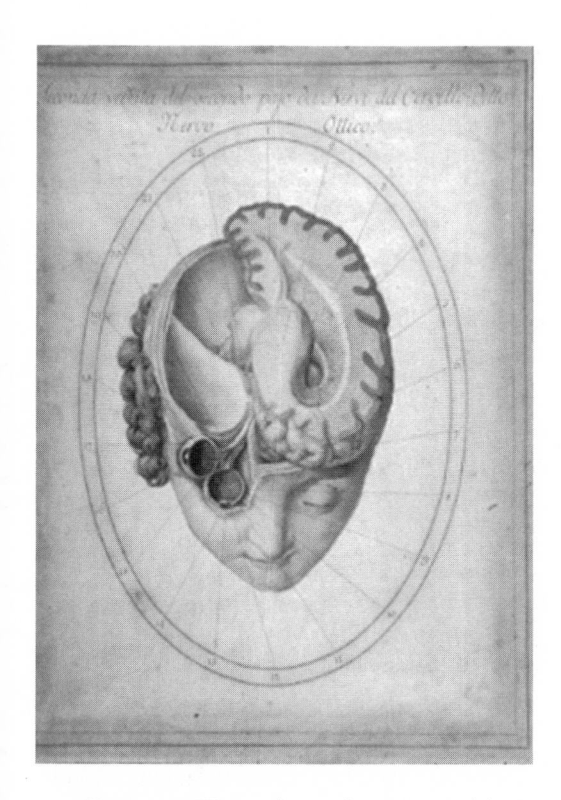

Figure 3.2 Watercolour of a wax model. Courtesy of Museo Zoologico 'La Specola', Florence.

illustrations; 2. the preparation, according to a chosen illustration, of an anatomical part; 3. the production of a model in coarse wax (*ceraccia*) or clay after the anatomical preparation; 4. the production of a demountable plaster mould of the model; 5. the spreading of one or more thin layers of fine coloured wax on the surfaces of the mould; and 6. the filling of the mould's cavity with wax-impregnated tow or with iron frames wrapped in tow. Blood vessels, nerves, lymphatics, and muscular striations were made on the finished model. At this stage a painter was asked to produce a watercolour of the model with numbers referring to each anatomical part (Fig. 3.2). The watercolour was generally framed and hung on the wall next to the model, which was placed in a glass case. Included in this case was a drawer containing sheets of paper listing the anatomical nomenclature corresponding to the numbers appearing on the watercolour.

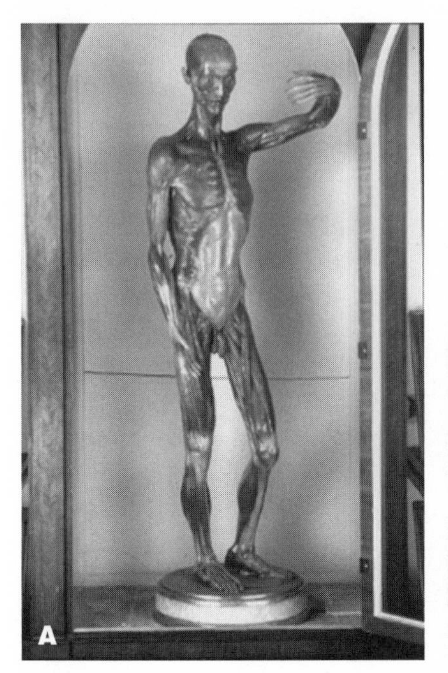

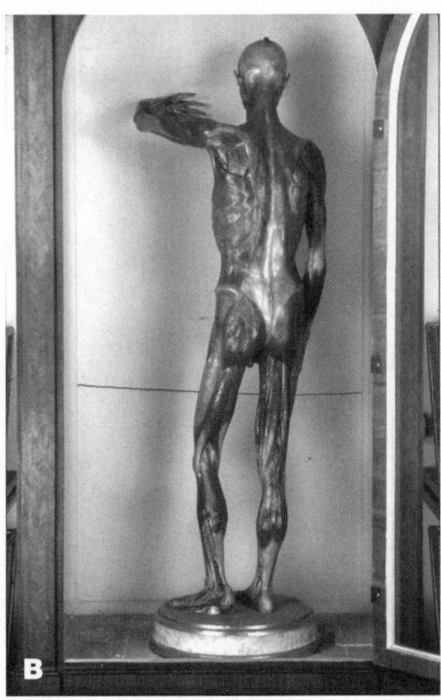

Figure 3.3 Life-size statue in wax mounted on a pedestal rotated by a lever, (A) front view, (B) rear view. Courtesy of Museo Zoologico 'La Specola', Florence.

Life-size statues were not cast in one piece, but resulted from the joining together of model parts that had previously been made separately. The standing life-size statues were placed on pedestals that could be rotated by a lever, thus enabling the visitor to observe the model clearly from all sides (Fig. 3.3).[8] Most of the models were life-size, but some were magnified in order to point up fine details. Although the entire collection was of gross anatomy, it also included, like many plates of the time, some microscopic anatomy modelled directly after illustrations (Fig. 3.4).[9]

For the first two years, Fontana himself dissected the parts he intended to be modelled, but later he instructed dissectors to relieve him of this task. After 1771 the main modeller in the museum was Giuseppe Ferrini, who from 1773 was joined by Clemente Susini (1754–1814), regarded as the institution's most distinguished artist. Apart from the modellers and painters employed on a regular basis, many artisans enrolled as day-labourers. In

Figure 3.4 Wax models of the dissected parts of an eyeball. The left-hand model of the middle row is of the ciliary processes; it was produced after illustrations made with the microscope by Johann Gottfried Zinn in 1755. Courtesy of Museo Zoologico 'La Specola', Florence.

order that their work could be better supervised, they were asked to keep a diary of their daily duties.

The phases in the production of the demountable anatomy in wood were less complicated, but the work required even greater skill. According to Fontana his prototype statue was made of 3000 independent wooden parts, each of which had to be carefully machined in perfect dimensions so that they could all fit in the statue.[10] It should be noted that most of the work in wood was not carried out at the museum, but in Fontana's own workshop in his home. There is evidence that this new project had already started in 1786 and that in later years much of it was conducted at Fontana's own expense. Preliminary to the work itself a number of experiments were carried out with many sorts of oriental woods in order to choose the best possible material. A dissector always prepared the anatomical part to be reproduced, a draftsman drew the preparations at different sizes, and a modeller fashioned them into a model in coarse wax. On various occasions preexisting wax models were used. The sculptor had then to reproduce the wax model in wood—which was the most difficult task—and finally a painter had to colour it correctly (Martelli 1977, 115–17).

FONTANA'S RELATIONS WITH THE MODELLERS

The display of the anatomical models in the Florentine museum suggested to Goethe an idyllic combination of sciences, arts, taste, and technique. But the backstage was different: fund-raising, difficult negotiations with the Crown, administration, dissection of corpses, analysis of anatomical plates, technical improvement of wax preparations, and tense relations between Fontana and the young artists he employed.

Working on the anatomical waxes was not the ideal job for a young sculptor, because he had no artistic liberty and was considered to be a simple executor of detailed instructions. Fontana regarded the young artists he employed as instruments. As he put it in 1791 to one of his correspondents: "I use them, as they use a plane, a saw, a chisel."[11] It was the same technique he had used to train instrument-makers when he discovered that the personnel available lacked "the most elementary notions of the art, of drawing, of arithmetic, or reading.... Then I resolved to use them as one would a hammer, a file, that is to consider them simply as tools, and to make them do everything without understanding anything."[12]

At an early stage of the enterprise Fontana himself dissected the corpses. Later he obtained a professional dissector and, intermittently for a few years (1782–85), the help of an able anatomist, Paolo Mascagni, professor of anatomy at the University of Siena, and a true innovator in the study of lymphatic vessels.[13] The quantity of work, the low salary, contradictory instructions, and the often abusive and outrageous manners of Mascagni and Fontana caused much animosity among the artists, and complaints of ill treatment were forwarded to the Grand Duke and supported by Giovanni Fabbroni, a protégé of Fontana who soon became the Grand Duke's informer at the museum.[14]

Fontana's correspondence with Sebastiano Canterzani, secretary of the Institute of the Sciences in Bologna, throws some light on the procedures followed by Fontana in enlisting new modellers. If possible the candidate would have won a prize in modelling at an art academy, and he had to be young, "able and docile".[15] On the other hand he did not need any knowledge of anatomy or special instruction in drawing. They are, he often said, very ignorant: "*ignorantissimi*".

Not all artists, however, were prepared to work for Fontana once they saw what was involved. In 1784 Canterzani found a candidate who had all the qualifications Fontana wanted. His trip from Bologna to Florence was organised and paid for, but as soon as the young man saw the kind of work he had to perform, he fled back to his home town without even greeting Fontana.[16]

The next year Canterzani found another sculptor, a Venetian named Agostino Ferrari. On 12 February 1785 Fontana reported to his Bolognese correspondent: "He does not have the frights of the other, and he will not flee for fear. He seems to be a good lad."[17] As a matter of fact, not much time passed before Ferrari led, in Fontana's own words, a "kind of sedition" of the modellers then working on the waxes that had to be sent to Vienna.[18] Since Ferrari dared to present the Grand Duke with a memorandum that Fontana deemed to be full of falsehoods, the anatomist decided he had to fire the sculptor notwithstanding Canterzani's petition for clemency.[19]

In order to oversee the work of his sculptors from early morning to evening, Fontana set up a workshop in his own house. Later he would pay some of the sculptors with his own money, especially those working on the wooden models, and recruit them in the Trentino, his native province, presumably because he felt they would be more faithful and obligated to him.[20]

PERSONAL AND ADMINISTRATIVE CONFLICTS

A surviving register indicates that 7,663 persons visited the museum during a period of just over thirteen months—from 13 September 1784 to 19 October 1785—and that they could choose between two guided tours a day, at 8.30 and 10.00 a.m.[21] From the names and towns of origin of the visitors, one notes that most of them were from Florence itself or from Tuscany, but also that there were foreigners from both the Italian states and across the Alps. The tour concentrated on the rooms devoted to experimental philosophy and those showing human anatomy. Among the men the fascination for the waxes fluctuated between admiration for the mimesis of nature and amazement at the artistic rendering, while—according to one source—on some occasions the ladies fainted (Ferber 1773, 85–86; Stendhal 1977, 2: 401; Otto 1825, 1: 105). Marguerite, Countess of Blessington (1789–1849) lamented the absence of restrictions "to preclude men and women from examining these models together" because, she argued, "although highly useful for scientific purposes" the waxes were "unfit for this promiscuous exhibition" (Knoefel 1984, 259–60).

The most eminent statesman to visit the museum in the early 1780s was the Emperor Joseph II, the Grand Duke's elder brother. He was so impressed by the waxes and their didactic value that he requested a similar collection for the military surgical academy then being built in Vienna (Schmidt 1996, 102). At first Fontana was reluctant to engage in such a large new project, but later accepted, even though Peter Leopold felt that it would retard completion of the exhibition halls of the Florentine museum. For this reason a compromise was struck. Fontana was allowed to oversee the project in a private workshop, and the use of the museum's moulds was permitted as long as it did not interfere with other activities. But while the collection for Vienna—consisting of 1,192 models packed in 368 cases and sent by mule between 1784 and 1788—was a success, probably also financially, relations between the Grand Duke and Fontana suffered. The former felt that Fontana had given priority to the Emperor's request over his obligations to the museum and himself. Furthermore, Peter Leopold could not accept the idea that Fontana considered the museum very much his own creation, more than that of the Crown, and that he showed little interest in the administrative needs of a public institution (Pasta 1989, 177).

The Grand Duke appreciated Fontana's merits as a savant, but by the end of the 1780s he was dissatisfied with him because of the large financial resources used up by the museum, the numerous complaints from the workers, the many changes in the display of the instruments, models, and natural objects in the collections, and because there appeared to be no end to the wax collection. It is true that Fontana frequently changed his mind. In the early 1780s he reported that the wax collection was perfect and unrivalled. But in the late 1780s he lamented, on the basis of the work he had carried out for Vienna, that a number of the Florentine models had to be improved and that others had to be supplemented with some variations. Both the Grand Duke and Fabbroni felt this was unnecessary, failing to realise that these changes were due to new observations made by Fontana himself during dissections, and to the fact that the collection needed to be updated to take account of anatomical discoveries then being made. Fontana executed the changes all the same. It is interesting to note that on visiting the collection in 1786 Antonio Scarpa, professor of anatomy at the University of Pavia, was astonished to see a model of discoveries on the olfactory nerves that he had published only the previous year: "There I have found my latest work on the olfactory mechanism so well done in wax that it seemed I had before my eyes the same part of the cadaver from which I drew my figures."[22]

In 1789 the Grand Duke decided that Fontana's activity should be limited. He named Fabbroni assistant director and gave him overall responsibility for the administration of the museum (Pasta 1989, 183). Fontana's liberty as a decision-maker was now reduced to a minimum and his projects of including in the wax collection exhibits showing human pathological conditions and animal anatomies were rejected (Fontana 1783; Knoefel 1980, 24–25). In the meantime Fontana embarked on a new enterprise: demountable models of the human body in wood. These were viewed with much scepticism by both the Grand Duke and Fabbroni.

FONTANA'S IDEAS CONCERNING ANATOMICAL MODELS

If one looks at Fontana's bibliography, one notes that his last major work—a collection of essays—was printed in 1783 and that after that date he published only papers and pamphlets.[23] This is somewhat surprising since various sources suggest that he had collected a great amount of experimental data in various research areas and that he intended to publish them. One of the reasons that might have prevented him from doing so is the fact that he became obsessed by 3-D demountable models of the human body in wood, to which he devoted most of his time and personal income. His ideas on these models are evidenced principally by three sources: his own correspondence, administrative records, and reports by a few of his contemporaries.

At the end of the 1780s Fontana became dissatified with wax anatomical models because, as he wrote in a report of 1789, "they are of no use when one wants to relate parts to parts" in the living body.[24] Furthermore, he argued, the human body cannot be known unless one assembles its constituent parts; one can only understand the entire bodily structure by analysis and synthesis, i.e., by disassembling and reassembling it. Fontana resorted to the idea of a statue that could be disassembled and rebuilt by hand. Here we see the influence of the theory of knowledge that Étienne Bonnot de Condillac expounded in his *Treatise of Sensations*, which was well known to Italian scholars of the time because Condillac had spent several years (1758–67) in Italy as tutor to Ferdinand of Bourbon, later Duke of Parma (Condillac 1754).

Condillac sought to explain the origin of ideas by conducting a thought experiment concerning an animated statue of a man that had never received any sense impressions and had no innate ideas. By endowing this imaginary statue with one sense at a time (first smell and then hearing, taste, sight, and touch) he analysed the ideas generated by the sensations produced in the statue. He then compared the interactions between the sensations produced by two senses, e.g., sight and touch. By using this method Condillac concluded that the sense of touch alone was responsible for the awareness of external objects, and for the ideas of extension, distance, magnitude, and solidity. The attribution of form and colour to a given surface or volume was due to the motions of the eyes and the hand. With no previous experience of touch, a globe would appear to be a flat surface and its convexity could not be apprehended. According to Condillac, learning to see entailed coordinating sight with touch.

It was also because of Condillac's theory that Fontana wanted a model that could not only be looked at, but also touched and handled. Since a model of this kind could not be made of wax because of its fragility, he turned to wood. First he worked on a huge model of a hemisected demountable

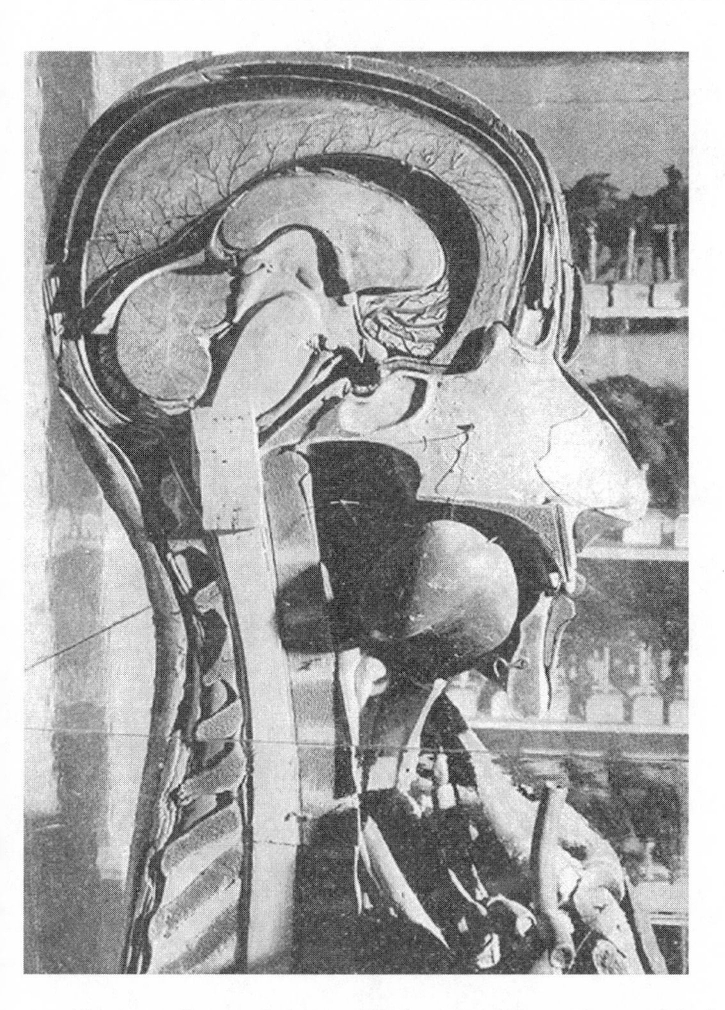

Figure 3.5 Fontana's large-scale demountable wooden model of a hemisected bust. Courtesy of Museo Zoologico 'La Specola', Florence.

bust (Fig. 3.5) made of coloured wood,[25] later on life-size human statues. According to one contemporary source he worked on six statues (Sonolet 1977, 450), but only two of them seem to have survived: one is a myological statue in bone and wood and is preserved in Florence,[26] the other is an unfinished demountable wooden statue, now to be seen in Paris (Sonolet 1977; Fontana 1980, 95–102)(Fig. 3.6). The fate of the other statues and especially of the prototype of which so much is said in the sources is unknown.

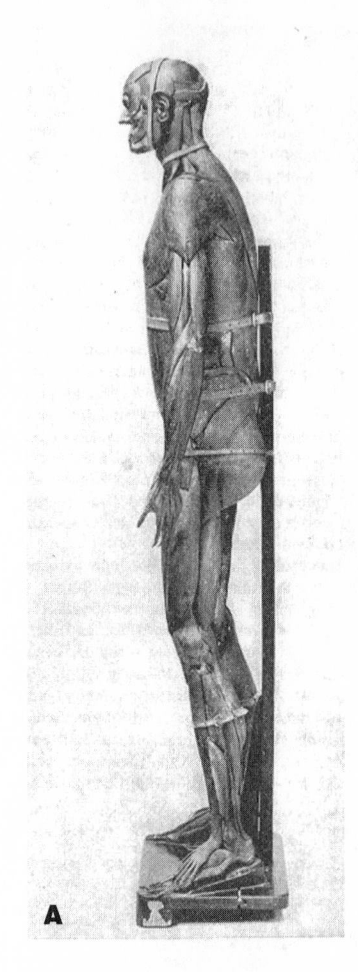

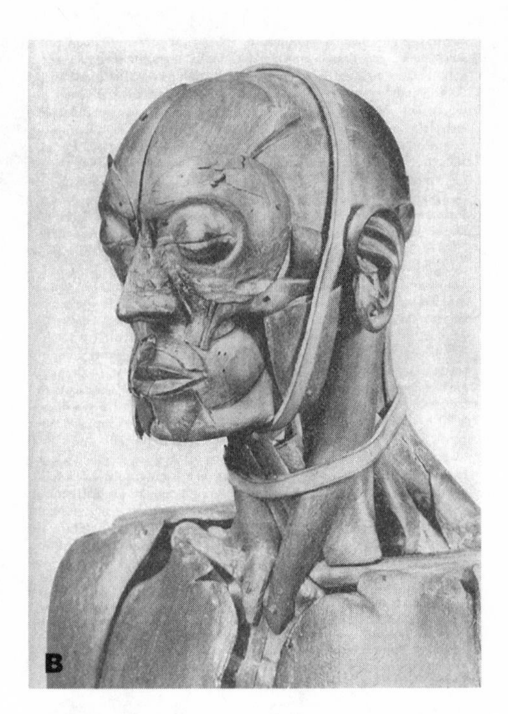

Figure 3.6 Fontana's demountable wooden statue, (A) side view, (B) detail of the head. Courtesy of Musée d'Histoire de la Médecine, Paris.

The Paris model in particular can be disassembled and reassembled, and the same applies to each organ model within the statue. In comparison to wax models and also to true anatomy, Fontana's model of the human body allows one to assemble all the individual parts in their mutual relations within the body and to multiply the views of an organ in relation to adjacent organs. This was very important to Fontana since the statue was intended to be a didactic tool replacing much of the teaching on the cadaver itself. In 1791 he wrote: "My opinion is . . . that a student who could make more progress in six months with the waxes than in six years with cadavers, would learn more in weeks with the anatomy in wood. Taking apart an entire man piece by piece and then putting him back together again, as he was, is the easiest and surely the most useful exercise for understanding the highly composite machine of the human body, for all this can be done in hours with the anatomy in wood but cannot be done at all with that in wax, still less with cadavers".[27]

The conflicts within the museum between Fabbroni and Fontana came to a head in the years 1793–94.[28] In the meantime Grand Duke Peter Leopold had moved to Vienna as Emperor Leopold II (1790–92), being succeeded in Tuscany by his son Ferdinand III. While Fabbroni was attempting to rationalise the entire management of the museum by submitting all its activities to rigorous bureaucratic control, Fontana was asking for more money, cadavers, and artisans to complete the wax collection and finish his wooden demountable statue. This meant that much material belonging to the museum was brought to Fontana's workshop in his private home and that Fabbroni could not be held responsible for its safety. Furthermore, Fabbroni thought that the waxes were complete and his own report quoted extensive passages from earlier written statements by Fontana in which the latter declared that the entire collection was perfect in all its parts and unrivalled;[29] so if changes or additions had to be made Fontana was contradicting himself. Fabbroni therefore did not see the necessity of any further works, which would have been expensive and would have required further alterations to the exhibition halls of the museum. He believed the statue a waste of time and money. Calling it a "sketch" (*abbozzo*), he wrote that, if it was not perfect, "the fault was neither of the director, nor of the artisans, but of the material", i.e., wood, which did not lend itself to being moulded like wax into all the minute details required in a work of that kind.[30]

The Crown asked Fontana to provide a list of what had still to be done and exact estimates of the time and money necessary to complete his work.[31] For his part Fontana asked that a competent external judge should be appointed to express his verdict, and proposed his friend Mascagni. Fabbroni could not but be ironic about the choice, which would inevitably favour Fontana.[32] To his own report he added a series of documents including the expert opinion of the anatomist Tommaso Bonicoli (1746–1802), since 1781 dissector at the museum.[33] This is a highly significant document since its content contrasts sharply with Fontana's ideas, showing that he was completely isolated in Florence and that support for his projects now came only from abroad.

Bonicoli mantained that "in imitating with some exactitude the elegance of animal nature, either in its colour or in its most delicate modifications and thinness" wood could not compete with wax.[34] He also stated that Fontana followed no consistent method in constructing the demountable statue and that he would say to his artisans: "Let us continue, we shall see what happens".[35] It was because of this unpredictable attitude, he argued, that many wooden demountable organs had been made and remade, and still lay about in a very crude state. But, above all, Bonicoli challenged Fontana's ideas concerning the didactic value of anatomy in wood. He mantained that at most it might be of some use to beginners, while the anatomical waxes were by far the best didactic device because they allowed medical students to save much of the time usually spent on the cadaver. He warned, however, that in the education of a skilful anatomist or surgeon waxes would never take the place of the cadaver. For this same reason Fabbroni wrote that if the cadavers provided to Fontana by the Hospital of the Innocenti and by the Hospital of Santa Maria Nuova were insufficient, he could well provide those of the Hospital of San Bonifazio, but added: "However, given all this, I do not think it is either useful or just to favour the museum, instead of the more important study that physicians and surgeons have to carry out on cadavers. If this happened it would entail the sacrifice of present instruction of undoubted value to a dubious future one."[36] It is possible that, careful not to trespass on the terrain of others, Bonicoli and Fabbroni were echoing criticism expressed by the hospital doctors, who were unwilling to share any of their teaching responsibilities in anatomy with the museum. But this is only a plausible supposition.[37]

In a long letter, dated 24 May 1794, addressed to his friend Leopoldo Marc'Antonio Caldani (1725–1813), professor of anatomy at the University of Padua, Fontana stressed the superiority of demountable anatomy over anatomical plates and wax models, maintaining however that "the cadaver alone is its original, and it is from the cadaver that demountable anatomy has derived" (Fontana 1980, 364). Nevertheless, he stressed that, as a didactic device, it could be superior to the cadaver itself: "Demountable anatomy represents all the organs of the human body as Nature formed them: that is to say with all the vessels and nerves that injections, the knife, and the microscope reveal. This state of affairs, this natural state of the human organs can be represented neither by plates nor by wax [models]. Not even the cadaver itself, as the anatomist well knows, can simultaneously display them all" (Fontana 1980, 364).

The possibility of handling the organs was another feature that made demountable anatomy preferable: "All the human organs have their particular curvatures, and so too do the muscles. Demountable anatomy represents them just as they are in Nature; they can be handled, rotated into any position, and inspected at leisure. Plates, wax [models], or a cadaver lend themselves to these operations only with difficulty and unsatisfactorily" (Fontana 1980, 365).

The relations of a single organ to its neighbouring parts could also be perceived globally: "Anatomical plates, wax [models], [and] the cadaver itself, given its rapid putrefaction and the time required for its dissection, cannot furnish complete and distinct ideas of the human viscera. The eye only views detached parts of them without ever seeing the whole created by Nature; nor can the imagination ever represent so many things together, which the eye sees only as separate, disconnected, haphazard. What idea would one have of a tree if one saw only its leaves, its flowers, its branches scattered here and there, divided and separated from each other?" (Fontana 1980, 365).

The properties of demountable anatomy as a teaching device for both medical students and the lay public are made clear in a few sentences: "I finally mention the great usefulness of being able to consult the demountable anatomy whenever one wishes, in whatever season, with no need for a cadaver, without the risk of morbid infections, without the vexation of noxious odours, in short without distastefulness of any kind, a device which can be studied by even the frailest persons not destined for medicine, and in consequence the object of research at once useful and enjoyable. There is no one who does not wish to know his own body, to discover how it is made: but the horror of the cadaver repels even the most curious and resolute persons" (Fontana 1980, 366).

What Fontana wanted to achieve with his wooden statues was an artificial anatomy which imitated the dissecting process with none of the limitations and perils of true dissection. In 1795 he informed the newly-founded *École de Santé* in Paris that he was constructing a "revolutionary" wooden anatomical model (Lemire 1990, 59). By 1798 the prototype of this model was almost finished, and on 27 September of the same year he received the approval of the Grand Duke to build a movable frame on which the statue, covered with real human skin, could be placed for public demonstrations.[38] In an enthusistic report sent to the *Société des Sciences et des Arts,* Georges Cuvier wrote that the skin had hairs and nails (Lemire 1990, 59). In the early years of the nineteenth century all important visitors to Florence had to see the statue;[39] it took one day to take apart and two days to reassemble.

Let me stress that what Fontana considered important in his model was the fact that the viewer could put his hands on the statue, take it apart, touch, and inspect the various organs, and, most of all, put it together again. This could not, of course, be achieved with fragile waxes. According to Fontana it was especially through this 'putting together' that knowledge of the structure of the human body could be acquired.

On 1 July 1796 General Napoleon Bonaparte visited the museum. The next day he informed the Directory that he had seen an anatomical wax collection certainly worth possessing because "of its great usefulness in the teaching of a science so essential to humanity" (Sonolet 1977, 446). He added that the museum's director was willing to duplicate the collection. Aware that, during their Italian campaigns, the French had seized many valuable artistic and scientific objects and sent them to France, Fontana's willingness to duplicate the collection was inspired not only by the desire to increase his reputation as an anatomist and to regain power over the museum, but also by the fear that the entire collection of waxes might be confiscated as the property of an enemy, namely the Grand Duke of Tuscany. In all the troubled years that followed the Grand Duke's flight, Fontana's general conduct was dominated by one thought alone: to save the various collections of the museum and to keep them in Florence.[40]

Negotiations over duplicates of the waxes and of the wooden demountable statue were held between Fontana and the French generals and administrators who succeeded each other in Florence.[41] They were, however, often interrupted by military events and sudden changes in the system of government. For instance, after the defeat of the French army by the Austrian–Russian coalition in June 1799, the Aretini—who had remained faithful to the Grand Duke—entered Florence in July 1799 and started persecution of all those inhabitants who had sympathised or collaborated with the French. Fontana was among them. His house was looted, many of his books and manuscripts were destroyed, he himself was mistreated, arrested, and imprisoned. While in jail, instructions to stop work on the wooden statue were delivered to the museum.[42] Later he was released but exiled.

When allowed to return to Florence, and when the French also returned, Fontana resumed his work on the models though physically weakened and embittered by what had happened. He nonetheless refused a pressing invitation to move to France in order to train modellers and carry out his work in that country (Sonolet 1977, 447), because he felt that his place was in Florence as guardian of the collections which he had assembled over so many years and with which he identified much of the meaning of his life.

CONCLUSION

In his vast literary production Goethe only once mentions the Florentine collection of anatomical models. He does so in the 1832 *Promemoria* on *Plastic Anatomy* quoted at the beginning of this paper. From this text alone it is not clear whether Goethe actually saw the Florentine waxes during his second visit to Florence from 29 April to 11 May 1788. However, on the basis of some indirect evidence (Mazzolini 1992, 265–67) and analysis of a long section of chapter 3 of Goethe's *Wilhelm Meisters Wanderjahre* (*Wilhelm Meister's Travels*) of 1829, it seems that he did (Goethe 1989, 602–7). In that section Goethe describes how, during a visit to a museum, Wilhelm is taken by the caretaker, an old savant and artist, into a secluded area of the museum, the description of which closely resembles the halls devoted to the wax models in Florence. The most striking aspect of the description is the fact that the old savant and artist takes Wilhelm also to his own home where he has set up a workshop and where not only wax models are produced, but also models in wood.

The temptation to identify the old caretaker with Fontana is very strong, because of both the contextual similiarities between Fontana's life and that described in the novel, and the ideas expressed by the old savant and artist: "You shall find out shortly that building up is more instructive than tearing down, linking is more instructive than separating, to enliven what is dead is more important than to kill further what has already been killed" (Goethe 1989, 604).

Although Fontana's wooden models had little direct impact upon contemporary teaching practices, it is historically significant that, by labouring for decades on volumetric representations in wax of the human body which aroused great public appreciation and admiration, Fontana came to recognise their theoretical and practical limitations. He consciously moved from one conception of the anatomical model as something only to be contemplated with the eye, to another conception of it as the object of a process of deconstruction and reconstruction by hand. In 1775, while describing the display of the wax models in their glass cases, he stated that "everything is seen without the need to open anything or to touch", but in the 1790s, when discussing his wooden model, he emphasised the sense of touch.[43] In short, he moved from a conception in which knowledge transfer was mainly accomplished by seeing, to one in which it was supplemented by doing. Rebuilding the statue became the cornerstone of his new enterprise. Wax models, he maintained, were like the scattered parts of a disassembled clock and would give no clue as to their meaning unless assembled.[44] In anatomy, rebuilding could be achieved only by using his wooden model. This shift brought with it a different conception of the functions of public scientific institutions like museums. But Fontana's projects to house the Florentine museum in a completely new building, to turn it also into an active centre of research, and to give the public greater access to demonstrations in natural philosophy and anatomy were pushed aside by Peter Leopold and his son Ferdinand III, not only because of the costs involved, but also because the grand dukes were satisfied with the museum as it stood, under their direct control, as a showcase for their policies of reform. All that Fontana could do was withdraw into his own workshop, and stubbornly continue with his experiments in physics and chemistry as well as his work on the wax models and his wooden statues, the importance of which was much better appreciated by the French generals than by his Tuscan contemporaries or fellow anatomists.

NOTES

1. For general reviews of waxworks and anatomical models in art and science, see Boschung 1980; Freedberg 1989, 192–243; and Lemire 1990. For histories of the Florentine wax collection, see Schiff 1928; Castaldi 1947; Belloni 1959–60; Knoefel 1978; and the catalogues by Lanza et al. 1979; and Düring et al. 1999. For the Viennese collection, see Allmer and Jantsch 1965; and Schmidt 1996 and the literature cited therein. For the models in Cagliari, see Castaldi 1947; and Cattaneo 1970; and for those in Pavia, Knoefel 1979.

2. For works by and on Fontana, see Knoefel 1980; and for biographies, Knoefel 1984; and Mazzolini 1997.

3. For histories of the Florentine museum, see Schiff 1928; Azzaroli 1977; and Contardi 1996.

4. Knoefel 1984, 252. Desgenettes (1793, 237) wrote that the full-size statues numbered twenty-four. Carl Otto wrote that during his visit in 1819 the models occupied fifteen rooms (Otto 1825, 1: 104).

5. For Fontana's proposal to Peter Leopold to hold public lectures and create a learned society, see Fontana's manuscripts in the Biblioteca di Casa Rosmini, Rovereto, 28 B. 110, file 2. On the Accademia del Cimento, briefly revived in 1801, see Knoefel 1984, 72.

6. For an overview of the reforms of Tuscan cultural institutions under Peter Leopold, see Pasta 1996.

7. The wax collection of comparative anatomy is now preserved in 'La Specola', of pathology in the Museo dell'Istituto di Anatomia Patologica, and of botany in the Istituto di Botanica and the Istituto Tecnico 'Galileo Galilei'. During the nineteenth century the most important modellers were Francesco Calenzuoli, Luigi Calamai, Giuseppe Ricci, Giovanni Lusini, and Egisto Tortori.

8. For a more detailed description of the techniques used in Florence, see Azzaroli 1977, 16–18; and Lanza et al. 1979, 50–53.

9. For example, Fontana's own illustrations of the structure of nerves (Fontana 1781, pls 3 and 4), and Johann Gottfried Zinn's illustration of the ciliary processes (Zinn 1755, pl. 2, fig. 3). See also Mazzolini 1991, 131–32. Wax models of microscopic structures were already mentioned by [Ceruti] 1775, 31.

10. Fontana to Francesco Chiusole, 16 December 1795, in Adami 1905, 10–11.

11. Fontana to Leopoldo Marc'Antonio Caldani, 30 October 1791, in Fontana 1980, 319.

12. Archivio di Stato, Florence, Imperiale e Reale Corte Lorenese, 119: "Memoria presentata in Vienna a S. M. l'Imperatore Leopoldo". English translation by Knoefel 1984, 344.

13. On Mascagni's contribution to the Viennese wax collection, see Schmidt 1996 and the literature cited therein.

14. For Fabbroni's biography and activity at the museum, see Pasta 1989.

15. Biblioteca Universitaria, Bologna, ms 2096.4, Fontana to Canterzani, 16 March 1784.

16. Ibid.

17. Ibid., Fontana to Canterzani, 12 February 1785.

18. Ibid., Fontana to Canterzani, [May] 1785.

19. Ibid., copy of Canterzani to Fontana of 17 January 1785.

20. At least two artists from the Trentino worked on wooden models in Fontana's private workshop and were therefore not recorded in the official documents. The first was the sculptor Giovanni Insom (Perini 1852, 2: 252); the second the modeller and painter Domenico Udine (Bibliothek des Tiroler

Landesmuseums Ferdinandeum, Innsbruck, ms 1104, "Del Pittore Dom[eni]co Udine", fol. 779–82).

21. Istituto e Museo di Storia della Scienza, Florence, "Segue il Giornale del R. Museo 1784 Tutto ottobre 1785".

22. For Scarpa's letter to Gregorio Fontana (Felice's brother) of 16 September 1786, see Scarpa 1938, 112; and Knoefel 1978, 333. Scarpa's own drawing of the olfactory nerves, and the plate of his *Anatomicarum annotationum liber secundus de organo olfactus praecipue* of 1785 (Scarpa 1785) are reproduced in Mazzolini 1991, 121–23.

23. Interestingly, these projects were resumed by scholars working in the museum in the early nineteenth century.

24. Istituto e Museo di Storia della Scienza, Florence, Filza di negozi, 1789, fol. 428r–31v (quotation from fol. 429v).

25. Museo Zoologico 'La Specola', Florence. See also Fontana 1980, 317–21.

26. At the Museo Zoologico 'La Specola', Florence. See also Martelli 1977, 117–18, 121.

27. Archivio di Stato, Florence, Imperiale e Reale Corte Lorenese 119: "Memoria del Cav. Fontana in sua giustificazione".

28. The main documents of the dispute are preserved in both the Istituto e Museo di Storia della Scienza, Florence, Filza di negozi, 1793 and 1794, and the Archivio di Stato, Florence, Imperiale e Reale Corte Lorenese, 378.

29. Fabbroni's report is dated 26 April 1794: Istituto e Museo di Storia della Scienza, Florence, Filza di negozi, 1794, fol. 95r–108r.

30. Ibid., fol. 104r.

31. On 25 October 1793: Istituto e Museo di Storia della Scienza, Florence, Filza di negozi, 1793, documents 130–35, fol. 357.

32. Mascagni's report is preserved in the Archivio di Stato, Florence, Imperiale e Reale Corte Lorenese, 378. On this report and Fabbroni's reaction, see Coturri 1977; and especially Ruffo 1996.

33. Istituto e Museo di Storia della Scienza, Florence, Filza di negozi, 1794, fol. 68r–70r.

34. Ibid., fol. 68r.

35. Ibid., fol. 68v.

36. Ibid., fol. 98v.

37. Before moving to the museum, Bonicoli had been dissector at the Hospital of Santa Maria Nuova. A number of anatomical waxes made by Fontana and showing pathological conditions were transferred by Fabbroni to that institution for the instruction of physicians and surgeons: Istituto e Museo di Storia della Scienza, Florence, Filza di negozi, 1792, fol. 111r.

38. Ibid., 1798, fol. 107. See also ibid., 1801, fol. 118.

39. Ibid., 1801, fol. 97, 113, 123, 125.

40. Biblioteca di Casa Rosmini, Rovereto, 28 B., file 110.

41. Martelli 1977, 122–23; Sonolet 1977. These historical reconstructions should be supplemented by Fontana's own account of the difficulties encountered in carrying out the work for the French. See Biblioteca Comunale, Trento, ms 910: "Ristretto della Condotta del Fontana relativa alle Commissioni Anatomiche date allo stesso in diversi tempi, da diversi Sovrani".

42. Istituto e Museo di Storia della Scienza, Florence, Filza di negozi, 1799, fol. 44r–v.

43. [Ceruti] 1775, 31. This anonymously published guide to the Florentine museum was chiefly drawn from Fontana's manuscripts by the abbé Ceruti; see Knoefel 1980, 41–42.

44. Fontana used this analogy both in his report of 1789 (Istituto e Museo di Storia della Scienza, Florence, Filza di negozi, 1789, fol. 428r–31v), and in the justificatory memoir cited in note 27.

REFERENCES

Adami, Casimiro. 1905. *Di Felice e Gregorio Fontana scienziati pomarolesi del secolo XVIII*. Rovereto: Ugo Grandi.

Albinus, Bernardus Siegfried. 1747. *Tabulae sceleti et musculorum corporis humani*. Leiden: Verbeek.

Allmer, Konrad, and Marlene, Jantsch,. 1965. *Katalog der Josephinischen Sammlung anatomischer und geburtshilflicher Wachspräparate im Institut für Geschichte der Medizin an der Universität Wien*. Graz: Böhlau.

Armaroli, Maurizio, ed. 1981. *Le cere anatomiche bolognesi del Settecento*. Bologna: Università degli Studi di Bologna.

Azzaroli, Maria Luisa. 1977. "La Specola: The Zoological Museum of Florence University." In *La ceroplastica nella scienza e nell'arte. Atti del I Congresso Internazionale, Firenze, 3–7 giugno 1975*, vol. 1: 1–22. Florence: Olschki.

Belloni, Luigi. 1959–60. "Anatomia plastica. I. Gli inizi. II. Le cere Bolognesi. III. Le cere Fiorentine." *Symposium Ciba* 7: 229–33; 8: 84–87, 129–32.

Bernoulli, Johan. 1777. *Zusätze zu den neuesten Reisebeschreibungen von Italien*..., vol. 1. Leipzig: Caspar Fritsch.

Boschung, Urs. 1980. "Medizinische Lehrmodelle. Geschichte, Techniken, Modelleure." *Medita* 10, no. 5: ii–xv.

Castaldi, Luigi. 1947. *Francesco Boi, 1767–1860, primo cattedratico di anatomia umana a Cagliari, e le cere anatomiche fiorentine di Clemente Susini*. Florence: Olschki.

Cattaneo, Luigi. 1970. *Le cere anatomiche di Clemente Susini dell'Università di Cagliari*. Cagliari: Fratelli Stianti.

[Ceruti] 1775. *Saggio del Real Gabinetto di Fisica e di Storia Naturale di Firenze*. Rome: Giovanni Zempel.

Cheselden, William. 1733. *Osteographia; or, The Anatomy of the Bones*. London.
Condillac, Étienne Bonnet de. 1754. *Traité des sensations*, 2 vols. London and Paris: de Bure.
Contardi, Simone. 1996. "Unità del sapere e pubblica utilità: Felice Fontana e le collezioni di fisica dell'Imperiale e Regio Museo di Firenze." In *La politica della scienza. Toscana e stati italiani nel tardo Settecento*, ed. Giulio Barsanti, Vieri Becagli, and Renato Pasta, 279–93. Florence: Olschki.
Coturri, Enrico. 1977. "Paolo Mascagni e i suoi rapporti con le cere della Specola." In *La ceroplastica nella scienza e nell'arte. Atti del I Congresso Internazionale, Firenze, 3–7 giugno 1975*, vol. 1: 41–48. Florence: Olschki.
Desgenettes, Nicolas-René. 1793. "Réflexions générales sur l'utilité de l'anatomie artificielle; et en particulier sur la collection de Florence, et la nécessité d'en former de semblables en France." *Journal de Médecine, Chirurgie et Pharmacie* 94: 162–76, 233–52.
Düring, Monika von, Georges Didi-Huberman, and Marta Poggesi, with photographs by Saulo Bambi. 1999. *Encyclopaedia Anatomica: A Complete Collection of Anatomical Waxes*. Cologne: Taschen.
Ferber, Johann Jakob. 1773. *Briefe aus Wälschland über natürliche Merkwürdigkeiten dieses Landes*.... Prague: Wolfgang Gerle.
Fontana, Felice. 1775. *Descrizione ed usi di alcuni stromenti per misurare la salubrità dell'aria*. Florence: per Gaetano Cambiagi.
———. 1781. *Traité sur le venin de la vipere*..., 2 vols. Florence. Et se trouve à Paris chez Nyon l'Ainé; à Londres chez Emsley.
———. 1783. *Opuscoli scientifici*. Florence: Gaetano Cambiagi.
———. 1980. *Carteggio con Leopoldo Marc'Antonio Caldani, 1758–1794*, ed. Renato G. Mazzolini and Giuseppe Ongaro. Trento: Società di Studi Trentini di Scienze Storiche.
Freedberg, David. 1989. *The Power of Images: Studies in the History and Theory of Response*. Chicago: University of Chicago Press.
Goethe, Johann Wolfgang. 1964. *Aufsätze, Fragmente, Studien zur Morphologie*, ed. Dorothea Kuhn (*Die Schriften zur Naturwissenschaft*, section 1, vol. 10). Weimar: Hermann Böhlaus Nachfolger.
———. 1989. *Wilhelm Meisters Wanderjahre*, ed. Gerhard Neumann and Hans-Georg Dewitz (*Sämtliche Werke*, vol. 10). Frankfurt am Main: Deutscher Klassiker-Verlag.
Haller, Albrecht von. 1743–54. *Iconum anatomicarum partium corporis humani*, 8 vols. Göttingen: A. Vandenhoeck.
Hunter, William. 1774. *Anatomia uteri humani gravidi tabulis illustrata*. Birmingham: John Baskerville.
Ingen-Housz, John. 1779. *Experiments upon Vegetables*.... London: P. Elmsly and H. Payne.
Knoefel, Peter K. 1978. "Florentine anatomical models in wax and wood." *Medicina nei Secoli* 15: 329–40.

———. 1979. "Antonio Scarpa, Felice Fontana, and the wax models for Pavia." *Medicina nei Secoli* 16: 219–34.

———. 1980. *Felice Fontana, 1730–1805: An Annotated Bibliography*. Trento: Società di Studi Trentini di Scienze Storiche.

———. 1984. *Felice Fontana: Life and Works*. Trento: Società di Studi Trentini di Scienze Storiche.

Lanza, Benedetto, Maria Luisa Azzaroli Puccetti, Marta Poggesi, and Antonio Martelli, with photographs by Liberto Perugi. 1979. *Le cere anatomiche della Specola*. Florence: Arnaud.

Lemire, Michel. 1990. *Artistes et mortels*. Paris: Chabaud.

Martelli, Antonio. 1977. "La nascita del Reale Gabinetto di Fisica e Storia Naturale di Firenze e l'anatomia in cera e legno di Felice Fontana." In *La ceroplastica nella scienza e nell'arte. Atti del I Congresso Internazionale, Firenze, 3–7 giugno 1975*, vol. 1: 103–33. Florence: Olschki.

Mazzolini, Renato G. 1991. "Schemes and models of the thinking machine (1662–1762)." In *The Enchanted Loom: Chapters in the History of the Neurosciences*, ed. Pietro Corsi, 68–143, 198–200. New York: Oxford University Press.

———. 1992. "L'opera del fisiologo Felice Fontana nella cultura tedesca del secondo Settecento." In *Deutsche Aufklärung und Italien*, ed. Italo Michele Battafarano, 251–78. Bern: Lang.

———. 1997. "Fontana, Gasparo Ferdinando Felice." In *Dizionario biografico degli italiani*, vol. 48: 663–69. Rome: Istituto della Enciclopedia Italiana.

Otto, Carl. 1825. *Reise durch die Schweiz, Italien, Frankreich, Grossbritannien und Holland*..., 2 vols. Hamburg: August Campe.

Pasta, Renato. 1989. *Scienza, politica e rivoluzione. L'opera di Giovanni Fabbroni (1752–1822), intellettuale e funzionario al servizio dei Lorena*. Florence: Olschki.

———. 1996. "Scienza e istituzioni nell'età leopoldina. Riflessioni e comparazioni." In *La politica della scienza. Toscana e stati italiani nel tardo Settecento*, ed. Giulio Barsanti, Vieri Becagli, and Renato Pasta, 3–34. Florence: Olschki.

Perini, Agostino. 1852. *Statistica del Trentino*, 2 vols. Trento: Perini.

Ruffo, Patrizia. 1996. "Paolo Mascagni e il Real Museo di Fisica e Storia Naturale di Firenze." In *La scienza illuminata. Paolo Mascagni nel suo tempo (1755–1815)*, ed. Francesca Vannozzi, 241–51. Siena: Nuova Immagine Editrice.

Scarpa, Antonio. 1785. *Anatomicarum annotationum liber secundus de organo olfactus*. Pavia: Monastero S. Salvatore.

———. 1938. *Epistolario*, ed. Guido Sala. Pavia: Società medico-chirurgica.

Schiff, Ugo. 1928. "Il Museo di Storia Naturale e la Facoltà di Scienze Fisiche e Naturali di Firenze." *Archeion* 9: 88–95, 290–324.

Schmidt, Gabriela. 1996. "Sul contributo di Paolo Mascagni alla collezione viennese delle cere anatomiche nel Josephinum." In *La scienza illuminata*.

Paolo Mascagni nel suo tempo (1755–1815), ed. Francesca Vannozzi, 101–9. Siena: Nuova Immagine Editrice.

Sonolet, Jacqueline. 1977. "À propos d'un mannequin anatomique en bois: Napoléon Bonaparte et Felice Fontana." In *La ceroplastica nella scienza e nell'arte. Atti del I Congresso Internazionale, Firenze, 3–7 giugno 1975*, vol. 1: 443–58. Florence: Olschki.

Stendhal [Marie-Henry Beyle]. 1977. *Diario*, ed. E. Rizzi, 2 vols. Turin: Einaudi.

Zinn, Johann Gottfried. 1755. *Descriptio anatomica oculi humani iconibus illustrata*. Göttingen: A. Vandenhoeck.

CHAPTER FOUR

Fish and Ships: Models in the Age of Reason

Simon Schaffer

> Emerson has no patience with analysis. A Ship to him is the Paradigm of the Universe. All the possible forces in play are represented each by its representative sheets, stays, braces, and shrouds and such—a set of lines in space, each at its particular angle. Easy to see why sea-captains go crazy—godlike power over realities so simplified (Pynchon 1997, 220).

This chapter is concerned with models of electric fish and of ships' hulls made in late eighteenth-century London to manifest rational principles in the artificial and the natural realms. The two stories told here relate to the establishment of new sciences in the decades around 1800. Models of electric fish were decisive in the construction of the electric pile, so helped spawn newfangled chemical physiology and electrodynamics. Elegant trials with ship models showed how rational mechanics could join a new alliance of military and academic sciences. Furthermore, both projects were entangled with important philosophical puzzles, newly pressing in the wake of the work of David Hume. One problem involved the possibility of inferring prescriptions from descriptions, another challenged the legitimacy of projecting from microscopic to macroscopic systems. Solutions to the scale problem—the puzzle of generalising from intimately manageable trials to full-size realities—might establish model-builders' own rights and skills. It was important to navigate between an overambitious claim that natural philosophy could simply govern commercial schemes and the worrying charge that natural philosophers were incompetent to judge the worlds of trade and industry. There was thus a politics implicated in modelling. Mimicry of artificial or natural systems often involved remodelling the social order of production and design. The chapter traces ways in which these imitative and normative structures allowed moves between the making of representations and the exercise of power.

The novelist Thomas Pynchon's recent fictionalised image of the eighteenth-century mathematician William Emerson rings true more generally. Ingenious simplification seemed then to offer "godlike power over realities". Enlightened measures required manageable spaces where otherwise unruly phenomena could be directed at will by expert reason often backed with force (Lindqvist 1990; Edney 1994; Alder 1997, 56–86). In Britain, such spaces included those run by industrial managers and the state, patrons of metropolitan trades in instrument design and modelling (Ogborn 1998, 194–98; Ashworth 1998b; Millburn 1988). Inside these institutions and on the public stages they sponsored, modelling might capture the artificial world of mechanics and the natural economy of creation. Mechanics knew these worlds were not orderly systems well mimicked by wood and string, but modelling made them seem so or suggested how they might become so. Warships were not designed by slavish imitation of principles of rational geometry and enlightened mechanics. Yet trying ship models showed how shipwrights might change their ways and become subservient to reason. So in the natural order too, the enlightened natural philosophy which saw electric action in lightning storms and stingrays got its authority from building viable models, then argued for lightning-rods or the truths of galvanism. When model-makers designed, tried, and showed their well-behaved models, they could claim the right to govern and represent the macroscopic systems these models represented.

British engineers of the period were keen on trial models, in contrast with French mathematical analysis or American exploratory projects (Smith 1977; Kranakis 1997). Test models were favoured by enlightened instrument-makers turned civil engineers, such as James Watt and John Smeaton. In the 1760s the entrepreneurial Watt conducted repairs at Glasgow University to "a workable model of a Steam Engine which was out of order". Such models were crucial for his engine trials, in the law suits which plagued Watt's patent claims, and in his industrial plans for rational maximisation of profit and discipline (Hills 1989, 51–54; Robinson and McKie 1969, 254, 364; Miller 2000, 5–7). Smeaton was equally entrepreneurial, ambitious for philosophical status, and legally embroiled. He used small-scale experimental machines to estimate the friction and efficiency of his great mill wheels, and applied the same model technique in 1762 for experiments run on a

London fishpond by the newly formed Society of Arts to assess the best design for ship hulls (Reynolds 1979; Smith 1973–74, 181, 184; Robinson and McKie 1969, 429; Harley 1991). Watt showed the originality of his steam-engine designs, Smeaton proved overshot water-wheels were more efficient, and the Society of Arts demonstrated the speed of conventional ships, using models which made natural principles manifest and their own skills authoritative (Morton 1993; Stewart 1998, 275–76; Miller 1999, 188–91). The establishment of such authority was not easy. In the 1770s, the Cambridge mathematician George Atwood commissioned a pulley machine, which by minimising friction could show the truth of Cambridge mechanics. Atwood used this demonstration model to evince the fallacies of modish ideas about self-moving matter, espoused by republican agitators such as Joseph Priestley and later Tom Paine. Atwood's loyalty was rewarded with a lucrative government job when his student, William Pitt, became prime minister in 1784. Atwood also reckoned that models like those of Smeaton should never be used to challenge the academic truths of Newtonian mechanics. The omnipresence of friction in real world mills would frustrate any inferences from their working to the cloistered principles of established natural philosophy (Schaffer 1994, 173–74).

Since mechanical reasoning was supposedly governed by scale-insensitive geometry, it seemed hard to understand why projection to larger systems ever failed. The Wearside mathematician William Emerson, Pynchon's exemplary protagonist, treated this scale problem in a best-selling mechanics textbook (1758), which reached a fourth edition by 1794 (Hamilton 1952). Emerson referred his readers to the origin of classical mechanics, Galileo's *Two New Sciences* (1638). Galileo had used the scale puzzle as the basis of his new science of the strength of materials, formulating principles of levers, pulleys, and beams which would explain the role played by materials' own weight in large-scale structures (Galileo 1974, 11–14). These problems of reasoning from scale models in new materials still haunted engineering in the late eighteenth century (Skempton and Johnson 1962). In 1798, inspired by the success of iron bridges in the Midlands and the North, plans were launched for a vast iron bridge across the Thames to allow larger ships into the crowded wartime London docks. Parliament set up a committee to investigate the plan. Atwood and Watt were members, as was the Royal Military School professor Charles Hutton. The engineer Thomas Telford constructed a model of an iron bridge spanning 600 feet for show at the Royal Academy and before the Committee (Skempton 1980). Atwood commissioned his own experimental model arch (Fig. 4.1), then derived numerical data to guide unprincipled engineers. This model of brass and wood convinced him his Newtonian mechanics could explain all the properties of complex stone and iron arches. He claimed engineers had struck on satisfactory designs by luck. Their "fortuitous arrangements" were helped by friction, a force aiding cohesion of the most rickety and untheorised structures. So models and rational mechanics demonstrated that "practical artists" had always followed "rules established by custom, without being referred to any certain principle" (Atwood 1801, vi–vii and supplement, vii–viii).

In these enterprises of 1790s engineers and analysts, the rival claims of custom and principle were fought out with experimental models. Some showed that artisan custom was a legitimate authority; others that rational principles could govern mere tradition. Hutton, for example, considered that Atwood's trials on bridge models, "though common and well-known to other writers, he unfortunately fancied to be new discoveries". The conservative philosopher Samuel Taylor Coleridge saw Atwood's bridge model on show at Kew Observatory. Reflecting on enlightenment's failings, Coleridge used the apparent weakness of the mathematician's elegant set-up to suggest the fallacies of rationalist analysis and to laud custom and inspiration (Hutton 1815, 1: 189–90; Coleridge 1956–59, 4: 588; Coleridge 1969, 496–97). These contrasts between customary conduct and rational principles were the stuff of antagonistic politics and philosophy, especially in the wake of Edmund Burke's *Reflections on the Revolution in France* (1791). Burke damned the revolutionaries as deluded utopians whose model societies violated custom and would fail in practice (Golinski 1992, 178). Those who countered Burke often appealed to the rationality of mechanical models. Immanuel Kant's 1793 essay on theory and practice argued against Burke that enlightened reason could indeed be applied to mundane affairs. Kant offered a critique of the Burkean account of "the British constitution, of which the people are so proud that they hold it up as a model for the whole world" (Kant 1991, 83). He used rational mechanics to show that the apparent feebleness of high theory was a consequence of incomplete theoretical work:

> All of us would ridicule the empirical engineer who criticised general mechanics or the artilleryman who criticised the mathematical theory of ballistics by declaring that while the theory is ingeniously conceived, it is not valid in practice, since experience in applying it gives results quite different

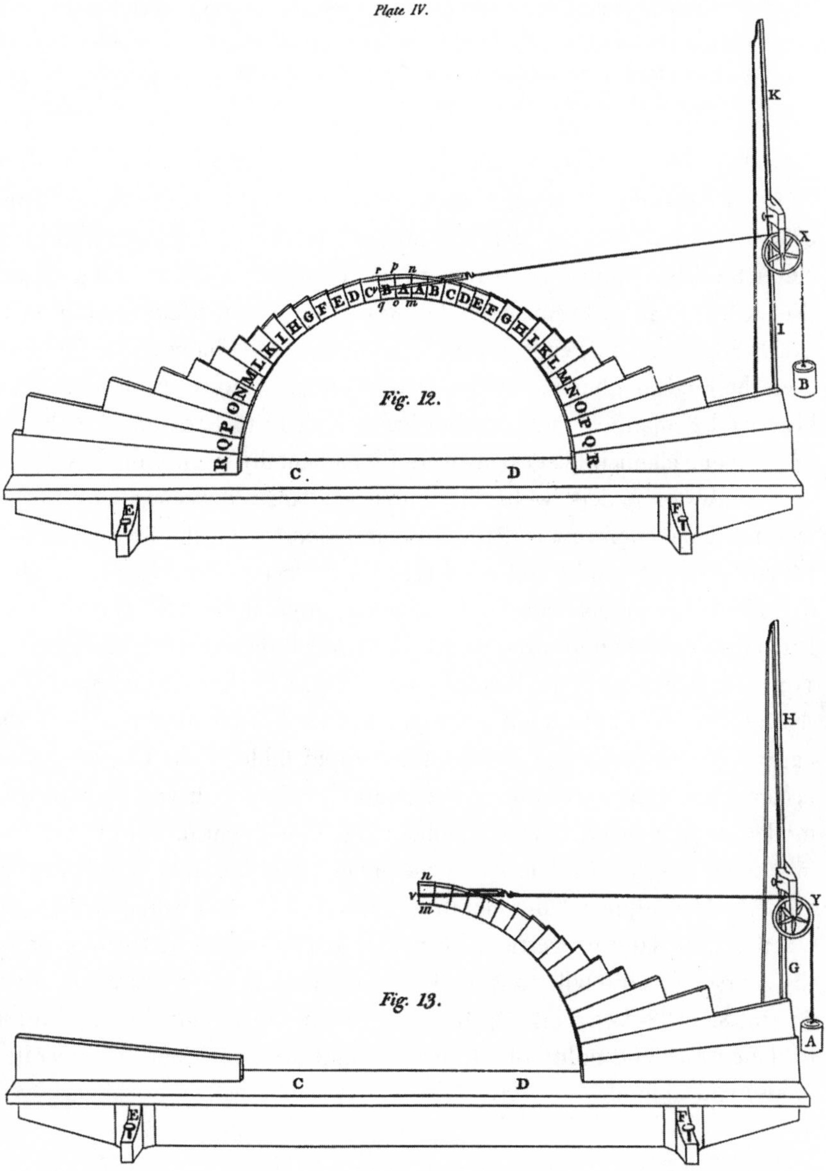

Plate IV.

Fig. 12.

Fig. 13.

Figure 4.1 George Atwood's experimental model of the load-bearing capacity of London Bridge, built in 1803 by Matthew Berge. Atwood used balance weights held over the pulley to estimate the forces at work in different arch designs. Samuel Taylor Coleridge saw this model at Kew Observatory in 1811. Source: George Atwood, *Dissertation on the Construction and Properties of Arches*, supplement (London, 1804), pl. 4. By permission of the Syndics of Cambridge University Library.

> from those predicted theoretically. For if mechanics were supplemented by the theory of friction and ballistics by the theory of air resistance, in other words if only more theory were added, these theoretical disciplines would harmonise quite well with practice (Kant 1991, 62).

"If only more theory were added"—the slogan of enlightened reason. Tom Paine's *Rights of Man* (1791) was the most radical answer to Burke. Paine called his pamphlet "my *political* bridge". He linked his development of better models of political order with his long-term work in bridge design (Keane 1995, 282). From the mid-1780s Paine sought backing for a novel iron bridge, first in Pennsylvania, then across the Thames, the Seine, or even the English Channel. He made spectacular demonstration models for his friend Benjamin Franklin, then brought them to Paris and London in 1787. Paine claimed his designs were fundamentally *natural* models, based on the laws of spiders' webs. He toured Midlands ironworks with Burke, met academic engineers in Paris, sought support from the Royal Society's president Joseph Banks and persuaded the Ordnance Board's gunfounders to build a huge model bridge in London for public show. The scheme failed. Patrons were more impressed by Telford's models than those of Paine (Kemp 1977–78; James 1987–88; Keane 1995, 267–82). But when he composed his *Age of Reason* in a Paris jail in 1793, Paine used mechanical models as the basis of his incendiary assault on the "stupid Bible of the Church". God became "the great mechanic of the creation" whose cult was embodied in models such as orreries, which Paine had seen with enthusiasm in London in the late 1750s. "We know that the greatest works can be represented in model", Paine argued, "and that the universe can be represented by the same means . . . The same properties of a triangle that will demonstrate upon paper the course of a ship will do it on the ocean" (Keane 1995, 42; Paine 1991, 73–76, 82, 188–89). This 'age of reason' involved the rational application to the cosmos and polity of allegedly natural principles first demonstrated in mechanical models.

FISH

For the London showmen who had so attracted Paine, electricity was a pre-eminent natural power. The aesthetics of electricity corresponded to those theorised by Burke in his *Philosophical Enquiry* on the sublime and the beautiful (1757). Models could show how sublime phenomena might be managed in the macrocosm. The invention of the Leyden jar in the mid-1740s, Benjamin Franklin's account of its operation, and his arguments for the possibility of protecting against the terrors of lightning with tall, pointed, well-grounded metal rods reinforced electrical modelling (Cohen 1990, 142). Franklin hung a pair of brass scales, one pan of which was electrified, from a wooden beam. This mimicked thunder clouds. Twist the balance so that its pans would slowly spin and descend. If hung over some vertical metal instrument, the array would spark violently. But if a pin were attached to the top of the instrument, or one's finger were slowly brought towards the array, the model cloud would slowly discharge without sparking (Franklin 1941, 212–36, 406). In early 1753 at St Johns, Antigua, a capital of the West Indian sugar plantocracy, one of Franklin's allies demonstrated how an electrical flash could be "made to strike a small house and dart towards a little lady sitting upon a chair, who will, notwithstanding, be preserved from being hurt; whilst the image of a negro standing by and seeming to be further out of danger will be remarkably affected by it". Thunder houses, electric models of churches, and of powder magazines, were also on show (Warner 1997). A London draper, William Henly, backer of Franklin's theory, performed model trials in spring 1772 using a thunder house "which I thought a nearer resemblance of the operations of nature on these occasions". To counter the worry that clouds move and float, while the prime conductors he used were static, he modelled clouds with a copper-coated bullock's bladder. Other electricians used metal-plated placentas for this trick (Franklin 1959–, 18: 229–31 and 19: 119–21; Henly 1774, 135, 142, 145–46). The ambitious society painter and electrical philosopher Benjamin Wilson contested Franklin's rod recipe. He built his own models to show that points were dangerous because they would pull electricity from any passing cloud (Wilson 1764).

Franklin was in the imperial capital as Pennsylvania's lobbyist and habitué of philosophical and commercial clubland. He explained why models of clouds with cotton rags and metal rods, and the imitation of high buildings by his own finger, should compel his confrères. "It has also been objected that from such electric experiments, nothing certain can be concluded as to the great operations of nature, since it is often seen that experiments which have succeeded in small, in large have failed. It is true that in Mechanics this has sometimes happened. But ... we owe our first knowledge of the nature and operations of lightning to observations on such small experiments" (Franklin 1959–, 19: 253). This knowledge became politically crucial. Between 1763 and 1768 the Ordnance Board arranged to shift its large powder magazines, whence the navy was supplied, from Greenwich down the Thames to Purfleet. In 1771 the government imposed new safety regulations on the magazines (Hogg 1963, 1: 108–9). In May 1772 both Wilson, hired as the Ordnance Board's artist and a well-supported royal client, and Franklin, the Royal Society's spokesman, advised on lightning-rods at Purfleet. The embarrassing and politicised conflict between Wilson's critique and Franklin's endorsement of high pointed rods forced the appointment of a committee dominated by Franklin's allies, including the aristocratic natural philosopher Henry Cavendish, who in winter 1771–72 presented the Royal Society with an important new paper on exact trials of electric forces (Randolph 1862, 1: 24–25; Heilbron 1979, 379–80; Franklin 1959–, 19: 153–56, 232–33; Cavendish et al. 1773). Debates centred on the right interpretation of the model experiments of Franklin, Henly, and Wilson. Wilson rightly claimed that points could not discharge distant thunderclouds. Franklinists rightly answered that protection was better the higher the conducting rod above the building. The conflict revealed the political fractures of metropolitan life in the epoch of the American war. It also raised important issues for model-makers (Franklin 1959–, 19: 424–30).

In his dissenting report, completed at the end of 1772, Wilson defended small-scale modelling precisely because real-world electrical phenomena were so obscure. He reflected that test models got status because they accurately represented the world, but then this status was used to test the representation's accuracy. Wilson reckoned the Franklinists erred because they confused experimental modelling with real-world needs. Points drew lots of electrical fire from heaven—just what experimenters wanted; balls were more restrained—just what the Ordnance managers needed. "Our intention is not to make electrical experiments but by the means of conductors to preserve buildings from the dangerous effects of lightning" (Wilson 1773, 52; Mitchell 1998, 314–16). Wilson's polemic brought into the domains of ministerial politics and theatrical showmanship the implicit reasonings on which such model-making depended. Model-makers claimed they already understood the system being modelled, then used their models to investigate it. But, as Wilson sagely argued, though it might be possible for experimenters to generate precision measures of electricity in cabinet, yet it was not obvious that this precision could be maintained when the experimenters projected their model results to large-scale systems.

During the same months as the first Purfleet crisis of 1772, Franklin and his colleagues were absorbed with another, equally complex, problem in electrical modelling of the natural economy. Natural historians had long known of the existence of cramp-fish, stingrays, eels, and their ilk. Most referred their stunning effects to mechanical action—this was, for example, the favoured theory of the eminent La Rochelle naturalist René-Antoine de Réaumur published in 1714. But such shocks were also candidates for absorption by the electrical cosmologies of the mid-eighteenth century. Linnaean naturalists, for example, were convinced such effects must be electrical (Ritterbush 1964, 33, 35–39). To show that stingrays were indeed electrical, it was necessary to model them with artificial electricity, then to reconcile these model trials with anatomy and natural history. Colonial connections supplied resources for dealing with electric fish (Walker 1937, 90–92). In spring 1772, Franklin encouraged a wealthy MP, John Walsh, veteran of the East India Company administration in Madras, to start work on the European torpedo. He went to Paris and La Rochelle in summer 1772 with Leyden jars and electrometers (Franklin 1959–, 19: 160–63).[1] As the Purfleet committee started its debates, Franklin gave Walsh detailed instructions on how to show the fish was electric by substituting it in electrical set-ups for a Leyden jar. Walsh showed the La Rochelle academy that the mechanistic views of their favourite son, Réaumur, could be destroyed with English equipment and the prestige of Franklinist natural philosophy. Gathering as many fish as he could from local boatmen, performing messy anatomies and startling electrical trials, Walsh "announced the effect of the Torpedo to be Electrical, by being conducted through Metals and intercepted by Glass and Sealing wax". The Rochellais were shocked into conviction by joining in the show (Franklin 1959–, 19: 189–90, 204–6, 233–35, 285–89; Torlais 1959, 119–20). Walsh told Franklin there was not a single shock "of the torpedo that we do not most exactly imitate with the Phial". But he conceded there were striking disanalogies between the fish and the jar: "the charged phial occasions attractive or repulsive dispositions in neighbouring bodies, its discharge is obtained through a portion of air, and is accompanied by light and sound, nothing of which occurs with respect to the torpedo" (Franklin 1959–, 19: 205 and 20: 266).[2] Since the stingray did not exert distant forces nor emit sparks, several London natural philosophers rather doubted Walsh's model argument. Walsh told Franklin that "as artificial electricity had thrown light on the natural operation of the torpedo, this might in return if well considered throw light on artificial electricity" (Franklin 1959–, 20: 260; Walsh 1773). Here was just the ingenious technique also involved in the debate about lightning-rods. By using the accepted Franklinist theory of the workings of the electric phial, it was possible to show that the fish mimicked an electrical instrument. Then any differences between the instrument and the animal would be referred not to the failure of the analogy but to the incompleteness of the theory.

Walsh and Franklin needed new resources to make the stingray look like the electric phial (Ingenhousz 1775; Franklin 1959–, 20: 75–79, 433; Hamilton 1773). It was not enough to use the fish as a substitute for electrical apparatus. Such apparatus also had to be arranged to mimic the fish. In 1773 Walsh supplied London's principal anatomist, John Hunter, with torpedos from Torbay, to show that "the Leyden phial contains all [the fish's] magic power" (Walsh 1774, 473). Hunter gave public demonstrations of the fishes' electric organs, "liberally supplied with nerves" and structurally similar to columnar phials (Hunter 1773). Franklin sent Walsh's reports to eminent natural philosophers. Walsh even received the prestigious blessing of Linnaeus himself. "I cannot tell you how delighted I was by your explanation of this phenomenon", the Swedish master wrote, "because it confirms the hypothesis which I have already adopted of an electrical force in the nerves" (Franklin 1959–, 21: 148, 150).[3] The Royal Society awarded Walsh its Copley Medal. In autumn 1774 Franklin's friend John Pringle, President of the Royal Society, lectured on the medallist's work. Pringle told the Society that "between lightning itself and the Leyden phial there is no specific difference, nay scarcely a variety, why then should we multiply species and suppose the torpedo provided with one different from that which is everywhere else to be found?" (Singer 1950, 251). So the authority of the lightning model was used to make the fish model convincing.

Just as in the Purfleet episode, so in the fish experiments, model-making helped because it let experimenters exert detailed control over an artefact, and thus legitimated a bolder philosophical claim about the behaviour of natural systems. Walsh conceded that the "veil of nature" obscured the electrical organs of the fish (Franklin 1959–, 20: 267). In early 1773 he turned to his friend Henry Cavendish to lift it. Cavendish's resources included the ability to measure very small electrical quantities and a theory which distinguished electrical quantity from the electrical intensity that could then be made manifest in a model (Walker 1937, 93–101; Jungnickel and McCormmach 1996,

189–90). A large array of Leyden jars in series could deliver a comparably large amount of electricity at an intensity so low that it could not produce discharge, nor easily travel across any air gap. By building a model fish, Cavendish could bring the unmanageable and 'veiled' animal's electricity within the scope of his own accurate electrometry. Most of Cavendish's decisive measures of conductivity and electrical quantity relied on his own idiosyncratic capacity to distinguish different shocks with his own bodily responses. His fish model let him distribute in public the account of electrical discharge which he had painstakingly developed backstage. In January 1775, therefore, Cavendish gave the Society a paper describing an "artificial torpedo" (Fig. 4.2). First he tried a wooden frame containing a number of

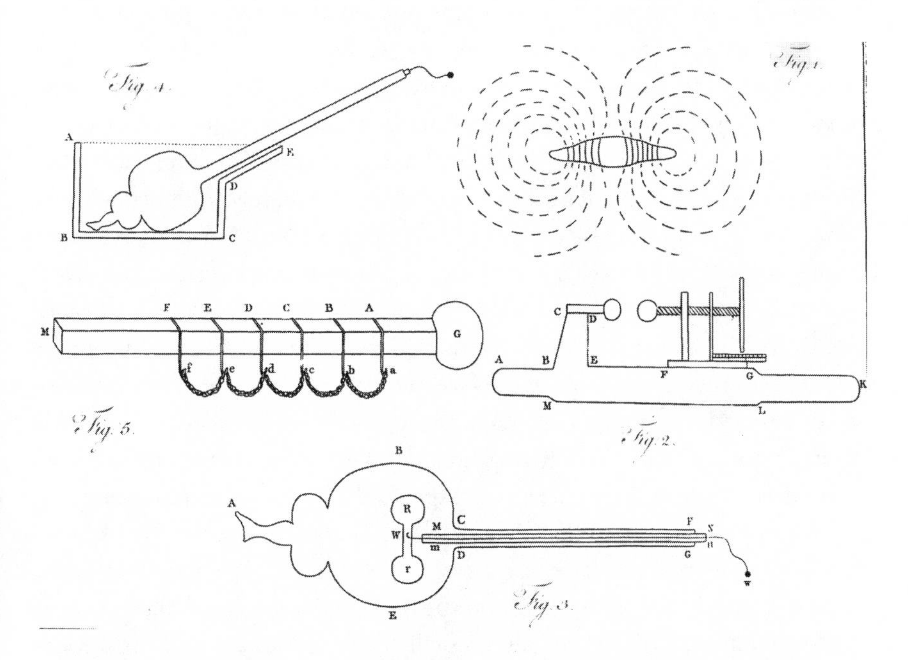

Figure 4.2 Henry Cavendish's model of an electric fish, shown in London in 1775. His figures show (1) the paths of electric fluid through water; (2) an electrometer to measure spark distances; (3) the fish model with a long wire passing through a glass tube to the internal artificial electric organ; (4) the model dipped in salt water; (5) a set-up to detect the transmission of the electric shock. Source: Henry Cavendish, "An account of some attempts to imitate the effects of the torpedo by electricity", *Philosophical Transactions* 66 (1776): 222, pl. 3. By permission of the Syndics of Cambridge University Library.

jars, then, when he reckoned this was a worse conductor than the fish itself, he built a frame of brine-soaked shoe-leather with thin pewter plates to mimic the electric organs. "The water used in this experiment was of about the same degree of saltness as that of the sea, that being the natural element of the torpedo and what Mr Walsh made his experiments with" (Cavendish 1879, 204). It was no argument against the electrical theory of the fish that it could not pass the shock through a short metal chain but that Cavendish's electric model still could. Rather, this showed that "the battery I used is not large enough" (Cavendish 1879, 211).

Cavendish's ingenious modelling did not sway his critics at once. Henly reckoned that no perfect conductor could ever behave the way the torpedo did. One Dublin commentator observed that if Walsh and Cavendish were right then lightning storms should happen underwater (Jungnickel and McCormmach 1996, 189). So in May 1775 Cavendish invited Henly, Hunter, and Priestley into the inner sanctum of his Great Marlborough Street laboratory. He showed them the model fish and let them experience its shocks. He hid the lights in his room to show the faint discharge. The implication was that previous observers of the fish had not been adequately equipped with the embodied skills and high-class instruments of Cavendish's armamentarium. Henly certainly changed his view. By 1777 he was mimicking Cavendish and Walsh in their instrumental manipulation of the electric fish. "The experiments by which Mr Walsh proves the existence and effects of this faculty are simple and elegant", the nabob's secretary recorded. "He lately repeated them publickly before numerous companies of the Royal Society and others to their great wonder and astonishment".[4] Once these witnesses had been convinced, Cavendish published his paper in the *Philosophical Transactions*. He by no means asserted that the fish's electric organs were exactly like Leyden jars, since that would violate the logic of the model argument. Instead, he used his model's success to insist on the truth of his theory of high-quantity and low-intensity electricity. By modelling the fish Cavendish institutionalised his own new account of electric agency and his practice of accurate electrometry (Cavendish 1879, 194–215, 313).

The new institutionalisation of precision and dramatic modelling was then used to win backing for pointed lightning conductors at the Purfleet magazine. Throughout 1774–77 Wilson kept this issue alive in the London press (Franklin 1959–, 21: 223). In May 1777 the Purfleet arsenal, defended according to the Franklinists' advice, was hit by lightning. In the midst of the

Figure 4.3 Benjamin Wilson's model of a lightning strike on the Purfleet arsenal demonstrated at the Pantheon in September 1777. Source: Benjamin Wilson, "New experiments and observations on the nature and use of conductors", *Philosophical Transactions* 68 (1778): 245–313, fig. 3. By permission of the Syndics of Cambridge University Library.

critical moment of the American war, some reported considerable damage to the navy's chief powder magazine (Boddington et al. 1778). Wilson seized his chance to get court backing for a huge new model on a scale of 1:36 (Fig. 4.3). The King was notoriously fascinated by ingenious models, whether of ships or buildings. "The plan I conceived to be the most proper for this purpose", Wilson explained, "was to have a scene represented by art as nearly similar as might be to that which was so lately exhibited at Purfleet by nature". Using royal cash and Ordnance Board equipment, Wilson modelled the arsenal and its points under a thunder-cloud with a vast prime conductor of 3,900 yards of wire and a tinfoil cylinder 155 feet long and 16 inches in diameter. He would win the fight by bulk. He used his new cylinder machine to charge the prime conductor, then recognised he would have to move the model of the arsenal under it along specially crafted grooves, since the model cloud itself was too vast to shift (Wilson 1778a, 246).

Wilson set up his model for public display at the Pantheon, opened in 1772 under crucial patronage of the King. The Pantheon was fashioned for masquerades, the modish London shows in which sublime entertainments allowed the seductive inversion of hierarchy. Wilson used his Pantheon play to seduce his audience, first the royal court, then the Ordnance Board, the Royal Society and, ultimately, the paying public with a show repeated every day during autumn 1777 (Chancellor 1925, 247–58; Brewer 1997, 60–63, 398–99; Randolph 1862, 1: 36; Mitchell 1998, 322–23). There were some puzzles due to the scale of Wilson's show. It was hard, for example, to estimate the size of vast discharges by mimicking Cavendish's celebrated sense of touch, "uncertain as it is in many cases to determine the different effects occasioned by the interposition of these different terminations", high points or low balls. Furthermore, because his enormous artificial cloud, unlike the real thing, discharged all at once and was recharged only by degrees, the model house moved very slowly under it. He put gunpowder inside the model arsenal to increase the similitude and the drama. The scale and thus theatricality were designed to show just how catastrophic the rapid and far-reaching effects of points really were. Only courtly power could afford, or engineer, such model work (Wilson 1778a, 257, 283, 306; Mitchell 1998, 323–24).

Wilson had in the late 1760s enjoyed Edmund Burke's personal support in some of his more popular artistic efforts (Randolph 1862, 1: 22). According to Burke's fashionable aesthetics, which reached a seventh edition by the end of the 1770s, events such as terrifying thunderstorms represented the height of nature's sublimity. To reproduce sublime effects through deliberate design, Burke recommended, "a true artist should put a generous deceit on the spectators . . . No work of art can be great, but as it deceives: to be otherwise is the prerogative of nature only" (Burke 1987, 76). Hence arose a puzzle for the electrical model-builders. Model-makers had to spell out their artifice to make this imitation work, but too much artifice might undermine the allegedly natural, artless status of the model. This was what divided the protagonists in the lightning-rod controversy—the relationship between the measured art of the enlightened philosopher and the histrionic artfulness of the courtier. So after a disputatious visit by Henly and his Royal Society colleagues to the Pantheon in September 1777, it was reported that "this is a very unfair and unsimilar experiment" which gained its authority not

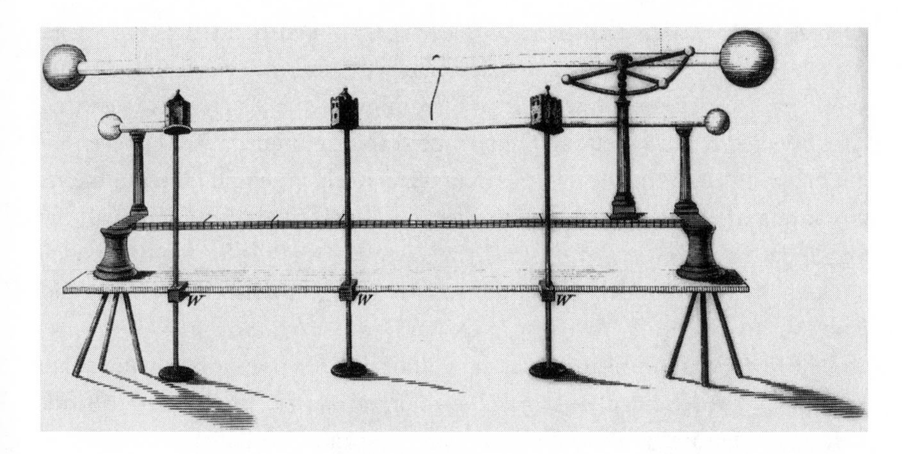

Figure 4.4 William Swift's version of the Wilson model of a thunder cloud above the Purfleet arsenal. Source: William Swift, "Account of some experiments in electricity", *Philosophical Transactions* 69 (1779): 454–61, fig. 2. By permission of the Syndics of Cambridge University Library.

from the validity of the model but the art of the show, Wilson's "apparatus of Drums and Princely Visitors".[5] Wilson's court interest tried to make the committee's views seem private, not those of the Society as a whole.

The Franklinists countered with new models (Benjamin 1896, 281–82; Franklin 1959–, 24: 163, 487–89 and 25: 5–6, 26). A Greenwich experimenter, William Swift, imitated Wilson's set-up on a much smaller scale, but with a pair of prime conductors, one positively and the other negatively charged, "to exhibit many experiments more analogous to the natural effects of lightning from the clouds . . . because in nature clouds are constantly flying in the air which are differently electrified and discharging themeselves in each other" (Fig. 4.4). Swift also added an insulated vessel of water interposed above the model arsenal to imitate rainfall. He backed the orthodox view that points could draw electricity at great distances, slowly, silently, and safely (Swift 1779). In summer 1778 the prestigious instrument-maker Edward Nairne showed that a compelling model of the arsenal could be constructed without Wilson's excessive scale. Nairne's professional skill lay in rendering such small-scale models into precision instruments. Then with carefully staged spectacle, and the decision to design clouds which could themselves be moved, Nairne convinced his fellows and customers that Wilson's confessedly unwieldy and inexact 'clouds' were imperfectly insulated. Hence they exaggerated the size and speed of discharges to high pointed rods (Nairne 1778). Wilson and his allies riposted that with accurate electrometers they could prove that points drew charge from great and dangerous distances. They denied he had cunningly interrupted the communication between his model points and the ground, a circumstance which would increase the fatal drama of electric explosions as the 'cloud' passed above. There was "no juggle in making his experiments". But juggling was the key to the model-makers' artistry. Models made the sublime into the artificial, measurable, and thus manageable (Wilson 1778b; Musgrave 1778, 805).

The history of the invention of the Voltaic pile can be read as an aftermath of these 1770s debates about lightning-rods and electric fish. In May 1782 the Lombardy natural philosophy professor Alessandro Volta came to London to publicise his new condenser, which could ease experimenters' difficulties in measuring charges gathered during thunderstorms. He met Walsh and discussed Cavendish's torpedo model (Heilbron 1979, 442, 457–58). The East India Company merchant and chemistry journalist William Nicholson soon designed a version of Volta's instrument which mechanised the collection of atmospheric electricity (Volta 1782; Nicholson 1788). Volta eventually substituted Nicholson's instrument with pairs of metal discs, and began to recognise that the residual electricity in the device could help him understand the nature of animal electricity too (Gill 1976, 355–58; Pera 1992, 104–16; Kipnis 1987, 118–25, 135). Nicholson then published an article which addressed Cavendish's fishy model-building. The disanalogy between the torpedo and the Leyden jar stayed impressive, so Nicholson instead proffered his multiplier as a better model of the electric fish. Thin discs of talc when pulled apart would produce just the large quantity of low-intensity electricity apparent in the torpedo (Nicholson 1797a; Cavallo 1797; Nicholson 1797b, 358). Within months, Volta designed his own model, pairs of zinc and silver discs separated by brine-soaked paper. In March 1800 he reported to Banks that the shock generated by this pile "has a perfect resemblance to that slight shock experienced from a torpedo in an exceedingly languishing state, which imitates still better the effects of my apparatus by the series of repeated shocks which it can continually communicate". Volta explained that torpedos must simply bring together otherwise separated conducting discs to generate all their electrical effects (Volta 1800, 431; Heilbron 1977; Pera 1992, 150–58). Nicholson reproduced Volta's new pile, criticised its author for ignoring its salient chemical effects, and advertised metropolitan shops where the components for this "artificial electric organ" could be purchased (Nicholson 1800; Kipnis 1987, 135–37; Sudduth 1980, 28–29; Golinski 1992, 203–7). The demonstration of Volta's model fish, now defined as the electrical pile, by no means settled the relationship between animal, atmospheric, and artificial electricity. Rather, it launched new enterprises for modelling electrical action and claiming the right to manage living behaviour. In Regency London philosophical materialists and electrical practitioners such as the radical Francis Maceroni repeated the lesson that brains and nerves were "real electrical machines, similar in principle, as they are similar in substance and structure, to the electrical discharging apparatus of the gymnotus and the torpedo" (Morus 1998, 131). The new institutions of public sciences took over the Enlightenment project to model nature's capacities and social realities with ingenious machines.

SHIPS

Workable models seemed to offer a way of managing systems that were otherwise hard for metropolitan institutions to control. Smarting from defeat in the American war, and facing new military threats from Spain and France, British naval experts were peculiarly aware of the virtues of model structures in reforming marine design. As at the Ordnance Board, so naval architecture could be remodelled, and hence dockyard and shipboard labour reorganised. In late 1789 a veteran publisher John Sewell, with close links to the East India Company and the Thames shipbuilders, began printing manifestos for an overhaul of naval construction. He stressed "the importance of the study of shipbuilding by philosophical as well as practical men" (Sewell 1791–92, part 1: vi). Sewell was a conservative opponent of what he saw as French and radical ideas. He printed attacks on the subversive Joseph Priestley and joined the Loyal Association formed in London to counter republicanism at the start of the Revolutionary War (Nichols 1812, 3: 738; Sewell 1791–92, part 2, appendix: xxi–xxiii; Gorham 1830, 202–3). Sewell's network established a Society for the Improvement of Naval Architecture, launched in a London pub in April 1791. Their avowed precedents included the Society of Arts' ship-model trials three decades earlier, and the more recent successes of London instrument makers' demonstrations, especially those of Nairne (Sewell 1791–92, part 1: 3, 63–66). The problem with naval design was that the "business was not studied as a science, but carried on more by precedent"

(Sewell 1791–92, part 1: iv). Members included Banks, mathematicians such as Hutton, several Thames shipbuilders, and some important naval administrators, including Charles Middleton, recently departed head of the Navy Board. By the following year membership had risen to over 300 (Sewell 1791–92, part 1: 66; Johns 1910, 29–31). The Society aimed at putting before the public new models in naval administration and science. Three strategies were proposed. They would "preserve for public exhibition" exemplary ship models, "with such improvements as have been adopted on the joint opinions of able mathematicians and skilful shipbuilders". They wanted a new academy in the naval dockyards to train cadets in the mathematics of navigation and shipbuilding. They would sponsor experiments with ship models in controlled settings (Sewell 1791–92, part 1: 2, 14–15; Martyn 1791, 4–5).

The theoretical problems these experiments confronted—ship stability, hull design, and frictional drag—were more obviously urgent concerns for national security than the problems of lightning protection at naval arsenals discussed by the Ordnance Board and the Royal Society. The Royal Navy was the nation's largest technical enterprise and a centre of the fiscal-military state. Speedier and stabler ships, capable of carrying larger and more guns, were essential to British military power. But Navy Board administrators were often sceptical of costly reforms to management and schemes for allegedly better ship design (Brewer 1989, 34–37; Ashworth 1998a, 65–71; Rediker 1987, 75; Morriss 1983, 15, 31–37). Block models of standard hulls were commissioned for the Admiralty and taken from captured enemy ships whose designs were considered especially admirable. It was hard to see how to scale up these established models to full-size vessels, though London instrument-makers marketed devices which apparently helped such scaling (Franklin 1989, 49–51; Lavery and Stephens 1995, 24–25, 94–95). Throughout the eighteenth century, administrators appealed to "the authority of practice" before agreeing to fund innovation (Rodger 1993, 137–40). This was why the Society for the Improvement of Naval Architecture was so impressed by the need to replace 'precedent' with 'science'—and their new models would generate this new reason.

The major conflicts over remodelling naval architecture took place around the labour process of the shipyards. A warship would be built from a few plans made at the traditional scale of one inch to four feet. Frames taken from the midship section were composed of arcs, which were scarcely optimal for design but easy to scale, and would be used to draw lines in the yards' mould lofts, where they would be applied to the timber (Lavery and Stephens 1995, 22–24). A moral economy existed in the vast shipyards to protect artisan skill and resist changes in wage rates. The very term 'strike' became widely used in 1768 when shipwrights paralysed the London fleet by striking its sails. Sudden mobilisation and re-equipment could easily disrupt this economy (Morriss 1983, 28–29, 60–1; Prothero 1979, 24–25; Rediker 1987, 110; Linebaugh 1991, 311). So could changes in power between private contractors and the major royal yards (Pollard 1953; MacDougall 1999). The war years saw a growth in labour militancy in yards whose military and economic value had correspondingly increased (Linebaugh and Rediker 2000, 219). In 1794 the Thames shipwrights founded their own friendly society to restrict labour supply, sustain wages, and secure workshop opacity. In 1797 Sewell established a Marine Voluntary Association to break up mutinies in the British Channel fleet. In 1798 the first Marine Police force was set up to oversee the docks (Prothero 1979, 33, 46–50; Nichols 1812, 3: 738; Linebaugh 1991, 430). Those who wished to remodel naval systems, such as Middleton at the Navy Board or Samuel Bentham as Inspector-General of Naval Works, spoke of "saving manual labour" and "ensuring greater despatch". Middleton planned a revision of the entire system of regulation of naval administration "to make each dockyard serve as a part only of a great machine" (Morriss 1983, 60, 186). In his own work for the Society of Naval Architecture, Middleton offered premiums for better models for ship design or naval administration. What might seem to some managers as rational reforms, such as steam engines in the Deptford yard in 1768, policing the loss of woodchips from Portsmouth in the 1790s, or the new well-fortified West India Docks in 1799–1802, were resisted by shipwrights as infringements of their moral traditions (Linebaugh 1991, 384, 425; Ashworth 1998a, 73–76).

The construction of ship models directly responded to the tensions of this naval system. Their everyday use in naval yards depended on collective skills judged impossible to render calculable (McGee 1999, 229). Use of models in trials and demonstrations was often taken as a sign of the incapacity of rational mechanics to describe or predict ship behaviour. The Society for the Improvement of Naval Architecture recognised the puzzle: "of two ships built by the same mould, and rigged exactly the same, one shall sail very well, and the other but indifferently" (Sewell 1791–92, part 1: vii). So the Society sought to turn "the poorer classes of workmen in the Yards" into paid experimenters (Sewell 1791–92, part 1: 3, 15). Model-building already formed a crucial element in shipwrights' traditional skilled work (Franklin 1989, 4, 177; Lavery and Stephens 1995, 35–36, 82–83). Commercial model-makers plied their trade alongside other marine instrument-makers. Models were important means of training by hands-on experience (Franklin 1989, 5; Lavery and Stephens 1995, 49; Morriss 1983, 37–38). Such models were not, therefore, so much a solution to the naval problem as a site where that problem was contested. Ingenious ship models and measured trials stood for this contest between the managerial reason of analysts and overseers and the tacit skills of the dockyards (Linebaugh 1991, 390; McGee 1999, 226).

Sewell's Society echoed a Newtonian theme in the programmatic writings of enlightened analysts of naval mechanics, the contrast between theoretical perfection and artisan ignorance. In the second book of the *Principia mathematica* (1687) Isaac Newton had set out the mechanical theory of fluid resistance which dominated academic theories of ship motion for a century. His doctrine, simplified by such writers as Emerson, implied that fluid resistance would vary as the square of the velocity of the moving body, its maximum cross-sectional area, and some constant dependent on the shape of the vessel. For a sphere, this constant was one-half. The analytic project was thus defined as the search for the highly-desirable solid of least resistance, a task well beyond and often inimical to the everyday work of shipwrights (Cohen 1974; Hall 1979, 164). Newton proposed that "by the same method by which we have found the resistance of spherical bodies in water and mercury, the resistance of other bodies can be found; and if various shapes of ships were constructed as little models and compared with each other, one could test cheaply which was best for navigation" (Newton 1972, 463; 1967–81, 6: 463). Though this proposal was omitted from all subsequent editions of the work, Newton maintained his interest in naval training as advisor to the Royal Mathematical School where boys were taught mathematics and navigation as preparation for apprenticeship at sea, and where the gradation from mechanic to mathematician was much in evidence (Newton 1959–77, 3: 359–60; Iliffe 1997).

Newton's prestigious hierarchy of mathematical reasoners and humbler artisans was an indispensable resource for enlightened theorists, especially in the French system of academic expertise and state regulation of the naval and merchant marine (Séris 1987, 132). In 1775, distinguished academicians such as Charles Bossut were commissioned by the French government to try resistance experiments on model boats at the Paris École Militaire. Instead of complex hulls, the experimenters relied on highly simplified geometrical shapes; they took for granted the dependence of resistance on the square of these models' speeds; and they claimed that frictional effects on the sides and stern of a model were negligible (Stoot 1959, 37; Wright 1989, 316). These academicians' experiments were closely studied in London. In 1776 an East India Company engineer, Henry Watson, also arranged an English edition of Leonhard Euler's *Theory of the Construction and Properties of Vessels*, the dominant academic text on the mathematics of shipbuilding. Like his colleagues, Euler endorsed the use of "good models in miniature which represent vessels exactly as they are"; but, like them, Euler relied on simplified geometrical solids to substitute for the behaviour of ships at sea (Euler 1776, 256). In summer 1790, Sewell publicised Euler's view that "it is not necessary that the model should exactly represent the whole vessel entirely", and that "the experiments which we might easily make would lead us without difficulty to a discovery of the good or bad properties which great vessels executed according to such models, ought to have with respect to resistance" (Sewell 1791–92, part 1: 27–28).

The Society's principal experimenter was a young, wealthy, and enthusiastic member of the Society's council, Mark Beaufoy, already expert in experiments on fluid resistance (Kerr 1974, 28; Beaufoy 1930, 163). In summer 1790, just made FRS, Beaufoy wrote to Sewell criticising the Paris academicians' trials on ship models, especially their claims about the solid of least resistance and the negligible effects of stern and side friction (Beaufoy 1834, xxiv; Sewell 1791–92, part 1: 24–26). Experiments should be done "by models drawn through the water by means of weights and pullies" (Sewell 1798–1800, part 1: 29–33). Beaufoy's trials along these lines eventually cost him almost £30,000 and lasted from spring 1793 until the extinction of the Society in late 1798. William Wells, another Society member, gave him use of the large Greenland Dock, and his collaborators included naval officers, the East India dock managers Randall and Brent, and Hutton, who performed the tedious calculations to derive precise results from the data accumulated in Beaufoy's almost 2,000 separate experiments (Stanhope 1914, 172, 182–83). Like Bossut and Euler, Beaufoy constructed idealised geometrical models, rectangular planks, spherical sections, and parallelepipeds. Applying Smeaton's designs already used in the Society of Arts trials, he used measured

Figure 4.5 Mark Beaufoy's apparatus for dragging ship models by horse-drawn pulleys at constant speed across Greenland Dock. Source: Mark Beaufoy, *Nautical and Hydraulic Experiments* (London, 1834). By permission of the Syndics of Cambridge University Library.

weights on horse-driven pulleys to drag these models across the surface of the dock (Fig. 4.5). The resisting force was computed from these weights, and as in Smeaton's set-ups, friction became the focus. Beaufoy also designed an ingenious system for measuring speed, using a horizontal rod on a smooth pulley which moved 12 times more slowly than the model itself. By marking the rod at known time intervals, a system comparable to that of Atwood's Machine, Beaufoy could measure the speed and acceleration of his geometrical ships. From the start of 1795, his attention turned to the effects on motion when the models were submerged. By winter 1795–96 he could demonstrate that friction was a major quantifiable factor in motion (Fig. 4.6). Against the orthodoxy of Newton, Bossut, and Euler, it seemed that bow shape was not the only variable affecting ship performance. Beaufoy's collaborators helped him define fluid resistance as a combination of the pressure effects at bow and stern, plus the friction along the surface of the hull. Beaufoy introduced what he called 'friction planks' to show that in general speed varied with resistance at a power of speed of between 1.71 and 1.82, well

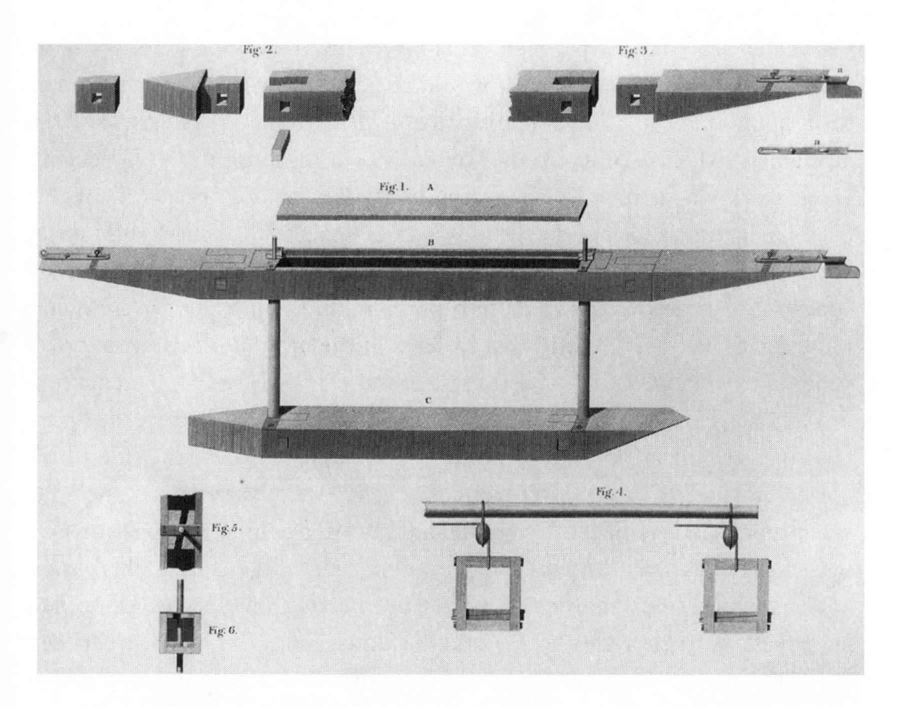

Figure 4.6 Mark Beaufoy's new friction planks to estimate the drag on ship models' motion through water. Source: Mark Beaufoy, *Nautical and Hydraulic Experiments* (London, 1834). By permission of the Syndics of Cambridge University Library.

below the Newtonian square law (Beaufoy 1834, xxvii–xxviii; Wright 1989, 317–21). Soon after Beaufoy completed his runs, his equipment was taken over by other Society members, with the aid of Wells' staff. Experimenters armed with trial data which showed that curved hulls were most stable, and that ships could be designed much longer than they were wide, hoped "that in time those absurd maxims which have so long governed the constructors of shipping will submit to refutation and be laid aside" (Gore 1799, 6). The Society advertised the extraordinary accuracy of Beaufoy's "curious and instructive" trials. "They clearly prove that experiments can now be made, by means of proper models, so as to ascertain the comparative advantages or disadvantages arising from the form . . . of all kinds of navigable vessels" (Society for the Improvement of Naval Architecture 1800, i–ii).

Yet to make Beaufoy's trial models count as exemplars of shipyard realities and then capable of changing shipwrights' ways, more was needed than the sterling efforts of the experimenters at Greenland Dock and Hutton's desk-top calculations at Woolwich. It would require an entire transformation of the labour conditions and administrative structure of the dockyards and the academies (McGee 1999, 229). Atwood's own response to the Greenland Dock work was telling. In early 1796 he finished a long survey of enlightened navigational science. He proposed large-scale tests to check differences between analytical theory and shipwrights' custom. Atwood was struck by Beaufoy's demonstrations that stern pressure and hull friction were quantifiable factors in ship motion, or at least in the motion of his geometrical solids. They showed frictional resistance must be at least a cubic equation in velocity. Atwood distinguished two senses of the term 'theory', the "pure laws of mechanics", known by academicians, and "a systematic rule which individuals form to themselves from experience and observation alone". He was prepared to credit the "experimental knowledge in naval constructions which has been transmitted from preceding times", the shipwrights' "skill and ingenuity". But analysts' difficulty in *inferring* shipyard tradition had no effects on their rights to *direct* the shipwrights. The alliance between experimental modelling and rational analysis was the only means through which the global reach of naval systems could be engineered (Atwood 1796, 125–29). Atwood was awarded the Royal Society's Copley Medal for his rules for calculating the metacentre at the large angles of heel neglected by previous analysts. Though needing as much as two years' work to apply, his unwieldy methods were adopted as guides. In early 1798, as Beaufoy's experiments were reaching completion, Atwood finished a treatise on ship stability. He claimed that the theory was "liable neither to ambiguity nor error". Nothing that happened in the shipyards or at sea had ever falsified the true theory of motion. The sole reason for the apparent inapplicability of theory was "that steady adherence to practical methods rendered familiar by usage which creates a disposition to reject, rather than to encourage, proposals of innovation in the constitution of vessels" (Atwood 1798, 202–4). Sewell published Atwood's mathematical methods, alongside Emerson's theory of the solid of least resistance, as part of the campaign to rationalise shipwrights' work (Sewell 1798–1800, part 1: 39–41 and part 3: 1–11).

Atwood's programme can be compared with contemporary attempts by Samuel Bentham and his allies at Portsmouth, where the principles of rational mechanics and precision engineering were applied to break the shipwrights' resistance and transform dockyard production lines. Bentham held that "the mode of putting together or fastening any of the component parts of that very complicated machine, a ship" would be useless without "a perfect knowledge of the principles of mechanics". From 1801 he tried to set up an apprenticeship system with teaching in these principles made compulsory and wages dependent on theoretical expertise (Ashworth 1998a, 68–69; Morris 1983, 110–12). Resistance in the dockyards was not limited to the Portsmouth strikes against Bentham's system. Atwood contacted the shipyard contractors, Randall and Brent, who had helped administer Beaufoy's trials. They supplied him with schemes of a fine East India ship to test his model of stability and loading (Atwood 1798, 287). Randall was the main supplier of ships to the East India Company. He planned his own treatise on naval architecture through the medium of the Society for the Improvement of Naval Architecture before being forestalled by Atwood and Beaufoy. In spring 1802, as the navy demobilised at the end of the Revolutionary War, Randall tried to impose a wage cut in his yards. His workers struck and Randall tried to get scab labour from the nearby Deptford royal dockyards. The Admiralty offered troops to guard Randall's yard and sacked their men who refused to work there. This was a major labour crisis of the age. During the summer Randall was hurt during a strikers' demonstration and died of his wounds. The shipwrights claimed victory: "how so large a body of men as we are could possibly resist the combinations of by far the greater part of our employers assisted by various departments of Government, civil, military and naval, is a mystery none but ourselves can develop" (Prothero 1979, 47–48; Morriss 1976; MacDougall 1999, 50–51). Unlocking the shipwrights' mystery was indeed the concern of yard managers, model experimenters, and academic theoreticians.

When Beaufoy's model data were published, they were prefaced with a hagiographic account of Randall's death during the 1802 strike (Beaufoy 1834, xxxviii). After the strike, Beaufoy continued his campaign for naval modelling. He used the popular natural philosophy journals to publicise new experiments on stability and hull design (Beaufoy 1817a, 9; Stanhope 1914, 183). Like Atwood, Beaufoy wanted the Admiralty to run model experiments using their own barges, transformed into more regularly shaped boats. Perhaps even Greenland whalers could be turned into scientific vessels. Beaufoy saw that unless ships were changed into forms more manageable by rational analysis, models could never generate useful data (Beaufoy

1817b, 257–58, 261; Beaufoy 1817c). By the end of the Napoleonic War, model-building had become more integrated into the formal system of naval training. Commissioners at the Navy Board, veterans of the Society for the Improvement of Naval Architecture, including Charles Middleton, agreed to establish a Portsmouth School of Naval Architecture. This realised the Society's plan for "a seminary at one of our principal royal dockyards" to expropriate the skills of artisans otherwise "totally incapable of paying that arduous and unremitting attention to the highest branches of mathematics" (Sewell 1791–92, part 1: 14–15). Under Cambridge-trained mathematicians, Portsmouth students studied Newtonian mechanics and algebraic geometry, read Emerson on the scaling problem in timber, and Atwood on ship stability. Special ship models were used to train the students in naval architecture. "In all the details of executing a draught from given dimensions, laying off in the mould loft and the actual building, they will not have much to learn" (Morriss 1983, 113–14, 221; Franklin 1989, 166).

But this was a controversial claim. One radical London magazine, *The Chemist*, edited by the socialist artisan Thomas Hodgskin, carried shipwrights' complaints that the School's models were designed to destroy traditional means of skill and advancement. Hodgskin's short-lived magazine was the organ of metropolitan artisan radicals fascinated by the materialist lessons of chemical philosophy and galvanism, and hostile to the pretensions of the professoriate at the Royal Institution and other homes of fashionable natural philosophy (Morus 1998, 15–16, 102; Golinski 1992, 244). The journal turned its guns on the new school of naval architecture and its model curriculum. The school's defenders might boast that "till the school's establishment *few persons* in any of our naval arsenals ever thought of guiding their practice by maxims drawn from the *legitimate* principles of science". Radical shipwrights countered that "legitimate principles" must mean "principles lawfully begotten in the cranium of some lawfully appointed professors of abstractions". The radicals directly named these detestable theorists, including Atwood himself. There was an obvious link between the legitimism of the professors and the legitimism of the reactionary regimes of postwar Europe. The shipwrights singled out for their bile "these *Holy Alliance* parts of science and the imputation thus to be cast on our best ship builders in all times, past and present". It was clear that "Calculus will not make them good ship builders". The lesson for the entire enterprise of model-building and rational science was damning.

> Let them show to the world that they possess that almost omnipotent power of determining with precision the various properties of that complex body called a ship, which even the immortal Newton thought impossible and that the person most likely in any degree to succeed was the man of practical experience.... To talk to an able practical shipwright of predicting the displacement, stability, weatherly qualities, and other essentials of a large ship by calculations made (as is the case in all calculations) from a drawing on a small scale, is only making a laughing stock of one's self. What makes the whole matter still more laughable is that the whole of their calculation of stability is founded on a centre of gravity which they assume and which may be almost any where but where they place it. What should be said of our modern theorists, who think it nowise absurd to attempt the construction of a whole navy upon nothing? (Working Shipwright 1825)

Hodgskin's radical magazine was short-lived, but so was the "laughable" Naval Architecture School. It shut in 1832, only to be replaced in the following decade by a system of shipyard schools which continued to suffer from the troublesome contrast between traditional artisan custom and formally trained expertise (Casey 1999). Training, modelling, and labour conflict stayed salient themes in the scientific regimes of the period. It has been convincingly argued that the physical sciences were established in early-nineteenth-century Britain through the combination of urban mathematical practitioners, the Cambridge professoriate, and scientifically expert members of the armed forces. The character of this coalition helped define the research agenda in astronomy, navigation, optics, meteorology, and geomagnetism. Cavendish, master of precision measures of gravitation and electricity, and Beaufoy, medallist of the new Astronomical Society, were among the heroes of this alliance (Miller 1986; Ashworth 1994). Stories of model-making in late-eighteenth-century Britain help illuminate the origins of these new sciences and their practitioners. Because models could be controlled and disciplined, they lent themselves to campaigns for precision management of society and cosmos. Institutions such as the Ordnance Board, the Royal Society, the military schools, the dockyards, and the Navy Board backed these campaigns. Inside and around these institutions, however, were groups who resisted and undermined the remodelling of practice, custom, and tradition. The evocative account of an eighteenth-century mathematician's universal paradigm and god-like power used as this essay's epigraph should be juxtaposed with that of the pseudonymous "working shipwright", who in early 1825 wrote in the London press against those who claimed "that almost omnipotent power of determining with precision the various properties of that complex body called a ship".

NOTES

1. John Walsh, Diary of Journey to France, 1772, 53–55, John Rylands University Library, Manchester, English MS 724.

2. John Walsh, "Experiments on the torpedo or electric ray at La Rochelle and l'Ile de Ré", 1772, Royal Society Library, MS 669, folio 72.

3. Carl Linnaeus to John Walsh, 16 September 1774 (copy), India Office Library MS Eur.E3.41.

4. David Davies to Francis Fowke, 26 November 1777, India Office Library, MSS Eur.E3.69.

5. J. H. de Magellan to Achille-Guillaume Lebègue de Presle, 15 September and 3 October 1777 (copies), Library of Congress, Franklin Papers.

REFERENCES

Alder, Ken. 1997. *Engineering the Revolution: Arms and Enlightenment in France, 1763–1815*. Princeton: Princeton University Press.

Ashworth, William. 1994. "The calculating eye: Baily, Herschel, Babbage, and the business of astronomy." *The British Journal for the History of Science* 27: 409–42.

———. 1998a. "System of terror: Samuel Bentham, accountability, and dockyard reform during the Napoleonic Wars." *Social History* 23: 63–79.

———. 1998b. "John Herschel, George Airy, and the roaming eye of the state." *History of Science* 36: 151–78.

Atwood, George. 1796. "Construction and demonstration of geometrical propositions determining the positions assumed by homogeneal bodies which float freely and at rest on a fluid's surface, also determining the stability of ships and other floating bodies." *Philosophical Transactions* 86: 46–130.

———. 1798. "Disquisition on the stability of ships." *Philosophical Transactions* 88: 201–310.

———. 1801. *A Dissertation on the Construction and Properties of Arches*. London: Bulmer [with "Supplement", separately paginated, published in 1804].

Beaufoy, G. 1930. *Leaves from a Beech Tree*. Oxford: Blackwell.

Beaufoy, Mark. 1817a. "Suggestions and remarks on naval subjects." *Annals of Philosophy* 10: 6–10.

———. 1817b. "Suggestions for building experimental vessels for the improvement of the Navy." *Annals of Philosophy* 10: 256–64.
———. 1817c. "On the Northwest Passage." *Annals of Philosophy* 10: 424–28.
———. 1834. *Nautical and Hydraulic Experiments*, ed. Henry Beaufoy. London: privately printed.
Benjamin, Park. 1896. "On a letter to Benjamin Franklin." *Cassier's Magazine* 9: 273–82.
Boddington, John, et al. 1778. "Sundry papers relative to an accident from lightning at Purfleet." *Philosophical Transactions* 68: 232–44.
Brewer, John. 1989. *The Sinews of Power: War, Money, and the English State, 1688–1783*. London: Unwin, Hyman.
———. 1997. *The Pleasures of the Imagination: English Culture in the Eighteenth Century*. London: Harper Collins.
Burke, Edmund. 1987 [1759]. *A Philosophical Enquiry into the Origin of Our Ideas of the Sublime and the Beautiful*, 2nd edn, ed. James T. Boulton. Oxford: Blackwell.
Casey, Neil. 1999. "Class rule: The hegemonic role of the Royal Dockyard Schools, 1840–1914." In *History of Work and Labour Relations in the Royal Dockyards*, ed. Kenneth Lunn and Ann Day, 66–86. London: Mansell.
Cavallo, Tiberio. 1797. "On the multiplier of electricity." *Journal of Natural Philosophy, Chemistry and the Arts* 1: 394–98.
Cavendish, Henry. 1776. "An account of some attempts to imitate the effects of the torpedo by electricity." *Philosophical Transactions* 66: 196–205.
———. 1879. *Electrical Researches*, ed. James Clerk Maxwell. Cambridge: Cambridge University Press.
Cavendish, Henry, William Watson, Benjamin Franklin, John Robertson, and Benjamin Wilson. 1773. "A report of the committee appointed by the Royal Society to consider of a method for securing the powder magazine at Purfleet." *Philosophical Transactions* 63: 42–47.
Chancellor, E. Beresford. 1925. *The Pleasure Haunts of London during Four Centuries*. London: Constable.
Cohen, I. Bernard. 1974. "Isaac Newton, the calculus of variations, and the design of ships." In *For Dirk Struik: Scientific, Historical, and Political Essays in Honor of Dirk J. Struik*, ed. R. S. Cohen, J. J. Stachel, and M. W. Wartofsky, 169–87. Boston: Reidel.
———. 1990. *Benjamin Franklin's Science*. Cambridge, Mass.: Harvard University Press.
Coleridge, S. T. 1956–59. *Collected Letters*, ed. Earl Leslie Griggs, 4 vols. Oxford: Clarendon Press.
———. 1969 [1818]. *The Friend*, ed. B. E. Rooke. Princeton: Princeton University Press.
Edney, Matthew. 1994. "Mathematical cosmography and the social ideology of British cartography, 1780–1820." *Imago Mundi* 46: 101–16.

Euler, Leonhard. 1776. *A Compleat Theory of the Construction and Properties of Vessels*, ed. Henry Watson. London: Elmsley.

Franklin, Benjamin. 1941. *Benjamin Franklin's Experiments: A New Edition of Franklin's Experiments and Observations on Electricity*, ed. I. Bernard Cohen. Cambridge, Mass.: Harvard University Press.

———. 1959–. *Papers*, ed. Leonard W. Labaree, William B. Willcox, Claude A. Lopez, and Barbara B. Oberg, 37 vols published so far. New Haven: Yale University Press.

Franklin, John. 1989. *Navy Board Ship Models, 1650–1750*. London: Conway Maritime.

Galilei, Galileo. 1974 [1638]. *Two New Sciences*, ed. Stillman Drake. Madison: University of Wisconsin Press.

Gill, Sydney. 1976. "A Voltaic enigma." *Annals of Science* 33: 351–70.

Golinski, Jan. 1992. *Science as Public Culture: Chemistry and Enlightenment in Britain, 1760–1820*. Cambridge: Cambridge University Press.

Gore, Charles. 1799. *Result of Two Series of Experiments towards Ascertaining the Respective Velocity of Floating Bodies*. London: James Hayward.

Gorham, George Cornelius. 1830. *Memoirs of John Martyn and Thomas Martyn*. London: Hatchard.

Hall, A. R. 1979. "Architectura navalis." *Transactions of the Newcomen Society* 51: 157–73.

Hamilton, S. B. 1952. "Historical development of structural theory." *Proceedings of the Institute of Civil Engineers* 3: 374–419.

Hamilton, William. 1773. "Account of the effects of a thunderstorm." *Philosophical Transactions* 63: 324–32.

Harley, Basil. 1991. "The Society of Arts' model ship trials, 1758–1763." *Transactions of the Newcomen Society* 63: 53–71.

Heilbron, John. 1977. "Volta's path to the battery." In *Proceedings of the Symposium on Selected Topics in the History of Electrochemistry*, ed. G. Dubpernell and J. H. Westbrook, 39–65. Princeton: Electrochemical Society.

———. 1979. *Electricity in the Seventeenth and Eighteenth Centuries*. Berkeley and Los Angeles: University of California Press.

Henly, William. 1774. "Experiments concerning the different efficacy of pointed and blunted rods in securing buildings against the stroke of lightning." *Philosophical Transactions* 64: 133–52.

Hills, Richard. 1989. *Power from Steam: A History of the Stationary Steam Engine*. Cambridge: Cambridge University Press.

Hogg, O. F. G. 1963. *The Royal Arsenal: Its Background, Origins, and Subsequent History*, 2 vols. Oxford: Oxford University Press.

Hunter, John. 1773. "Anatomical observations on the torpedo." *Philosophical Transactions* 63: 481–89.

Hutton, Charles. 1815. *Philosophical and Mathematical Dictionary*, 2 vols. London: Hutton.

Iliffe, Robert. 1997. "Mathematical characters: Flamsteed and Christ's Hospital Royal Mathematical School." In *Flamsteed's Stars*, ed. Frances Willmoth, 115–44. Woodbridge: Boydell and Brewer.

Ingenhousz, Jan. 1775. "Experiments on the torpedo." *Philosophical Transactions* 65: 1–4.

James, J. G. 1987–88. "Thomas Paine's iron bridge work, 1785–1803." *Transactions of the Newcomen Society* 59: 189–221.

Johns, A. W. 1910. "An account of the Society for the Improvement of Naval Architecture." *Transactions of the Institution of Naval Architects* 52: 28–40.

Jungnickel, Christa, and Russell McCormmach. 1996. *Cavendish*. Philadelphia: American Philosophical Society.

Kant, Immanuel. 1991 [1793]. "On the common saying: 'This may be true in theory but it does not apply in practice'." In *Kant: Political Writings*, 2nd edn, ed. Hans Reiss, 61–92. Cambridge: Cambridge University Press.

Keane, John. 1995. *Paine: A Political Life*. London: Bloomsbury.

Kemp, E. L. 1977–78. "Thomas Paine and his 'pontifical' matters." *Transactions of the Newcomen Society* 49: 21–40.

Kerr, B. 1974. *The Dispossessed: An Aspect of Victorian Social History*. London: John Baker.

Kipnis, Naum. 1987. "Luigi Galvani and the debate on animal electricity." *Annals of Science* 44: 107–42.

Kranakis, Eda. 1997. *Constructing a Bridge: An Exploration of Engineering Culture, Design, and Research in Nineteenth-Century France and America*. Cambridge, Mass.: MIT Press.

Lavery, Brian, and Simon Stephens. 1995. *Ship Models: Their Purpose and Development from 1650 to the Present*. London: Zwemmer.

Lindqvist, Svante. 1990. "Labs in the woods: The quantification of technology during the late Enlightenment." In *The Quantifying Spirit in the Eighteenth Century*, ed. T. Frängsmyr, J. L. Heilbron, and R. Rider, 291–314. Berkeley and Los Angeles: University of California Press.

Linebaugh, Peter. 1991. *The London Hanged: Crime and Civil Society in the Eighteenth Century*. Harmondsworth: Penguin.

Linebaugh, Peter, and Marcus Rediker. 2000. *The Many-Headed Hydra: Sailors, Slaves, Commoners, and the Hidden History of the Revolutionary Atlantic*. Boston: Beacon Press.

MacDougall, Philip. 1999. "The changing nature of the dockyard dispute, 1790–1840." In *History of Work and Labour Relations in the Royal Dockyards*, ed. Kenneth Lunn and Ann Day, 41–65. London: Mansell.

Martyn, Thomas. 1791. *Address of the Society for the Improvement of Naval Architecture*. London: Sewell.

McGee, David. 1999. "From craftsmanship to draftsmanship: Naval architecture and the three traditions of early modern design." *Technology and Culture* 40: 209–36.

Millburn, John R. 1988. "The Office of Ordnance and the instrument-making trade in the mid-eighteenth century." *Annals of Science* 45: 221–93.
Miller, David Phillip. 1986. "The revival of the physical sciences in Britain, 1815–1840." *Osiris* 2: 107–34.
———. 1999. "The usefulness of natural philosophy: The Royal Society and the culture of practical utility in the later eighteenth century." *The British Journal for the History of Science* 32: 185–201.
———. 2000. "'Puffing Jamie': The commercial and ideological importance of being a 'philosopher' in the case of the reputation of James Watt." *History of Science* 38: 1–24.
Mitchell, Trent. 1998. "The politics of experiment in the eighteenth century: The pursuit of audience and the manipulation of consensus in the debate over lightning-rods." *Eighteenth-Century Studies* 31: 307–31.
Morriss, Roger. 1976. "Labour relations in the royal dockyards, 1801–5." *Mariner's Mirror* 62: 337–46.
———. 1983. *The Royal Dockyards during the Revolutionary and Napoleonic Wars*. Leicester: Leicester University Press.
Morton, Alan Q. 1993. "Men and machines in mid-18th-century London." *Transactions of the Newcomen Society* 64: 47–56.
Morus, Iwan Rhys. 1998. *Frankenstein's Children: Electricity, Exhibition, and Experiment in Early Nineteenth-Century London*. Princeton: Princeton University Press.
Musgrave, Samuel. 1778. "Reasons for dissenting from the report of the Committee, including remarks on some experiments exhibited by Mr Nairne." *Philosophical Transactions* 68: 801–21.
Nairne, Edward. 1778. "Experiments on electricity being an attempt to shew the advantage of elevated pointed conductors." *Philosophical Transactions* 68: 823–60.
Newton, Isaac. 1959–77. *Correspondence*, ed. H. W. Turnbull, J. F. Scott, A. R. Hall, and Laura Tilling, 7 vols. Cambridge: Cambridge University Press.
———. 1967–81. *Mathematical Papers*, ed. D. T. Whiteside, 8 vols. Cambridge: Cambridge University Press.
—— 1972 [1726]. *Philosophiae naturalis principia mathematica: Third Edition with Variant Readings*, ed. Alexandre Koyré and I. Bernard Cohen. Cambridge: Cambridge University Press.
Nichols, John. 1812. *Literary Anecdotes of the Eighteenth Century*, 6 vols. London: Nichols.
Nicholson, William. 1788. "A description of an instrument." *Philosophical Transactions* 78: 403–7.
———. 1797a. "Description of an instrument which renders the electricity of the atmosphere very perceptible." *Journal of Natural Philosophy, Chemistry and the Arts* 1: 16–18.

———. 1797b. "Observations on the electrophore tending to explain the means by which the torpedo and other fish communicate the electric shock." *Journal of Natural Philosophy, Chemistry and the Arts* 1: 355–59.

———. 1800. "Account of the new electric or galvanic apparatus of Sig. Alessandro Volta." *Journal of Natural Philosophy, Chemistry and the Arts* 4: 179–87.

Ogborn, Miles. 1998. *Spaces of Modernity: London's Geographies, 1680–1780.* London: Guilford.

Paine, Thomas. 1991 [1795]. *The Age of Reason*, ed. Philip Foner. New York: Carol Publishing.

Pera, Marcello. 1992. *The Ambiguous Frog: The Galvani–Volta Controversy.* Princeton: Princeton University Press.

Pollard, Sydney. 1953. "Laissez-faire and shipbuilding." *Economic History Review* 5: 98–115.

Prothero, Iorwerth. 1979. *Artisans and Politics in Early Nineteenth-Century London.* London: Methuen.

Pynchon, Thomas. 1997. *Mason and Dixon.* London: Cape.

Randolph, Herbert. 1862. *Life of General Sir Robert Wilson*, 2 vols. London: John Murray.

Rediker, Marcus. 1987. *Between the Devil and the Deep Blue Sea: Merchant Seaman, Pirates, and the Anglo-American Maritime World, 1700–1750.* Cambridge: Cambridge University Press.

Reynolds, T.S. 1979. "Scientific influences on technology: The case of the overshot waterwheel, 1752–1754." *Technology and Culture* 20: 270–95.

Ritterbush, Philip. 1964. *Overtures to Biology: The Speculations of Eighteenth-Century Naturalists.* New Haven: Yale University Press.

Robinson, Eric, and Douglas McKie. 1969. *Partners in Science: Letters of James Watt and Joseph Black.* Cambridge, Mass.: Harvard University Press.

Rodger, Nicholas. 1993. *The Insatiable Earl: A Life of John Montagu, Fourth Earl of Sandwich.* London: Harper Collins.

Schaffer, Simon. 1994. "Machine philosophy: Demonstration devices in Georgian mechanics." *Osiris* 9: 157–82.

Séris, Jean-Pierre. 1987. *Machine et communication: Du théâtre des machines à la mécanique industrielle.* Paris: Vrin.

Sewell, John, ed. 1791–92. A *Collection of Papers on Naval Architecture*, vol. 1 (part 1, 1791 and part 2, 1792, separately paginated). London: Sewell.

———. 1798–1800. *A Collection of Papers on Naval Architecture*, vol. 2 (part 1, 1800; part 2, 1798; part 3, 1800, separately paginated). London: Sewell.

Singer, Dorothea Waley. 1950. "John Pringle and his circle." *Annals of Science* 6: 248–61.

Skempton, A. W. 1980. "Telford and the design for a new London Bridge." In *Thomas Telford: Engineer*, ed. Alastair Penfold, 62–83. London: Telford.

Skempton, A. W., and H. R. Johnson. 1962. "First iron frames." *Architectural Review* 131: 175–86.
Smith, Denis. 1973–74. "The professional correspondence of John Smeaton." *Transactions of the Newcomen Society* 46: 179–88.
———. 1977. "The use of models in nineteenth-century British suspension bridge design." *History of Technology* 2: 169–214.
Society for the Improvement of Naval Architecture. 1800. *Report of the Committee for Conducting the Experiments of the Society for the Improvement of Naval Architecture*. London: Sewell.
Stanhope, Ghita. 1914. *The Life of Charles, Third Earl Stanhope*. London: Longmans.
Stewart, Larry. 1998. "A meaning for machines: Modernity, utility, and the eighteenth-century British public." *Journal of Modern History* 70: 259–94.
Stoot, W. F. 1959. "Some aspects of naval architecture in the eighteenth century." *Transactions of the Institution of Naval Architects* 101: 31–46.
Sudduth, William. 1980. "The voltaic pile and electro-chemical theory in 1800." *Ambix* 27: 26–35.
Swift, William. 1779. "Account of some experiments in electricity." *Philosophical Transactions* 69: 454–61.
Torlais, Jean. 1959. "L'Académie de La Rochelle et la diffusion des sciences au 18e siècle." *Revue de l'Histoire des Sciences* 12: 111–25.
Volta, Alessandro. 1782. "Del modo di render sensibilissima la più debole elettricità sia naturale, sia artificiale." *Philosophical Transactions* 72: 237–80 (English translation in separately paginated appendix).
———. 1800. "On the electricity excited by the mere contact of conducting substances of different kinds." *Philosophical Transactions* 90: 403–31.
Walker, W. Cameron. 1937. "Animal electricity before Galvani." *Annals of Science* 2: 84–113.
Walsh, John. 1773. "Of the electric property of the torpedo." *Philosophical Transactions* 63: 461–80.
———. 1774. "Of torpedos found on the coast of England." *Philosophical Transactions* 64: 464–73.
Warner, Deborah Jean. 1997. "Lightning-rods and thunder houses." *Rittenhouse* 11: 124–27.
Wilson, Benjamin. 1764. "Considerations to prevent lightening from doing mischief to great works, high buildings, and large magazines." *Philosophical Transactions* 54: 247–53.
———. 1773. "Observations upon lightning and the method of securing buildings from the effects." *Philosophical Transactions* 63: 49–65.
———. 1774. *Further Observations upon Lightning*. London: Nourse.
———. 1778a. "New experiments and observations on the nature and use of conductors." *Philosophical Transactions* 68: 245–313.

———. 1778b. "New experiments upon the Leyden phial respecting the termination of conductors." *Philosophical Transactions* 68: 999–1006.

[Working Shipwright]. 1825. "Naval architecture." *The Chemist* 2: 349–51, 428–30.

Wright, Thomas. 1989. "Mark Beaufoy's nautical and hydraulic experiments." *Mariner's Mirror* 75: 313–27.

PART TWO

Disciplines and Display

CHAPTER FIVE

Modelling Monuments and Excavations

Christopher Evans

Reflecting general developments in the social sciences, over the last decade-and-a-half archaeology has been much concerned with the past as text, and has explored diverse readings and the nature of authorship (e.g., Hodder 1986; 1989). Yet, in the presentation of its information, the subject has historically been closely allied with architecture and (particularly military) engineering, and much of its reportage is graphic. Archaeological fieldwork reports use illustrations much more than, for example, those from ethnography. Stemming from an emphasis upon the standardised documentation of excavation as an unrepeatable experiment, the discipline's 'language' is certainly as much visual as textual.

However essential they may be to the communication of the discipline's prime activity, digging, archaeological graphics were not really conventionalised until as late as the 1920s and 1930s.[1] More than a century of field practice preceded the establishment of this graphic-to-text formula, and amongst the media experimented with were models. These have received almost no study, and museum stores must be frequented and excavation reports closely read to stumble upon them. Though time has lent them anachronistic charm, in their day they were a 'modern' vehicle for technical demonstration and, for popular amusement, edification, and professional display, models were a common form of representation in the later eighteenth and nineteenth centuries. They featured in international exhibitions, with more than 70 displayed in the Great Exhibition (Physick and Darby 1973, 13), and when Richard Lucas' models of the Parthenon first went on show in the British

Museum in 1846 they, like other major display models of the age, drew much public attention.[2]

Generally not requiring disciplinary initiation, it is the visual directness of most models that still makes them an appropriate means of presentation to the broader museum-going public. They are usually not symbolic in the way that archaeology's technical illustrations have become, and do not, for example, require a hachured line to render the slope of a ditch or the sides of a pit. Features can be appreciated in miniature without conventions, and specialised graphic codes need not stand between viewer and object. This appeal would have been all the greater before the twentieth century, when general literacy—and probably with it the ability to read complex map-based information—could not be presumed for the population at large (Thompson 1968, 447, 782–83).

This survey is restricted to archaeology in Britain, but it would be foolhardy and even misleading to forge a unified tradition of model construction. Paucity of citation, fragility, and neglect make it impossible to estimate the models' original frequency, and the subject's practitioners were too few and far between. Instead, we seem to have multiple origins through small networks, individual initiatives, and sporadic influence; the broader narrative that links them is the development of archaeological practice. Three main kinds of modelling can nevertheless be identified: (1) antiquarian, (2) demonstrative/presentational, and (3) reconstructive. Each predominated in a different period, with the first prior to circa 1850 (see also Baker, this volume) and demonstration models holding sway in the second half of the nineteenth century, while reconstructions were largely twentieth-century products. Yet there was much overlap between these categories.

The earlier antiquarian pieces displayed standing 'architectural' monuments, variously building ruins and large stones, or 'megaliths'. After reviewing these, this chapter concentrates on the demonstration models, that is, renderings of sites that to a greater or lesser degree illustrate excavation itself. Having affinities with the scientific models of the period—and opposed to the later 'peopled' reconstructions of past life that are primarily a phenomenon of museum display to laypeople—these models relate to the establishment of archaeological methods and tell more directly of specialist ways of seeing the past. Reflecting upon the drawn-out processes by which the discipline eventually came to visualise its practice, I also consider the changing institutional domains of its display and 'performance'—from private residences and society meeting rooms to museums (e.g., Levine 1986; Moser 1999).

THE PRESENTATION OF MONUMENTS

Amongst the most renowned model collections of the nineteenth century was that of the architect Sir John Soane at his No. 13 Lincoln's Inn Fields residence (Richardson 1989). Aside from those of his own buildings, the models were largely of classical ruins but also included a Stonehenge and a generic prehistoric house (Richardson 1989; Wilton-Ely 1969). Apart from his personal interest and inspiration as an architect and collector, these served to illustrate his public lectures, and appropriate examples could be viewed in his house on the days preceding and after their delivery (Thornton and Dorey 1992, 38; Watkin 1996).

Pride of place in Soane's collection was given to an enormous model of the buildings exposed by excavation at Pompeii (Fig. 5.1), and this influenced Soane's representation of his own buildings (e.g., Gandy's "The Bank of England" of 1830; see Evans 2000, fig. 1). Although Soane probably acquired this piece directly from Neapolitan model-builders, it may alternatively have been obtained from Dubourg's fee-charging museum of cork models. Soane is known to have visited this establishment, which operated in Pall Mall from 1778 until it succumbed in 1785 to a fire that resulted from an overly successful demonstration of the Vesuvius effect (Altick 1978, 392–96, fig. 140; Richardson 1989, 225–26; Thornton and Dorey 1992, 67), and apparently his own model room duplicated several of its exhibits.

Amongst the earliest surviving archaeological models in Britain is that of the megalith at Mont St Helier, Jersey in the collection of the Society of Antiquaries of London (Fig. 5.2; Hibbs 1985). Discovered during levelling operations for a military parade ground in 1785, the Island's Assembly presented it to the retiring Governor, Field Marshall Conway, who by 1788 had it re-erected within the grounds of his estate at Park Place, Henley-on-Thames. Dubbed the 'Little Master's Stonehenge', this gave rise to considerable public interest and was, for example, reported in the *Gentleman's Magazine*. Conway donated a wooden scale model of his 'Druidical Temple' to the Society of Antiquaries, to whom he also communicated a description of its discovery (Conway 1787; Molesworth 1787). Three, possibly four, other early models are known to have been made of the site. Conway gave one to his cousin, the

TEMPLE of JUPITER

Figure 5.2 The Society of Antiquaries of London model of the 'Druidical Temple' at Mont St Helier, Jersey (catalogue no. 57). Reproduced with permission of the Society of Antiquaries of London.

renowned author, man of letters, and connoisseur, Horace Walpole (1717–97), who apparently displayed it in his neo-Gothic mansion at Strawberry Hill.[3] So we have a model of a site destined to become a folly (the monument re-built at Park Place) itself residing within a folly (Strawberry Hill).

Probably intended to illustrate the architect's public lectures, the Soane archives contain a large watercolour of the same site (Evans 2000, fig. 4). Given the megalith's displacement, what is remarkable about this illustration is its rendering as if from life. For it obviously cannot show the monument *in situ* at St Helier, since this had left Jersey some 20–30 years previously. Much simplified, the site is seen in a slight bird's-eye perceptive and shown perfect without the clutter of fallen stones, bushes, or humans. This fabricated setting begs the question, from what source the watercolour was taken. While one cannot be absolutely certain that it does not derive from an engraving, it

←

Figure 5.1 The Soane model room, Lincoln's Inn Fields, with the Pompeii excavation model displayed on the lower stand. Acquired by Sir John Soane in 1826, it is thought to be the work of Dominico Padiglione, a model-maker in a Naples museum (Richardson 1989, 224–25). Photograph by M. Cyprien.

was most likely taken from a model. A ready source would have been that of the Society of Antiquaries. Yet on the upper left side the Soane figure lacks one of the stone cists, or chambers, which is present in Conway's original model. This stone setting was also absent in Richard Molesworth's 1787 description (and accompanying plan) and so conforms to the engravings of the monument which are published in Francis Grose's *Antiquities of England and Wales* of 1787 and thought to have been taken from Walpole's non-extant model (Hibbs 1985, 57–58, pl. 4). Since the only other surviving model, that in the collection of the Société Jersiaise, includes this feature, the Soane watercolour is most likely after Walpole's model or one that he acquired or commissioned for himself, but of which the provenance and history are otherwise unknown.

Whatever the exact source of this watercolour, the potential relationship between a model and as-if-from-life figures has implications for other 'topographic' site renderings. How many derive from models? This is especially likely for some of the overtly rusticated bird's-eye views of Stonehenge (e.g., Piggott 1978, figs 5 and 7). Could not the perspective come from looking down upon models, and the gnarled rustication of their stones be the result of depicting cork?

Stonehenge was an obvious subject for modelling. William Stukeley—the person most responsible for introducing the Druids into British antiquity—records the making of a model in 1716 and describes another chain of convoluted translations between three- and two-dimensional media, from a view to a model to plans: "Happening to fall into a set of thoughts about Stonehenge in Wiltshire, by a prospect of Loggan's which I met withall, I undertook to make an exact Model of that most noble and stupendous piece of Antiquity, which I have accomplish'd, and from thence drawn a groundplot of its present ruins, and the view of it in its pristine State . . . and propose from thence to find out . . . its design, use, Founders etc." (Stukeley, as quoted in Piggott 1989, 125).

Later in the century the natural philosopher and polymath, John Waltire, toured with cork models of the site to accompany his popular lectures (Higgins 1827, xviii; Peck and Wilkinson 1950, 160–61) and Henry Browne, the first 'guardian' of the great henge, is known to have produced models of it since 1807.[4] Although, as discussed below, Browne later became involved in London-based model networks, these are unlikely to have been his original source of inspiration. While other unknown eighteenth-century exemplars must lurk in the background, it is conceivable that Waltire provided the impetus; he and Browne shared similarly extraordinary theories concerning the Stonehenge's antediluvian origins (Chippindale 1985, 128; Piggott 1989, 146).

The most renowned of Browne's models accompany John Britton's Celtic Cabinet in Devizes Museum in Wiltshire (Fig. 5.3). Believed to date from circa 1824, the cabinet is thought to have been assembled on commission for a gentleman antiquarian to house his collections. (He refused it.) Topped by 'as it is' and 'as it was' cork models of Stonehenge, the front of this heavy furnishing carried framed watercolours of the most renowned megaliths of the time, a large Stonehenge roundel being matched by an aerial view of nearby Avebury (models of Avebury were in a drawer). A friend of Soane's, the antiquarian, topographer, and church historian associated with the Gothic Revival, Britton was the author of both the series, *Beauties of England and Wales*, and the first published account of Soane's collections, *The Union of Architecture, Sculpture and Painting* of 1827. The closeness of the perspective style of the cabinet's roundel to that of surveys by Soane's students indicates that either the architect must have given a version of the illustration to Britton (which was then subsequently cut down) or that Britton commissioned the piece in the manner of Soane's studio (Evans 2000, figs 7 and 8).

The Stonehenge models which top the cabinet were set within a glass box, of which the variously tinted sides effected the monuments' lighting as at different times of day. When demand for Browne's models increased, Britton explored the idea of establishing a 'Druidical Antiquarian Company' to promote their sale and also a display in London to exhibit the monument in pictures and models with accompanying magic-lantern effects (Chippindale 1985, 129). In archaeological circles this venture and the Britton Cabinet itself have been considered highly eccentric. However, recognising the linkage between Britton and Soane recasts the context of these enterprises, relating them to model exhibitions such as Dubourg's and Soane's Lincoln's Inn Fields collection. What, after all, is Britton's cabinet but a 'union of architecture' (at least furniture), 'sculpture' (models), and painting involving the main display media for both amusement and edification of the time?

The Britton cabinet also has parallels in the mixed media works of Richard Tongue of Bath—'painter and modeller of megaliths'—and is evident in his donations to the British Museum (Evans 1994). Tongue shared an interest,

Figure 5.3 John Britton's Celtic Cabinet of circa 1824. Reproduced with permission of the Wiltshire Archaeological and Natural History Society.

too, in the 'atmospheric' presentation of sites that Britton had sought to achieve with the tinted panes. The pictures the artist-antiquarian gave the museum were rendered to show three times of day: dawn (at the chambered tomb at Plas Newydd), noon (at Stonehenge), and dusk (at The Tolmen, a great natural boulder with druidical associations). Signing himself 'Honorary Curator of Models of the Bath Scientific and Literary Institute' in a letter announcing his presentation of six models to the Museum in 1834, Tongue declared that he had shown his models to Britton; his own earliest examples date from eight years before. This correspondence makes clear that Tongue envisaged the paintings and models being exhibited together as a unified group, with the latter to be shown below the three canvases. Although never realised, this was intended as a multi-media display in a manner reminiscent of Britton's cabinet or the unfulfilled ambitions of the Druidical Antiquarian Company.

Though it may be tempting to see this idea for the multi-media display of ancient monuments as having been transmitted from Soane to Britton to Tongue, there is no direct evidence that Tongue was familiar with Britton's cabinet. Similarly, although Britton was obviously aware of the layout of Soane's collection, other dioramas and panoramic displays of the day may have been equally influential (Altick 1978), and he apparently had immediate knowledge of Philippe-Jacques de Loutherbourg's magic-lantern variant, the Eidophusikon (Chippindale 1985, 129).

While also relating to traditions of architectural and craft-based model construction (e.g., ship-building; see Schaffer, this volume), the specific context of these early efforts can be further appreciated by considering other such holdings in the collection of the Society of Antiquaries of London. Aside from a plaster model of the amphitheatre at Dorchester given in 1854 (catalogue no. 433), these were of classical ruins. Catalogue entry 8, presented by Captain W. H. Smyth in 1830, is "A model of the ancient thermal baths discovered in the Island of Lipari" (1830), and in 1767 the Society received a cork model of the Temple of the Sibyl at Tivoli (catalogue no. 16):

> That a Man of very singular Talents having lately appeared at Rome, whose merit consists in making Models of the Antiquities, it raised in Mr Jenkins a Desire to have one of the most respectable ones executed by him & chose that usually called the Sibyls Temple at Tivoli; being one of the most elegant & pittoresque Objects in the Country, as well as the most singular point of Architecture. In order to render this work as compleat as possible, Mr Jenkins got the Assistance of Signr. Gio: Stern, an eminent Architect, to inspect the Proportions & every singularity of the Building: And to compleat the Imitation, Mr. James Forrester, a Countryman, and an ingenious Landscape Painter, has been so obliging as to colour the various Parts of it in its proper tints; which has so good an Effect, as to become the Admiration of all who have seen it.[5]

As in the Soane model collection, these Italianate benefactions—mementoes of travel and testimonials to taste—attest to the appreciation of classical prototypes and the impact of the Grand Tour.

Depicting monuments and not excavations, the early models of 'rude stone' megaliths are similarly architectonic and, I have argued, are essentially expositions of megalithic 'first architectures', i.e., standing constructions rather than earthen sub-surface traces (Evans 2000). (Excavation was then in its infancy and largely restricted to burial mounds.) As we saw for Tongue, the representation of monuments was a defining activity in constituting antiquarianism, but it did not relate only to developments within the fledgling discipline. Soane was interested in Stonehenge, and both Turner and Constable painted it (e.g., Chippindale 1983; Edmonds and Evans 1991). By depicting monuments archaeology participated in a broad and eclectic cultural discourse. Ancient monuments increasingly began to contribute to the 'new' national cultural landscape that was engendered by the unification of the United Kingdom and the further forging of national identity through the Napoleonic Wars (e.g., Colley 1992): "What man is there, in whose breast glows a spark of patriotism, who does not view the monuments of his country . . . which connect the present with the remote past . . . with feelings deeper and nobler than any exotic remains" (Babington 1865, 8).

This is the context of the Society of Antiquaries' Dorchester Amphitheatre model dating to the middle years of the nineteenth century. Its whereabouts are now unfortunately unknown, but its donor is recorded as Charles Warne, who was responsible for another model of the same site that still survives in Dorset County Museum. This one is catalogued as having been presented by Brighton Museum in 1913. Perhaps Warne distributed multiple versions of the same model in an attempt to bring Dorchester's archaeology to extra-regional notice, or maybe the Society of Antiquaries' model was transferred to Brighton and eventually returned to its Dorset 'home'. Author of *Celtic Tumuli of Dorset* (1866) and *Ancient Dorset* (1872), Warne was a renowned excavator of barrows and, as such, can be considered this study's first true 'archaeologist' (Levine 1986, 31–32; Piggott 1989, 158). His Dorchester Amphitheatre, a simple relief piece, represents both the topographic tradition of early fieldwork and an increased interest in national and regional antiquities from the middle decades of the nineteenth century.

PERFORMANCE AND SITE WORKINGS

Display models evidently sit in the 'background' of a number of later-nineteenth-century site reports, many of which were published in the form of the originally oral communications (Hodder 1989). This is apparent, for example, in Stephen Stone's "Account of certain (supposed) British and Saxon Remains, recently discovered at Standlake in the county of Oxford" to the Society of Antiquaries of London in May 1857:

> On removing the soil, the workmen first came to a circular pit whose diameter was 5 feet 6 inches. The pit, marked 'a' in the sketch I have prepared, I subsequently cleaned out, the gravel diggers not choosing to be at the trouble themselves.... Having prepared a model, as well as a ground plan, of the whole from measurements carefully made and repeated, I need not enter into a minute description of *each excavation* (Stone 1857, 93, 95; my emphasis).

By today's conventions the accompanying plan in the published text is abstract and highly stylised (Fig. 5.4A). Shaded, its features appear almost sculptural, like floating geometric solids. Typical of the time the text is difficult to follow; a disjointed chronicle of discovery with much use of 'We next came to'. Although the speaker evidently lacked a magic lantern, he clearly refers to a model somewhere in the room, which in this instance miraculously survives (Fig. 5.4B). It is a scale reproduction of all the features open at the end of excavation and is, in effect, a three-dimensional base-plan.

In Stone's account each feature is 'an excavation'. Not interpreted as part of a settlement cluster, each is a thing unto itself with little attention given to stratigraphic and plan interrelationships. Pits, by definition negative features, are ascribed (positive) architectural qualities, and incidental elements, evidently the by-product of their intercutting, assigned structural attributes: "A ledge of hard rock was found in No. 17, the situation and extent of which is shown in the model.... The question now arises, how were these thin walls of gravel, in some instances only 6 inches in thickness, made to stand?" (Stone 1857, 95). A comparable ethos (and means of presentation) still pervades, for example, R. C. C. Clay's account of the Fyfield Bavant Down settlement excavated some 70 years later: "A series of models of the more interesting pits, made to scale, to illustrate the various types, communicating pits, pits with recesses in their walls, pits with steps in their walls, pits with seats, and pits with flint shaft, have been made" (Clay 1924, 462) (Fig. 5.8A).

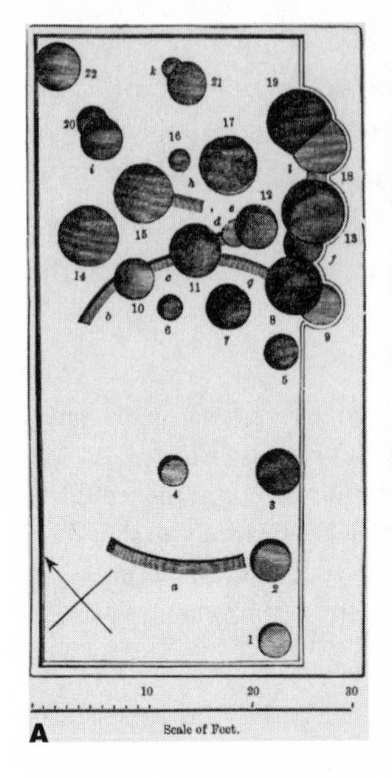

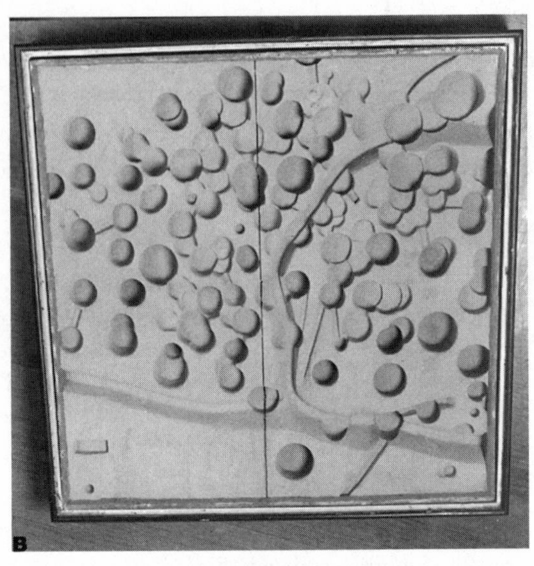

Figure 5.4 (A) Plan of Standlake. From Stephen Stone, "Account of certain (supposed) British and Saxon remains", *Proceedings of the Society of Antiquaries of London*, 1st ser., 4 (1857): 94. (B) Model of Standlake; plaster with sand coating (85 × 84 × 10 cm; scale 1/2": 1'). Reproduced with permission of Oxfordshire County Museums.

This is not the 'grammar' of excavation through which archaeologists envisage sites today: the subtle interrelationship of negative form, horizontal strata, and the context of finds. Site models, which apparently lurk behind many early reports, are clearly a factor standing between us and earlier practice. At a basic level they were a key accompaniment to presentation and an item of 'performance'. Colour slide film did not come into common usage until the 1920s and 1930s and magic-lantern slides were evidently rare in the discipline's academic circles. Illustrations could either be pinned up or held aloft by assistants,[6] but model displays were an integral part of archaeological congresses and conferences until well into the last century.

Modelling clearly also influenced the perception of sites. The above accounts relate to a renowned red herring of nineteenth- and early-twentieth-century practice, the pit dwelling: the notion that during later prehistory people actually lived in the pits that were readily evident in the ground (and often full of refuse). This view was much influenced by the small scale of excavation, which was to prove a drawback until the 1940 and 1950s, when earthmoving machinery was first regularly employed; it allowed large-scale exposure of sites and detection of the traces of free-standing post-built roundhouses (Evans 1988; 1989). Before then, site models granted architectural solidity to negative features, and reinforced these pit-dominated modes of interpretation. While villas, for example, could still be viewed in Italy, timber-set round buildings could not be seen directly and had no precedent in the immediate experience of most excavators. By contrast, models gave a sense of tactile reality to the 'pit house'.

Lieutenant-General Augustus Pitt Rivers, the nation's first Inspector of Ancient Monuments and a great innovator of fieldwork technique (Thompson 1977; Bradley 1983; Bowden 1991), is the figure most closely associated with archaeological models in Britain. There were more than 100, of both solid wood and 'hollow' (plaster on wire frame) construction, in his museum at Farnham in Dorset, where they were arranged centrally down rooms, along the sides of which artefacts were displayed (Dudley Buxton 1929, pl. VIII; Thompson 1977, fig. 14). Made by his estate carpenters under strict instruction, he was assured that their format would appeal to the general public: "Since the models of my excavations were exhibited at the Society of Antiquaries, the Excavators at [the Roman town of] Silchester have pursued the same course with very great success, and I think the use of such models promises to become general. *There is nothing that conveys such a correct idea of the work done in investigations of this nature*" (Pitt Rivers 1892, 298, my emphasis).[7]

Pitt Rivers' models fall into distinct types. Relating to his tenure as Inspector of Ancient Monuments, some simply document megaliths in a manner akin to earlier antiquarian efforts (e.g., Bowden 1991, fig. 26), and there is also a number of Celtic crosses. Most, however, are of his excavations, and while some only record individual pits or burials, many were remarkably sophisticated. Such models he considered the culmination of the site record.[8] In the case of the Wor Barrow—a Neolithic long barrow in Dorset—the locations of finds within the ditches flanking the mound are shown by pins (Figs 5.5 and 5.6). Their heights indicate the depth of the artefact, and the different colours of their heads show its period (Roman or Neolithic); drawings of skeletons, elevated on small pedestals, depict the positions of burials.[9] The model, therefore, contains an enormous amount of information, interrelating

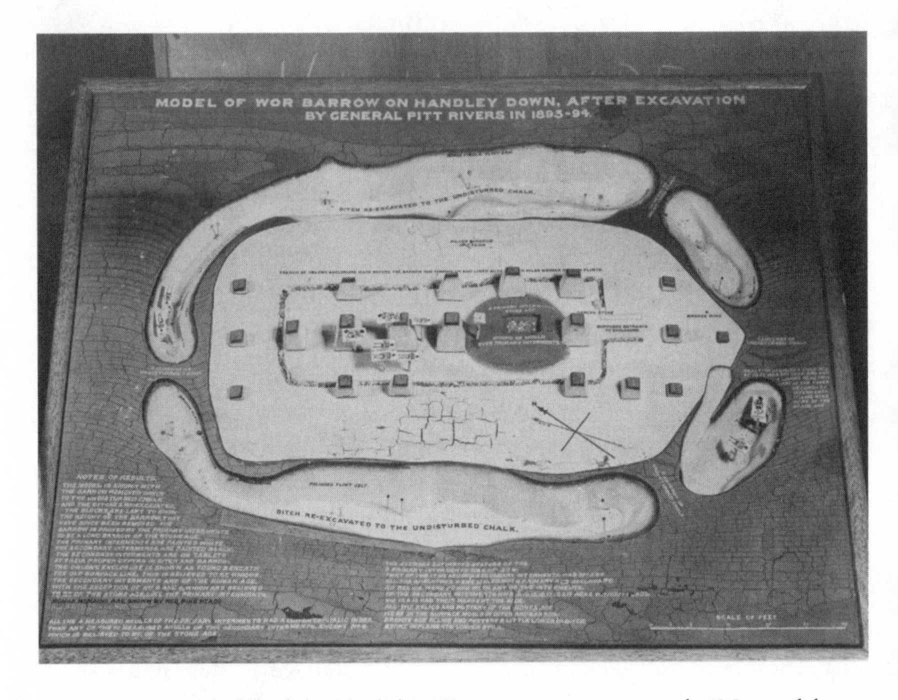

Figure 5.5 Model of the Neolithic Wor Barrow as excavated. "My models are of well seasoned mahogany and are carved from the contoured plans. Carpenters should be trained to the work; my estate carpenters are so used to it, that I have only to put a contoured plan before them on a proper scale, and they will cut it out with the utmost precision, but of course I supervise the construction of models very closely" (Pitt Rivers 1897, 333). Reproduced with permission of Salisbury and South Wiltshire Museum.

the evidence of section drawings, contour maps, base-plans, and artefact locations (Pitt Rivers 1898, pls 249–52). But a viewer must be able to 'read' its conventions to know what a pin's length and colour convey. Similarly, in other models, the ground-slope is moulded and painted contours have also been applied.

Another example, the model of the Iron Age hillfort of Cissbury in West Sussex, seems superficially less detailed—it does not portray the location of artefacts. It is nevertheless complicated in its demonstration of stratigraphic sequence (Bowden 1991, pl. 20). Carved from wood, it is hinged to allow viewers to appreciate the galleries of the Neolithic flint mine beneath the ramparts of the fort. Such models have marked affinities to scientific demonstration pieces of the age, like Thomas Sopwith's series of geological

Figure 5.6 Augustus Pitt Rivers' 1894 excavation at the Wor Barrow with one of the site models proudly propped on the front edge of the mound, with detail inserted left. Photograph, from a private album, provided courtesy of Prof. M. Jones.

models of 1841. Employing laminated woods, these show the basic principles of stratigraphic tilting, folding, and faults (Sopwith 1841; Whipple Museum catalogue, no. 1581). Here it is relevant that General Pitt Rivers' use of models was not confined to archaeology. In its spirit of public edification, his Farnham Museum also included agricultural collections, amongst which were miniature farming implements (Bowden 1991, fig. 47).[10]

While Pitt Rivers discussed his models only in relationship to public presentation, they may also have played a role in his fieldwork. Here a photograph of the Wor Barrow excavations in progress becomes telling (Fig. 5.6). Taken during the second 'campaign' of 1894, propped in the foreground is a site model indicating the results of the first season, when the surrounding ditch had been dug out. Was it there for show—or for *showing*, and possibly to demonstrate plan information to labourers whose map literacy may have been limited? Most revealing perhaps is the very fact that the General would have had the model built while fieldwork was still in progress. This suggests that modelling was part of the process of excavation and not just a museum display tool. Did the models aid his conceptualisation of sites and the complexities of deposition? To a degree unique for the time, Pitt Rivers evidently conceived of sites in 3-D and his long interaction with model formats surely helped him do this.

Through his military background the General would have been familiar with a range of fortification tableaux and topographic models. Volume 3 of his publications on the excavations at Cranborne Chase in Dorset fails to acknowledge any archaeological precursors for his models (Pitt Rivers 1892, 297–98), but in the first of this series he admitted to having seen Stone's Standlake model (Pitt Rivers 1887, 20), which pre-dated his own earliest models by some 15 years. Stone's rendering appears to have fostered other early examples. P. Stevens' delivery of a lecture concerning Highfield settlement at Fisherton, Salisbury to a meeting of the Wiltshire Archaeological and Natural History Society had apparently been accompanied by a model of a group of pit dwellings (Awdry 1870, 148). Unfortunately this cannot be located and was not mentioned when the excavations were finally published some 60 years later (Stevens 1934).

The Silchester models to which Pitt Rivers refers in the quotation above, while much more detailed than that of the Standlake site, are still basically just a record of excavated features. Made in 1890–93 by the sites' excavators, William Henry St John Hope and George Fox (and housed today in Reading Museum), these straightforwardly depict building foundations exposed within the Roman town (Fig. 5.7). 'Peopled' by archaeologists (and labourers), they are remarkably accurate and even show the ragged trench edges produced through a 'wall-chasing' excavation technique (Wheeler 1954, 127). Although applauded by the General and certainly demonstrative of the appearance of excavation, they lack the sophistication of Pitt Rivers' more complicated models. In effect, the Silchester tableaux reconstruct site work without demonstrating principles of excavation (e.g., distribution or sequence) or interpretation.[11]

It was Harold St George Gray, one of Pitt Rivers' assistants from 1888 until 1900, who most directly continued his legacy of modelling. Excavating a number of important sites in his own right and co-directing others (Bradley 1975; Bowden 1991, 163–64), upon leaving the General's service he went on to dig the stone circle and henge at Arbor Low in Derbyshire. He duly made a model of the site and was commissioned by the journal *Man* to provide a description of its manufacture (Gray 1903). Estimating that the model took 450 hours' labour, the note is illustrated with photographs showing the

Figure 5.7 George Fox and William Henry St John Hope's Silchester models (both approx. 1 × 2 m). Top: House 1, Insula VIII (1893). Bottom: House 2, Insula II (1891). Copyright Reading Museum Service (Reading Borough Council). All rights reserved.

sequence of its construction. The resulting piece is an unremarkable relief model showing the circle as an earthwork with the outline of the excavation trenches painted on (Gray 1903, fig. 1). It tells of the effort that went into it that its scaled standing stones were apparently carved in soapstone on the site itself. Dorset Museum also holds a similar relief model of Gray's 1903–13 excavations of the Neolithic henge and Roman amphitheatre at Maumbury Rings, Dorchester (Bradley 1975). Of the same site as Warne had modelled earlier, in this case the dug features and trenches have been moulded white to mimic chalk.[12]

To construct accurate site models presupposed the existence of 2-D scale plans. Yet although sympathetic to the conventions of engraved reproduction, the lunar-like abstraction of the published Standlake plan suggests that it was itself influenced by the site's model. Standlake is held to be amongst the first 'proper' excavations of a settlement site of this type; how pits and ditches were to be depicted was not then conventionalised and model representations may well have been influential. Hope and Fox's Silchester plans are equally simplistic and essentially consist of hard-edged renderings of building foundations. While in both sites artefacts would have been attributed to discrete features, finds were not individually plotted.

Appropriate to his advocacy of a 'fuller' and more nuanced archaeology, Pitt Rivers brought a wider range of evidence to bear in his excavations and practised more detailed recording. Consequently, many of his graphics are complicated. The stylised ditch sections from Wor Barrow or South Lodge show the vertical plotting of different artefact types; with section figures impinging on plans, some of his published Cissbury illustrations are, in fact, overloaded with information. The General was producing a dense and multi-faceted record, and his figures strive to incorporate more than one aspect of the data. In effect portraying the 'workings' of sites, his more dynamic and complex archaeology was, in many respects, better suited to the more 'mechanical' means of model depiction. Accordingly, amongst early practitioners, he alone pushed and explored the possibilities of model formats beyond straightforward map-based representation.

WORLDS IN SMALL: THE LIVED PAST

The majority of extant archaeological models date to the first half of the twentieth century, and excavation models continued to be built up until the Second World War. Unlike those discussed in the previous section, many of these more recent models seem to have been specially built for museums and usually not by the excavators themselves. This tells of the growing public interest in archaeology and that museums wished to document local fieldwork. But there are exceptions, notably Stuart Piggott, later Abercromby Professor of Archaeology at Edinburgh, who made plan models of sites with which he was engaged. These include a Neolithic 'pit dwelling' from a ditch segment of E. C. Curwen's excavations of the Trundle, a hillfort in West Sussex (the model is in the Barbican House, Lewis Collections) and a portion of Gerhard Bersu's renowned Iron Age settlement at Little Woodbury in Wiltshire (Bersu 1940, acknowledgments; Stone 1958, pl. 68). Stylistically the latter closely matches the Standlake model and was, I suspect, inspired by it. Equally, variants of demonstration models continued to be constructed; among these are Clay's 'casting mould-like', Iron Age pit 'trays', and Devizes Museum's

Figure 5.8 (A) R. C. C. Clay's model of Iron Age pits from Fyfield Bavant Down (1924). (B) The Cunningtons' "Woodhenge Restored" (1929). Reproduced with permission of the Wiltshire Archaeological and Natural History Society.

"Woodhenge Restored" (Fig. 5.8B). Depicting a later Neolithic timber circle (a monument type ancestral to Stonehenge), the latter was excavated by Maude and Benjamin Cunnington in 1926–28, and the model was built in the following year for the museum by the husband of this team. In the stark abstraction of its concentric post setting, it can be considered as bridging reconstruction and demonstration formats.[13]

Over the course of the century full reconstruction modelling came increasingly to the fore. 'Bringing to life' the results of excavations, this not only involved the reconstruction of sites in whole or in part, and individual buildings and room interiors, but also generic scenes of, for example, typical prehistoric flint mines or houses. Frequently peopled, it is these endearing reconstructions that captured attention in the museums of our youths. Certainly their appeal is markedly different than that of the comparatively 'lifeless' demonstration models of the Standlake and Pitt Rivers' excavations. Akin to model railway displays and dolls' houses, these 'worlds shown small' conceptually straddle a divide between play and instruction, and combine a sense of grand Lilliputian overview with a hobbyist's fine detail. A fair critique is their degree of invention from the ground up and their normative (e.g., gender stereotypic) assignation of activities. Sophus Muller, for example, as Director of the National Museum of Denmark at the turn of the nineteenth century, banned the inclusion of reconstruction models on the ground of their extreme 'fictionalisation' (Sørensen 1999, 205).

While this emphasis upon 'as lived' scenes obviously has parallels in the development of museum habitat dioramas (Moser 1999; Nyhart, this volume), the portrayal of archaeology within the popular press contributed too. Crucial was the *Illustrated London News* coverage of the great discoveries of the day, and particularly the accompanying dramatic reconstruction paintings by their 'special artist', Amedee Forestier. In 1911, for example, the paper devoted a section to Arthur Bulleid and Harold St George Gray's excavation of the Iron Age lake settlement at Glastonbury, Somerset. Waterlogged, this extraordinary site produced a wealth of material that encouraged 'full-bodied' reconstruction figures and, for the first time, allowed a convincing picture of life in later prehistoric Britain to be portrayed.[14] Effectively setting the scene, reworked images of Glastonbury also featured in Marjorie and Charles Quennell's hugely popular *Everyday Life in the New Stone, Bronze, and Early Iron Ages*, first published in 1922 and subsequently reprinted three times prior to 1960 (Quennell and Quennell 1922).[15]

Encouraged by the recovery of major settlement sites, from the 1930s archaeology itself began to don a 'reconstructionalist' ethos, both redefining its scope and seeing amongst its roles the putting of 'flesh onto the bare bones' of the past (Evans 1989). Through this emphasis on daily-life settlement archaeology and particularly the recovery of distinctly archaeological building types (e.g., the later prehistoric post-built roundhouse), the reconstruction of houses was by the middle of the twentieth century felt to be the job of excavators. Influenced by his German background and the 'Haus und Hof' tradition of ethnographic house studies, Bersu championed this approach:

> It is the duty of every excavator to attempt a reconstruction of the building he has found, despite the many uncertain factors involved. Speculation of this kind often leads us to notice features in the soil, which we should otherwise have overlooked. Many a building, believed to be completely excavated, would be recognised as not so if only the excavator had tried to reconstruct its mode of building (Bersu 1940).

In this capacity modelling, both on paper and in solid format, was, effectively, a matter of structural soundness: the archaeological construct had to be able to stand.[16] Not only did Bersu supervise the making of a reconstruction model of one of his Manx Iron Age houses (Bersu 1977, 23, pl. iii), but also various timber-frame models of the main Woodbury building (Hawkes 1946; Evans 1989; 1998, figs 6 and 7). The ultimate test, life-size building reconstruction, was undertaken for the main Woodbury house that was erected on a backlot at Pinewood Studios for the Ministry of Information's wartime film, *The Beginning of History* (Hawkes 1946).

It is relevant here that archaeology never really developed experimental models applied to building reconstruction and destruction or general depositional principles. Instead, miniaturisation related solely to visualisation. There was no sense of trying to scale down building stress (e.g., duplicate materials) or processes of weathering and decay. The full-scale Woodbury House was itself only an elaborate stage set, with the internal posts of concrete moulded on steel girders. It is only really in the landscape-scale projects since the 1960s—the Overton Down Earthwork or the Buster Experimental Farm—that, focusing much more on long-term processes than on structural form, an experimental component was incorporated into reconstruction.

MODEL NETWORKS AND OPERATIONAL DOMAINS

This chapter has been about how what is portrayable in different media has been linked to what is conceivable. Through experimentation with models in the nineteenth and earlier twentieth centuries archaeology explored how to visualise its own activity. Yet this also occurred within a broader milieu, involving both the representation of the past to the public at large and the employment of models as a means of technical demonstration within the sciences as a whole.

The application of modelling in British archaeology seems essentially to have hinged upon two networks. While successive, there is no evidence that the first inspired the second. From a broader background of eighteenth-century model-building, including those few known antiquarian exemplars (Stukeley and Waltire's Stonehenges, and Conway's Mont St Helier 'temple'), the first coalesced around the Soane–Britton axis, by whom Tongue was probably inspired. Within this, Britton's association with Browne may have had links to eighteenth-century practitioners, particularly Waltire. Pitt Rivers was clearly pivotal to later developments; possibly inspired by Stone's Standlake, he obviously fostered both Gray's and Fox and Hope's model-making efforts. Although few continuities exist between these networks—primarily the affinities between Pitt Rivers' simple rendering of megaliths for his official recording of ancient monuments and the earlier 'rude stone' constructions of Tongue and Browne—any bridging of the two must relate to what lay behind Stone's Standlake base-plan model. I suspect a geological source, but its direct inspiration remains unknown.

Another underlying strand is the specific impact of displays within, and benefactions to, the Society of Antiquaries of London. This was especially important in the latter half of the nineteenth century and is probably how Pitt Rivers was exposed to Stone's Standlake, just as it was there that the excavators of Silchester apparently first saw his own models. Involving the presentation of finds and site models, various Society meetings (both in London and the regions) were evidently the key venue of the discipline's interaction. It was in the hubbub of face-to-face *conversazione* that results would be adjudicated through the collective verification of a select audience. By the cumulative 'weight' of excavations, and shifts towards various schools of approach and developments within fieldwork and museum practice, over the first half of the twentieth century this small group dynamic began to fragment. This reflects the rise of professional domains and the growth of the subject's audiences. Establishing its own 'models' of methodological practice and codes of specialist reportage (i.e., graphic conventions and textual formulae), model-building largely ceased within the discipline itself. 'Professional' archaeologists increasingly groomed their output towards publication alone, whereas model construction was embraced by museums as a key display tool.[17]

Just as with the advent of off-set lithography and colour slide film (and now interactive computer graphics), this legacy attests that the 'archaeology' of archaeology should not just be intellectual history, but should also encompass its technologies and media of communication. The subject's histories have ignored this practical knowledge, of which models were part; for the practical discourse of archaeological representation has always involved *seeing*, as much as 'reading', the past.

NOTES

For the provision of 'model' information, I am particularly grateful to C. Anderson (Oxfordshire County Museum), R. Boast and C. Chippindale (University of Cambridge, Museum of Archaeology and Anthropology), A. Deathe (Salisbury and South Wilts. Museum), S. Fox (Roman Baths Museum & Pump Room, Bath),

J. Frisby and D. Morgan Evans (Society of Antiquaries of London), M. Fulford (Department of Archaeology, University of Reading), J. Greenaway (Reading Museum), A. Gwilt (National Museum of Wales), S. Jones (National Army Museum, London), F. Marsden (Sussex Archaeological Society, Barbican House, Lewes), S. C. Minnitt (Somerset County Museums Service), S. Needham (British Museum), M. O'Hanlan (Pitt Rivers Museum, Oxford), S. Palmer and M. Richardson (Sir John Soane Museum), P. Robinson (Wilts. Heritage Museum, Devizes), A. Sheridan (National Museum of Scotland), A. Sherratt (Ashmolean Museum), D. Thorold (St Albans Museum), H. S. Torrens (University of Keele), C. West (Whipple Museum) and P. Woodward (Dorset County Museum). M. Jones, Pitt Rivers Professor of Archaeological Science, Department of Archaeology, University of Cambridge kindly provided access to the General's photographic album. M. Bowden, R. Bradley, G. Lucas, I. Kinnes, J. Nordbladh, M. Rowlands, and M. L. S. Sørensen were generous with advice and critique.

1. See Bradley 1997 on archaeology's visualisations and 'learning to see' in fieldwork.

2. For example, the exhibition of Captain William Siborne's great model of the battle of Waterloo within the Egyptian Hall, Piccadilly, between 1838 and 1841 attracted over 100,000 fee-paying visitors (Siborne [circa 1838]; Anon. 1997). See also Altick 1978, 213 on Brunetti's equally renowned model of ancient Jerusalem and other Holy Land displays during the 1840s.

3. See Morrissey 1999, 108–30, concerning the false proportions of follies, and particularly the miniaturisation in Walpole's 'little castle', which was effectively a scale version of its larger Gothic sources.

4. Here is Dr William Withering writing to James Norris in July, 1797: "You will be pleased to learn that Mr Waltire has a most accurate model made of cork, to a pretty large scale, of Stonehenge in its late state, and to the exactness of which I can bear testimony. He has also made another model of what he thinks was its perfect state" (Peck and Wilkinson 1950, 160–61). Apart from the Cabinet models, Browne is known to have sold at least four other 'before and after' Stonehenge pairs. These went for seven guineas each and, anticipating greater demand, he even had moulds made so that plaster copies of the stones could be quickly produced (Chippindale 1985, 128–29). Conveying a sense of a 'heavy' architectonic record and an almost geological solidity of representation, the Ashmolean Museum's versions of Browne's Avebury and Stonehenge models were respectively accompanied by specimens of "earth from its centre" and "stones from the circles and altar" (catalogue nos 329–31).

5. Society of Antiquaries of London Minutes 1767 (vol. 10: 393–94). While the "Man of very singular Talents" may have been one of Jenkins' own assistants, he was more likely the Neapolitan model-maker, Giovanni Alteri; Giovanni Stern (1734–94) was the architect to the papal palaces (*Antiquaries Journal* 1965, 222, notes 1 and 2). See Vogt 1998, 260–66 on the modelling of classical buildings during the eighteenth century. Upon receipt of Conway's St Helier model, when

recording their gratitude in the Minute Books, the Society of Antiquaries remarked upon the uniqueness of this form of presentation (vol. 12: 81–84). Presumably this refers to the modelling of a 'native' monument, since they had received the model of the Sibyl temple some 20 years earlier.

6. The Ashmolean Museum has a nineteenth-century cranked roller-blind system on which illustrations could be advanced to accompany lectures (A. Sherratt, personal communication). The 1877 engraving of Heinrich Schliemann's evidently slideless, audience-packed lecture concerning his work at Mycenae to the Society of Antiquaries of London in the *Illustrated London News* in March of that year (reproduced in Bacon 1976, 55, fig. 30) conveys a sense of the circumstances of these early 'performances'. A colleague experienced what must be comparable lecture conditions whilst attending an archaeological symposium in the final years of the Soviet Union. Lacking slides and with only limited publication opportunities, the speakers hung their figures around the hall. When referred to, members of the audience would cross the room to examine these, interjecting comments and queries. In short, a state of theatrical pandemonium! (M. L. S. Sørensen, personal communication.)

7. See also Pitt Rivers 1898, 23–24; and Thompson 1977, 99. Here is Pitt Rivers' assistant, Harold St George Gray, on the General's Farnham models: "The plaster models, which are heavy and on the whole less satisfactory than the wooden models, are built up on boards bearing traces of plans of the ancient sites. Vertical wires of brass are then stuck into the board at all places at which levels have been taken, and they are cut off at the ascertained height of each spot above a low datum. The wire-tops mark the surface of the plaster model when completed. Such additions as gates, hedges, trees and graves containing human skeletons, etc. have to be added to scale, and the model finally painted and lettered" (Gray 1929, 36–37).

8. Christopher Hawkes' reassessment and publication of the Iwerne villa (Pitt Rivers' one unpublished Cranborne Chase site) 50 years after its excavation, owed much to the surviving model (Hawkes 1947; Dudley Buxton 1929, pl. XI). Equally, in their recent Cranborne Chase investigations Barrett and colleagues note that the South Lodge Camp contour model so accurately reproduced the original mapping that it shows earthwork features which the General did not recognise and were identified only in the later excavations (Barrett, Bradley, and Green 1991, 14–15). Similarly, during the later campaign an extant model provided the only known plan of a site (Handley Barrow 26) and—telling of the limitations of graphic depiction—in other instances models proved a more sensitive record than the published excavation plans. Examples are the Late Neolithic cemetery around the Wor Barrow and the irregular ditch at Handley Barrow 27 (Barrett, Bradley, and Green 1991; R. Bradley, personal communication).

9. There is also a pre-excavation contour model; in its 'after'-companion the height of the removed mound is indicated by the 'monolithic' baulks employed in its excavation.

10. Upon his return to Britain from Java, Sir Stamford Raffles (the founder of Singapore) brought back a series of doll-like figures illustrative of local 'social types', and also miniature sets of contemporary weapons and the instruments for gamelan orchestras. Used as the basis for 'as-if-from-life' illustrations in his *History of Java* of 1817, these were commissioned as a means to transmit information—this being held to be the prime goal rather than any intrinsic value of the original object (Barley 1999). Ethnography would develop its own range of reconstruction models, such as of the African or South Pacific village (M. O'Hanlon, personal communication). Although not involving miniaturisation, the archaeological equivalent of Raffles' pieces are casts of individual artefacts. Ranging from flint axes to Bronze Age swords, the British Museum, for example, has some 90 in its stores. Pitt Rivers also commissioned casts as copies; the only extant record of the West Buckland Bronze Age hoard is, for example, from such a set (Taylor 1982). Mounted on sticks like giant lollipops, the General also had casts made of excavated stakehole impressions.

11. Generally having much more tangible remains of 'formal buildings' (and closer professional architectural ties) than prehistory, there has been a strong tradition of reconstruction modelling in the archaeology of the classical and ancient worlds. On a scale similar to Soane's Pompeii model, in the Roman Baths Museum, Bath, there is a large wooden model showing the base of the ground-storey walls and other features of the great baths complex. Although the precise date of its construction is unknown, it was part of the City of Bath's displays in the British Empire Exhibition of 1923 (illustrated in Fletcher 1945, 143). London, Colchester, and St Albans museums also have a number of models reconstructing their respective Roman towns. See Bernbeck 2000 for models in the Pergamon Museum, Berlin, and Kemp 2000 on the large model of Tell el-Amarna, Akhenaten's ancient Egyptian capital, recently commissioned for the Boston Museum of Fine Arts.

12. Accredited to O. C. Vidler of Dorchester, Gray would, at the very least, have overseen the construction of this piece. It cannot be coincidental that the transfer of Warne's earlier model to Dorset corresponds with the dates of Gray's excavation of the 'Rings' and that Herbert Toms (Pitt Rivers' other active later assistant) was then curator of Brighton Museum and apparently still in contact with Gray (R. Bradley, personal communication). Aside from two topographic models of the great hillfort of Maiden Castle, excavated by Wheeler in the 1930s, the Dorset Museum also holds a three-stage sequence model of an imaginary Roman excavation, inspired by the recently excavated town house at Colliton Park. Made of painted plaster, card, and sawdust (each 40 × 51 cm), they progressively show a nineteenth-century estate cottage setting into which has been cut a grid of survey trenches, and the subsequent exposure of a Late Roman corridor building. It was apparently made in the 1930s by the daughters of the museum's Curator, Colonel Drew.

13. The roofed depiction of this monument—on paper—had to wait for Piggott's overview of such structures (Piggott 1940).

14. At least in Britain, Forestier arguably established the 'rules' for subsequent domestic reconstruction which, until recently, have remained rather formulaic. Although claiming to be based on a detailed study of the Glastonbury results, as a professional illustrator his settings—for example, males returning home from the hunt poised at the roundhouse door with expectant females huddled around the hearth—basically follow the conventions of nineteenth-century historical interiors (e.g., "The Life of Alfred" or "Scenes from Shakespeare"). Lacking the drama of Forestier's pictures, Bulleid, the site's director, had two 'unpopulated' models made to show the portions of the settlement as reconstructed (Coles, Goodall, and Minnitt 1992, fig. 21). Vogt 1998 discusses the broader social and architectural impact of the prehistoric Swiss lake village discoveries and its models during the later nineteenth century (Vogt 1998, 260, figs 192–94).

15. These figures are reproduced in Coles and Minnitt 1995; see also Bacon 1976, 135–36.

16. Bersu's archives reveal that his thinking and recording processes were predominately visual, with 'text' a secondary concern (Evans 1998); see Bradley 1997 for the visual arts backgrounds of many leading archaeologists.

17. Some archaeologists have participated with museums in the making of major site-reconstruction models over the last 30 years, for example, Francis Pryor's Iron Age Fengate compounds (Pryor 1984, pls 27 and 28) and the Museum of London's 'Roman waterfront' (Milne 1985), and several were commissioned for the 1986 British Museum exhibition, "Archaeology in Britain since 1945".

REFERENCES

Altick, Richard. 1978. *The Shows of London*. Cambridge, Mass.: Belknap Press.

Anon. 1997. "Siborne's Waterloo." *Military Illustrated* 111: 47.

Awdry, J. W. 1870. "President's address." *Wiltshire Archaeological and Natural History Magazine* 12: 133–52.

Babington, Charles. 1865. *An Introductory Lecture on Archaeology Delivered before the University of Cambridge*. Cambridge: Deighton, Bell.

Bacon, Edward, ed. 1976. *The Great Archaeologists*. London: Secker & Warburg.

Barley, Nigel. 1999. "Introduction." In *The Golden Sword: Stamford Raffles and the East*, ed. Nigel Barley, 11–15. London: British Museum.

Barrett, John, Richard Bradley, and Martin Green. 1991. *Landscape, Monuments, and Society*. Cambridge: Cambridge University Press.

Bernbeck, Reinhard. 2000. "The exhibition of architecture and the architecture of an exhibition." *Archaeological Dialogues* 7: 98–145.

Bersu, Gerhard. 1940. "Excavations at Little Woodbury, Wiltshire. Part I: The settlement as revealed by excavation." *Proceedings of the Prehistoric Society* 6: 30–111.

———. 1977. *Three Iron Age Round Houses in the Isle of Man*, ed. C. A. Ralegh Radford. [Douglas, Isle of Man]: Manx Museum and National Trust.

Bowden, Mark. 1991. *Pitt Rivers: The Life and Archaeological Work of Lieutenant-General Augustus Henry Lane Fox Pitt Rivers, DCL, FRS, FSA*. Cambridge: Cambridge University Press.

Bradley, Richard. 1975. "Maumbury Rings, Dorchester: The excavations of 1908–1913." *Archaeologia* 105: 1–97.

———. 1983. "Archaeology, evolution, and the public good: The intellectual development of General Pitt Rivers." *Archaeological Journal* 140: 1–9.

———. 1997. "'To see is to have seen': Craft traditions in British field archaeology." In *The Cultural Life of Images: Visual Representation in Archaeology*, ed. Brian L. Molyneaux, 62–72. London: Routledge.

Chippindale, Christopher. 1983. *Stonehenge Complete*. London: Thames & Hudson.

———. 1985. "John Britton's 'Celtic Cabinet' in Devizes Museum and its context." *Antiquaries Journal* 65: 121–38.

Clay, R. C. C. 1924. "An early Iron Age site on Fyfield Bavant Down." *Wiltshire Archaeological Magazine* 42: 457–96.

Coles, John, Armynell Goodall, and Stephen Minnitt. 1992. *Arthur Bulleid and the Glastonbury Lake Village, 1892–1992*. Somerset County Museums Services.

Coles, John, and Stephen Minnitt. 1995. *"Industrious and Fairly Civilized": The Glastonbury Lake Village*. Somerset County Museums Service.

Colley, Linda. 1992. *Britons: Forging the Nation, 1707–1837*. New Haven: Yale University Press.

Conway, Henry S. 1787. "Description of a druidical monument in the island of Jersey." *Archaeologia* 8: 386–88.

Dudley Buxton, L. H., ed. 1929. *The Pitt Rivers Museum, Farnham: General Handbook*. Farnham: Farnham Museum.

Edmonds, Mark, and Christopher Evans. 1991. "The place of the past: Art and archaeology in Britain." *Excavating the Present*, part 2. Cambridge: Kettle's Yard.

Evans, Christopher. 1988. "Monuments and analogy: The interpretation of causewayed enclosures." In *Enclosures and Defences in the Neolithic of Western Europe*, ed. Colin Burgess, Peter Topping, Claude Mordant, and Margaret Maddison, 47–73. Oxford: British Archaeological Reports, International Series 403.

———. 1989. "Archaeology and modern times: Bersu's Woodbury 1938 & 1939." *Antiquity* 63: 436–50.

———. 1994. "Natural wonders and national monuments: A mediation upon the fate of The Tolmen." *Antiquity* 68: 200–8.

———. 1998. "Constructing houses and building context: Bersu's Manx roundhouse campaign." *Proceedings of the Prehistoric Society* 64: 183–201.

———. 2000. "Megalithic follies: Soane's 'druidic remains' and the display of monuments." *Journal of Material Culture* 5: 347–66.
Fletcher, Banister. 1945. *A History of Architecture on the Comparative Method.* London: Batsford.
Gray, Harold St G. 1903. "Relief model of Arbor Low Stone Circle, Derbyshire." *Man* 3: 145–46.
———. 1929. "Models of ancient sites in Farnham Museum." In *The Pitt Rivers Museum, Farnham: General Handbook*, ed. L. H. Dudley Buxton, 36–37. Farnham: Farnham Museum.
Hawkes, Christopher, F. C. 1947. "Britons, Romans, and Saxons round Salisbury and in Cranborne Chase." *Archaeological Journal* 104: 27–81.
Hawkes, Jacquetta. 1946. "The beginning of history: A film." *Antiquity* 20: 78–82.
Hibbs, James. 1985. "Little Master Stonehenge: A study of the megalithic monument from Le Mont de la Ville, Saint Helier." *Annual Bulletin of the Society Jersiaise* 110: 49–74.
Higgins, Godfrey. 1827. *The Celtic Druids*. London: R. Hunter.
Hodder, Ian. 1986. *Reading the Past: Current Approaches to Interpretation in Archaeology*. Cambridge: Cambridge University Press.
———. 1989. "Writing archaeology: The site report in context." *Antiquity* 63: 268–74.
Kemp, Barry. 2000. "A model of Tell el-Amarna." *Antiquity* 74: 15–16.
Levine, Philippa. 1986. *The Amateur and the Professional: Antiquarians, Historians, and Archaeologists in Victorian England, 1838–1886*. Cambridge: Cambridge University Press.
Milne, Gustav. 1985. *The Port of Roman London*. London: Batsford.
Molesworth, Richard. 1787. "Description of the Druid Temple lately discovered on the top of the hill near St Hillary in Jersey." *Archaeologia* 8: 384–85.
Morrissey, Lee. 1999. *From the Temple to the Castle: An Architectural History of British Literature, 1660–1760*. Charlottesville: University Press of Virginia.
Moser, Stephanie. 1999. "The dilemma of didactic displays: Habitat dioramas, life-groups, and reconstructions of the past." In *Making Early Histories in Museums*, ed. Nick Merriman, 95–116. Leicester: Leicester University Press.
Peck, Thomas W., and Kenneth D. Wilkinson. 1950. *William Withering of Birmingham, M.D., F.R.S., F.L.S.* Bristol: John Wright & Sons.
Physick, John, and Michael Darby. 1973. *"Marble Halls": Drawings and Models for Victorian Secular Buildings*. London: Victorian and Albert Museum.
Piggott, Stuart. 1940. "Timber circles: A re-examination." *Archaeological Journal* 96: 193–222.
———. 1978. *Antiquity Depicted: Aspects of Archaeological Illustration*. London: Thames & Hudson.
———. 1989. *Ancient Britons and the Antiquarian Imagination*. London: Thames & Hudson.
Pitt Rivers, Lt-Gen. Augustus H. L. F. 1887. *Excavations in Cranborne Chase, near Rushmore on the Borders of Dorset and Wilts*, vol. 1. Privately printed.

———. 1892. *Excavations in Cranborne Chase, near Rushmore on the Borders of Dorset and Wilts*, vol. 3. Privately printed.
———. 1897. "Presidential address to the Dorchester meeting of the Archaeological Institute." *Archaeological Journal* 54: 311–39.
———. 1898. *Excavations in Cranborne Chase, near Rushmore on the Borders of Dorset and Wilts*, vol. 4. Privately printed.
Pryor, Francis M. M. 1984. *Excavation at Fengate, Peterborough, England: The Fourth Report*. Leicester: Northamptonshire Archaeological Society; Toronto: Royal Ontario Museum.
Quennell, Marjorie, and Charles Henry Bourne Quennell. 1922. *Everyday Life in the New Stone, Bronze, and Early Iron Ages*. London: Batsford.
Richardson, Margaret. 1989. "Model architecture." *Country Life* 183 (September 21): 224–27.
Siborne, William. [circa 1838]. *Guide to the Model of the Battle of Waterloo*. London: P. Dixon.
Smyth, W. H. 1830. "Account of an ancient bath in the Island of Lipari." *Archaeologia* 23: 98–102.
Sopwith, Thomas. 1841. *Description of a Series of Geological Models*. Newcastle upon Tyne: Printed for the author.
Sørensen, Marie Louise S. 1999. "Sophus Otto Muller." In *Encyclopaedia of Archaeology: The Great Archaeologists*, vol. 1, ed. Tim Murray, 193–209. Oxford: ABC-Clio.
Stevens, Frank. 1934. "The Highfield pit dwellings, Fisherton, Salisbury." *Wiltshire Archaeological Magazine* 46: 579–624.
Stone, J. F. S. 1958. *Wessex before the Celts*. London: Thames & Hudson.
Stone, Stephen. 1857. "Account of certain (supposed) British and Saxon remains . . . ." *Proceedings of the Society of Antiquaries of London*, 1st ser., 4: 92–100.
Taylor, Robin J. 1982. "The hoard from West Buckland, Somerset." *Antiquaries Journal* 62: 13–17.
Thompson, Edward P. 1968. *The Making of the English Working Class*. Harmondsworth: Pelican.
Thompson, M. W. 1977. *General Pitt-Rivers: Evolution and Archaeology in the Nineteenth Century*. Bradford on Avon: Moonraker.
Thornton, Peter, and Helen Dorey. 1992. *A Miscellany of Objects from Sir John Soane's Museum*. London: Laurence King.
Vogt, Adolf Max. 1998. *Le Corbusier, the Noble Savage: Toward an Archaeology of Modernism*. Cambridge, Mass.: MIT Press.
Watkin, David. 1996. *Sir John Soane: Enlightenment Thought and the Royal Academy Lectures*. Cambridge: Cambridge University Press.
Wheeler, Mortimer. 1954. *Archaeology from the Earth*. Oxford: Clarendon Press.
Wilton-Ely, John. 1969. "The architectural models of Sir John Soane: A catalogue." *Architectural History* 12: 5–38.

CHAPTER SIX

Monsters at the Crystal Palace

James A. Secord

Travelling by rail from London in the summer of 1854, the journalist Harriet Martineau overheard a conversation as her train neared the Crystal Palace. The great iron and glass structure had recently been rebuilt on the southern outskirts of the metropolis after the triumphant success of the Great Exhibition. Just before arriving, the passengers were presented with an astonishing sight: a lake with islands inhabited by models of huge beasts, some over thirty feet long:

> "What *are* those?" exclaimed a passenger in the railway carriage, as it ran along the embankment above the gardens. "The antediluvian animals, to be sure", a comrade informed him. "Why antediluvian?" "Because they were too large to go into the ark; and so they were all drowned" ([Martineau] 1854, 540).

Over a million people a year for the next half-century saw these remarkable full-scale models of extinct reptiles and mammals—more than had ever viewed a scientific work of any kind. Although ancient forms of life had previously been presented to a wide audience in panoramas and books, they had never been shown in three dimensions. And this was to be done as part of a commercial enterprise supported by thousands of small shareholders.

The Crystal Palace models continue to appear on postcards, mugs, games, and in countless books, articles, and television programmes. Even in historical writings, they have come to serve as iconic substitutes for an understanding of Victorian attitudes towards science. The New Year's eve dinner held inside the mould of one of the creatures is among the most celebrated meals in history (certainly the most famous involving a scientific model) and a potent symbol of the modern fascination with dinosaurs. Still surviving on their islands in Crystal Palace Park, the models possess a 3-D solidity that appears to offer a direct link with the past, so that the spectacle renews itself every time images of them are reproduced.

This chapter explores a different approach, by taking seriously the character of the displays as models. Like several other essays in this book, it places the models within a triangular relationship between patrons, producers, and public: in terms of projected and actual use. It suggests that large-scale displays intended for public consumption need, at least in the first instance, to be investigated using techniques developed in art history for examining the commissioning, construction, and viewing of monumental buildings and sculpture. The Crystal Palace exhibits can then be understood as the apotheosis of a short-lived conjunction between commercial capitalism and rational education in the early 1850s. It is a monument to this utopian vision—which differs from our own commodity spectacle—that 3-D scientific models were commissioned and built at vast expense. To understand how hard-headed investors expected to profit from an improving display, we need to recapture the moment when the very notion of the spectacle was under construction.

A UTOPIA OF THE EYE

The success of the Great Exhibition, it was anticipated, could continue forever. The response of the crowds that had thronged to central London in 1851 suggested that decades of class conflict and religious strife had come to an end. The huge structure that had housed the exhibition, re-erected south of the capital at Sydenham, could become the 'People's palace' (Fig. 6.1). The Hyde Park show, displaying the useful and the new, had been the experiment; Sydenham, founded on the ideal, would be the realisation, pointing towards "one immortal striving for the goal of moral and intellectual and physical advancement" (Cockle 1854, 33). The displays would make a profit by reforming the public's habits of observation, so that a programme of visual education would bring all the people together.

These grandiose projections were matched by the gardener Joseph Paxton's plans for the rebuilt palace, which was twice as tall as the original and covered almost one hundred more acres. (Cynics said it looked like a giant hotel.[1]) The interior was organised as a series of 'courts'. Some of these, as

Figure 6.1 The Crystal Palace and gardens. This view from the railway line was sketched by Benjamin Waterhouse Hawkins and printed in colour by George Baxter. Like most of the material relating to the exhibition of extinct animals, it was published before the display was completed. Note the inclusion of human figures to give a sense of (exaggerated) scale. Reproduced by permission of the Trustees of the British Museum, Department of Prints and Drawings, 1901-11-5-53.

at Hyde Park, were devoted to modern manufactures, from grand pianos and clothing to books and stationery. The great novelty was a sequence of exhibits displaying the progress that had led to modern civilisation. Visitors could marvel at plaster models of the peoples of Africa, India, North America, and the Pacific; they could stroll through replicas of a mediaeval cloister, a Renaissance court, the palaces of Assyria, and the pyramids of Egypt.

If the interior could be read as a model of progressive order, then the avenues and waterworks of the gardens extended this order into the surrounding landscape. The largest fountains ever built were to play high into the air. The whole site was designed as a visual treat, a school for the eye, and a factory for improving public taste. The scheme was grounded in the theories of the Swiss educational theorist Johann Heinrich Pestalozzi, with the aim of conveying knowledge directly through the senses. As the architect M. Digby Wyatt wrote, the sciences would be put to "the test of vision". The Palace, made of glass, was a prism through which the properly placed eye would see into the nature of things, in an understanding that went beyond words. Appreciating the ethnological exhibits, Wyatt noted, required "the simple use of the organs of sight, and little more than the natural instincts of a child" (Wyatt 1854, 10).

In an address to the Society of Arts, Benjamin Waterhouse Hawkins, the sculptor of the extinct animal models, stressed the philosophy of visual education upon which the Crystal Palace Company was founded. Not just the common objects of life, he observed, but the conclusions of the highest science, could be communicated:

> This direct teaching through the eye has been recognised as a principle and a facility of education for some years past, even in the limited sphere of schools; and I believe the name of Pestalozzi deserves the most honourable mention in connection with its first enunciation as a recognized facility on principle. His, and his followers', lessons on objects were urged upon the public some twenty years ago, and a writer who was quoted at the time ... shrewdly observed, that "we daily call a great many things by their names, without even inquiring into their nature and properties, so that in reality it is only their names, and not the things themselves, with which we are acquainted" (Hawkins 1854, 444).

In building the antediluvian animals, Hawkins said, the order would be reversed, so that knowledge of things would come before the names. Even for the educated viewer, the size, life-like character, and three-dimensionality of the restorations would appear as a revelation.

The educational aims were underlined by other aspects of the geological displays, which included much more than the models. There was a replica lead mine and a limestone cavern with stalagmites and stalactites. Huge quantities of stone were imported to create a succession of strata, including coal-bearing rocks and a sequence of faults. Rocks of the appropriate era formed the islands on which the extinct animals stood (Doyle and Robinson 1993). The aim was to create a 3-D panorama of life, from water-dwelling dragons of the older Primary rocks, through the massive reptiles of the Secondary, to the lumbering mammoths and megatheria of the Tertiary. On each island an attempt was made (as in the museums discussed in Lynn Nyhart's chapter in this book) to show realistic life-groups; for example, some of the extinct deer were moved apart when it turned out that in the wild they

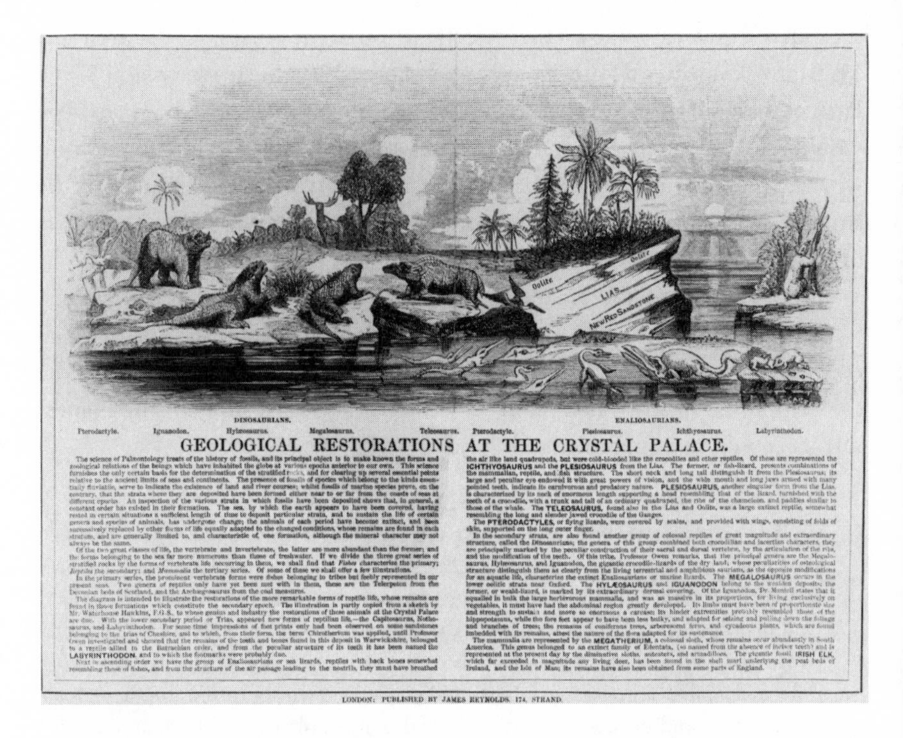

GEOLOGICAL RESTORATIONS AT THE CRYSTAL PALACE.

The science of Palæontology treats of the history of fossils, and its principal object is to make known the forms and zoological relations of the beings which have inhabited the globe at various epochs anterior to our own. This science furnishes the only certain basis for the determination of the stratified rocks, and for clearing up several essential points relative to the ancient limits of seas and continents. The presence of fossils of species which belong to the kinds essentially fluviatile, serve to indicate the existence of land and river courses; whilst fossils of marine species prove, on the contrary, that the strata where they are deposited have been formed either near to or far from the coasts of seas at different epochs. An inspection of the various strata in which fossils have been deposited shows that, in general, a constant order has existed in their formation. The sea, by which the earth appears to have been covered, having rested in certain situations a sufficient length of time to deposit particular strata, and to sustain the life of certain genera and species of animals, has undergone change; the animals of each period have become extinct, and been successively replaced by other forms of life equally adapted to the changed conditions, whose remains are found in each stratum, and are generally limited to, and characteristic of, one formation, although the mineral character may not always be the same.

Of the two great classes of life, the vertebrate and invertebrate, the latter are more abundant than the former; and the forms belonging to the sea far more numerous than those of freshwater. If we divide the three great series of stratified rocks by the forms of vertebrate life occurring in them, we shall find that *Fishes* characterise the primary; *Reptiles* the secondary; and *Mammalia* the tertiary series. Of some of these we shall offer a few illustrations.

In the primary series, the prominent vertebrate forms were fishes belonging to tribes but feebly represented in our present seas. Two genera of reptiles only have yet been met with in them, these are the Telerpeton from the Devonian beds of Scotland, and the Archegosaurus from the coal measures.

The diagram is intended to illustrate the restorations of the more remarkable forms of reptile life, whose remains are found in those formations which constitute the secondary epoch. The illustration is partly copied from a sketch by Mr. Waterhouse Hawkins, F.G.S., to whose genius and industry the restorations of those animals at the Crystal Palace are due. With the lower secondary period or Trias, appeared new forms of reptilian life,—the Capitosaurus, Nothosaurus, and Labyrinthodon. For some time impressions of foot prints only had been observed on some sandstones belonging to the trias of Cheshire, and to which, from their form, the term Chirotherium was applied, until Professor Owen investigated and showed that the remains of the teeth and bones found in this deposit in Warwickshire, belonged to a reptile allied to the Batrachian order, and from the peculiar structure of its teeth it has been named the LABYRINTHODON, and to which the footmarks were probably due.

Next in ascending order we have the group of Enaliosaurians or sea lizards, reptiles with back bones somewhat resembling those of fishes, and from the structure of the air passage leading to the nostrils, they must have breathed the air like land quadrupeds, but were cold-blooded like the crocodiles and other reptiles. Of these are represented the ICHTHYOSAURUS and the PLESIOSAURUS from the Lias. The former, or fish-lizard, presents combinations of the mammalian, reptile, and fish structure. The short neck and long tail distinguish it from the Plesiosaurus; its large and peculiar eye endowed it with great powers of vision, and the wide mouth and long jaws armed with many pointed teeth, indicate its carnivorous and predatory nature. PLESIOSAURUS, another singular form from the Lias, is characterized by its neck of enormous length supporting a head resembling that of the lizard, furnished with the teeth of a crocodile, with a trunk and tail of an ordinary quadruped, the ribs of the chameleon, and paddles similar to those of the whale. The TELEOSAURUS, found also in the Lias and Oolite, was a large extinct reptile, somewhat resembling the long and slender jawed crocodile of the Ganges.

The PTERODACTYLES, or flying lizards, were covered by scales, and provided with wings, consisting of folds of skin, supported on the long outer finger.

In the secondary strata, are also found another group of colossal reptiles of great magnitude and extraordinary structure, called the Dinosaurians; the genera of this group combined both crocodilian and lacertian characters, they are principally marked by the peculiar construction of their sacral and dorsal vertebræ, by the articulation of the ribs, and the modification of the teeth. Of this tribe, Professor Owen remarks, that the principal genera are the Megalosaurus, Hylæosaurus, and Iguanodon, the gigantic crocodile-lizards of the dry land, whose peculiarities of osteological structure distinguish them as clearly from the living terrestrial and amphibious saurians, as the opposite modifications for an aquatic life, characterize the extinct Enaliosaurians or marine lizards. The MEGALOSAURUS occurs in the lower oolitic strata near Oxford. The HYLÆOSAURUS and IGUANODON belong to the wealden deposits; the former, or weald-lizard, is marked by its extraordinary dermal covering. Of the Iguanodon, Dr. Mantell states that it equalled in bulk the large herbivorous mammalia, and was as massive in its proportions, for living exclusively on vegetables, it must have had the abdominal region greatly developed. Its limbs must have been of proportionate size and strength to sustain and move so enormous a carcase; its hinder extremities probably resembled those of the hippopotamus, while the fore feet appear to have been less bulky, and adapted for seizing and pulling down the foliage and branches of trees; the remains of coniferous trees, arborescent ferns, and cycadeous plants, which are found imbedded with its remains, attest the nature of the flora adapted for its sustenance.

The mammalia are represented by the MEGATHERIUM, a colossal sloth, whose remains occur abundantly in South America. This genus belonged to an extinct family of Edentata, (so named from the absence of incisor teeth) and is represented at the present day by the diminutive sloths, anteaters, and armadilloes. The gigantic fossil IRISH ELK, which far exceeded in magnitude any living deer, has been found in the shell marl underlying the peat beds of Ireland, and the Isle of Man; its remains have also been obtained from some parts of England.

LONDON: PUBLISHED BY JAMES REYNOLDS, 174, STRAND.

Figure 6.2 The panorama of geological progress at the Crystal Palace. From right to left, early amphibious forms are succeeded by dinosaurs, the crown of reptilian creation. Creatures from the age of mammals are shown in the background, including a megaloceros (the 'Irish elk', actually a giant deer) and a megatherium (a large ground sloth) grasping a tree. This card, published by James Reynolds, provided a textual supplement to the otherwise unlabelled visual display. Copies presumably could be purchased at the site. Source: Hawkins Papers, by permission of the Natural History Museum, London.

probably come together only during the mating season to fight.[2] The goal, though, was to model not only individual creatures or static tableaux, but the progressive history of life (Fig. 6.2). Because the display was in 3-D, the succession of living beings was experienced spatially, with different epochs revealed as one passed by on foot or by rail.

Geologists had included 2-D pictorial displays of ancient life in what one practitioner called their "raree show" for over two decades. In 1832, the *Penny Magazine* published a full-page woodcut of restored fossil remains as part of a series of introductory articles on geology (Rudwick 1992, 60–62).

Later in the same decade, the London Colosseum announced that a sequence of four scenes illustrating "The Geological Revolutions of the Earth" could be seen in its basement. These included the Creation, the Age of Iguanodon and Megalosaurus, the Garden of Eden, and the Deluge (Secord 2000, 440).

The most important precedents were in the work of the painter John Martin, whose visionary landscapes from the Bible and *Paradise Lost* provided the template for Paxton's overall designs for the rebuilt palace and gardens. Martin was also known for his illustrations of ancient life for geological treatises, scenes which combined lurid drama with the latest palaeontological research. One of these, "The Country of the Iguanodon", had shown a megalosaur feasting upon an iguanodon in a carefully researched landscape with appropriate vegetation (Fig. 6.3A). The original painting had hung in the Brighton museum of the surgeon and geologist Gideon Mantell, who also used it as the frontispiece to his introductory *Wonders of Geology* (Mantell 1838; Dean 1999; Rudwick 1992, 78–80). Such images provided a pictorial glimpse of what the Crystal Palace would attempt to realize in 3-D.

To fulfil this grandiose programme, the extinct animal models, like everything else at Sydenham, were made as large as possible, scaled up from the proportions of the biggest known individual bones. The projectors claimed that these were the largest sculptural models from which castings had ever been made. The cost was substantial: £13,729 for the mine, cave, and extinct animals alone (Sotheby 1855, 18–19). The publicity stressed their size, giving precise dimensions and the quantities of brick, mortar, and iron that went into the construction of each creature. To make an iguanodon, for example, took 600 bricks, 650 five-inch half-round drain tiles, 900 plain tiles, 38 casks of cement, 90 casks of broken stone, 100 feet of iron hooping and 20 feet of cube inch bar (Hawkins 1854, 447). Early pictures of the creatures show human figures walking nearby to give a sense of scale. Cartoonists brought the models into terrifying life, in illustrations that stressed their character as gigantic neogothic dragons.

The 1840s and early 1850s witnessed a taste for theatrical spectacle and extravagant stage sets. This was an age of the gargantuan in science and technology: the Great Eastern steam liner, Lord Rosse's Leviathan Telescope, James Nasymth's gigantic steamhammer. The Crystal Palace displays, and especially the exhibition of prehistoric life, appealed to this monstrous aesthetic. Geologists, palaeontologists, and the general public alike could wonder at the monsters that science had brought into being. As the *Quarterly*

A

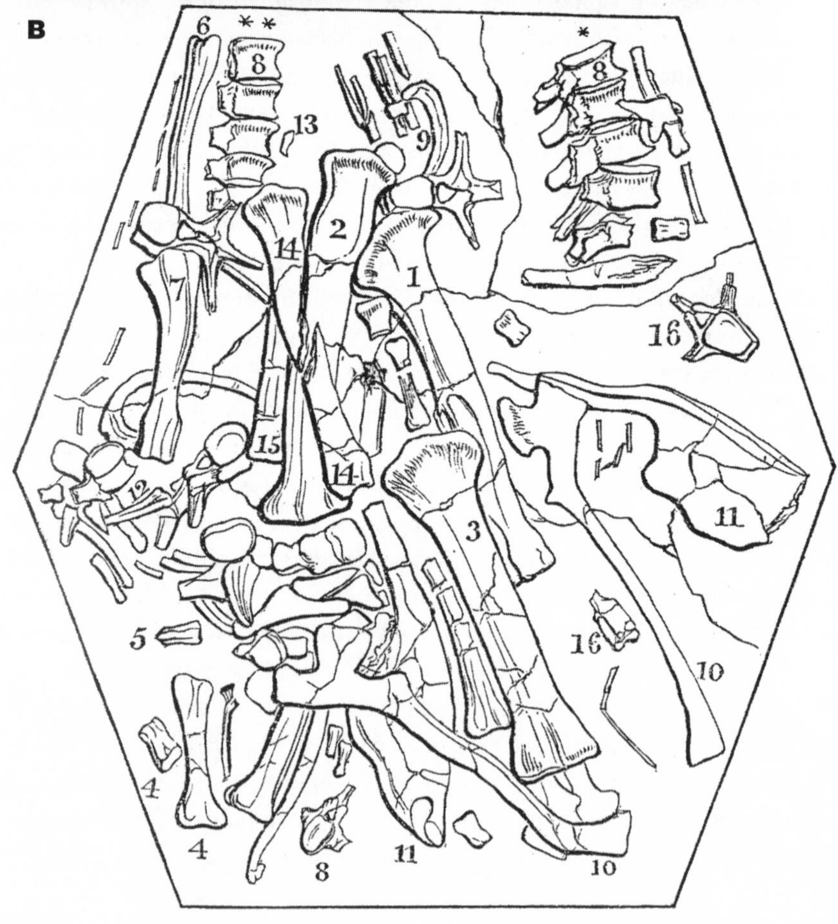
B

Review noted, reconstructions on this scale could attract both the unknowing multitude and the learned expert. "This is one of the most successful hits of the Company", the review observed, "for ignorance and knowledge will be alike gratified here" ([Eastlake] 1855, 353).

The educational value of the models depended on their size, for only in this way could they be seen by so many people at once. The creatures were not to be touched, but viewed from a distance on their islands. As in the case of the recently-completed statue of Lord Nelson at Trafalgar Square, this was monumental sculpture at the centre of the public eye. Contemporary reviewers stressed that the Crystal Palace monsters taught their most instructive lesson when the observer's gaze turned towards the crowd. Commentators from all points on the political spectrum claimed that the throngs could not be easily classified: it took a practised eye to distinguish working men, traders, and gentry when so few seemed destitute or poorly clothed ([Eastlake] 1855; [Martineau] 1854). "The comforts of life," the Nonconformist *London Quarterly Review* agreed, "must be widely spread in a land which can send out such a *mob*".[3] Such rhetoric extended the ideology of social cohesion embodied in the 1851 Exhibition (Auerbach 1999). Thus the models were not only designed to be seen by a large public but they were also intended to define what that public was.

From the moment exhibition-goers began the railway journey from London (which was included in the price of the entrance ticket), they were engaged in a process intended to appeal simultaneously to the senses and to reason. All of this, the projectors believed, could be accomplished through the economic mechanisms of commercial capitalism working according to the laws of nature. As the review in the radical *Westminster* noted, "The most timid politicians and economists need have no fear of any countenance to socialism from this experiment. It is property, in the shape of hard money,

Figure 6.3 (A) John Martin, "The Country of the Iguanodon", mezzotint frontispiece to Gideon Mantell's *Wonders of Geology* (London, 1838). The original painting hung in Mantell's Brighton museum above (B) the Maidstone iguanodon, one of the most complete specimens known at that time (Mantell, *Petrifactions and Their Teachings* [London, 1851], 306). Mantell encouraged displays of pictorial reconstructions in connection with the fossils on which they were based; this was not done at Sydenham.

every shilling earned by some individual, which has built this palace, and laid out these grounds" ([Martineau 1854], 542). The 'people', as members of a harmonious British nation, would be paying customers for their own visual education.

THE MODERN PYGMALION

That the Crystal Palace monsters were models and replicas, rather than actual fossil specimens, seems not to have bothered the projectors. As in nineteenth-century art exhibitions, the cult of the original had not yet taken hold, and casts were thought suitable for intellectual contemplation (Orvell 1989). The interior courts of the reconstructed Crystal Palace were filled with copies too—plaster portrait busts, replicas of famous sculptures, reconstructions of ancient ruins. Besides the modern manufactures, the only 'real' objects to be seen were the stuffed animals and potted plants in the natural history displays, and the original weapons, tools, and clothing in the ethnology galleries. On the geological islands, the strata on which the animals stood, and the antlers of the Irish elk, were specimens dug up from the ground. Everything else was a reconstruction.

From this perspective, the fragment, even if original, was dead: it was through the restored whole that the mind could best appreciate the life of past ages. This was thought to be especially true when the originals appeared to the non-expert eye as incomprehensible slabs of rock. The wonderfully preserved Maidstone Iguanodon was on show in the British Museum; but to the ordinary visitor it was a confused mass of bones (Fig. 6.3B). A true artist, like Martin, could infuse such remains with the appearance of life and forge a natural bond of sympathy with the observer. And for the people at large, an effective model could convey what the *Art-Journal* called "living action, with a vigour and truth fearful almost to look upon".[4]

Moving the Martinesque vision from two to three dimensions presented considerable obstacles. Full-scale models were expensive and effectively permanent. A sketch, wall painting, or panorama could be replaced in the light of new interpretations, but a model was set in stone. Financial speculation heightened the risks of scientific speculation. These were minimal if the restorations were unlikely to be challenged—no one doubted the fidelity of the model limestone cave or lead mine. Creating convincing full-sized versions of extinct forms of life, never seen by human eyes, was a very different matter. In 2-D art there were strong precedents for caricature and exaggeration, which had often been used in reconstructing hypothetical scenes of ancient life; but the Victorians expected 3-D sculptures to be realistic. There were some simple ways around this: close examination could be discouraged by placing the display on islands; reconstructing the hypothetical bits of sea-dwelling creatures could be avoided if these were hidden under water.

The fundamental criterion for contemporary sculpture was 'truth to life'. Accustomed to skilled taxidermy and living animals in zoos, viewers expected realistic poses, balanced proportions, and vivid expressions. In this respect, the problems posed at the Crystal Palace were similar to those faced by Edwin Landseer in fulfilling the commission for the lions at Trafalgar Square (Read 1982). Animal sculptures had to give the appearance of inner vitality, which created a bond of sympathetic perception. One American visitor commented that the model of the Irish elk had only one fault: "it did not actually breathe".[5] Great works of sculpture had to give the impression of life communicated by artistic genius.

For this reason, the Crystal Palace Company's directors realised that the right combination of skills in the model-maker would be the key to success. Waterhouse Hawkins, born in 1807, was an artist trained by William Behnes, a well-known painter and sculptor interested in natural history. Hawkins began his career as an oil painter in the early 1830s, exhibiting portraits, landscapes, and other works at the Royal Society of British Artists, the British Institution, and the Royal Academy. His skill in zoological drawing led to commissions to illustrate specimens in the Earl of Derby's menagerie at Knowsley Park from 1842 to 1847.

Hawkins developed an interest in educational models through his experience as a drawing master, which convinced him that traditional practices for teaching the art of sketching were unsuitable for all but the gifted. Too much time was spent on line, not enough on form. In 1843 Hawkins published a box of wooden models to help pupils master what he saw as the foundations of art in 3-D structure. The set consisted of six cubes, for the cube (as he explained in the accompanying pamphlet) was "the simplest of geometrical solids, and the archetype or parent of all forms". Each cube was composed of pieces that could be assembled in many different ways. By drawing these combinations, the student created the foundations for a knowledge of form. This was essential not only for drawing, but also for what Hawkins termed the "scientific training of the perceptive faculties of the mind" (Hawkins 1843, 4–5).

Convinced of the importance of 3-D form in relation to ordinary perception, Hawkins also turned his talents to animal sculpture. His work included commissions for the gardens of Biddulph Grange in Staffordshire and for the Zoological Society of London, including a "model from life" of a jaguar presented to Queen Victoria. He superintended several classes of the displays at Hyde Park in 1851, and exhibited a group of European bison in bronze (McCarthy 1994, 13).

The Sydenham project would never have been contemplated without Hawkins. He stressed, both at the Society of Arts and to those who visited his studio, how the sculptures were formed out of clay with his own hands, with attention to the finest details. Anatomical truth required the shaping hand of art. Conversely, as he had explained in connection with his models for teaching drawing, true works of art could only be created by the individual who united artistic skill with the knowledge of the naturalist. Only through this combination could the illusion of inner life be conveyed (Hawkins 1843, 5–6). Hawkins' own abilities were widely praised, and he presented letters of reference to the Crystal Palace Company from Charles Darwin and other leading naturalists.[6] One author hailed him as a "modern Pygmalion", a reference to the classical sculptor who breathed life into his own creations (Buckland 1858, 39). Hawkins was, as an enthusiast said, one of "God's scintillating stars of genius".[7]

Constructing "Frankensteinic"[8] models in 3-D posed an intricate combination of aesthetic, scientific, and practical problems. Hawkins began by consulting the writings of specialists who had figured and described the remains of extinct creatures. He also examined fossils in the British Museum, the Royal College of Surgeons, and the Geological Society. Hawkins then drew freehand sketches, often of great vividness, showing what the animals might have looked like in life. On this basis, just as for any sculpture, he prepared a model in clay at one-sixth or one-twelfth of the original size. Once this was corrected and adjusted, Hawkins instructed his workmen to make a full-scale model. This was a large undertaking, requiring in some instances 30 tons of clay and internal wooden supports. Hawkins then personally worked carefully over the detailing, to give the appearance of living skin and underlying muscle. The various stages, which are those typically found in Victorian monumental sculpture (Read 1982), are illustrated in Figure 6.4.

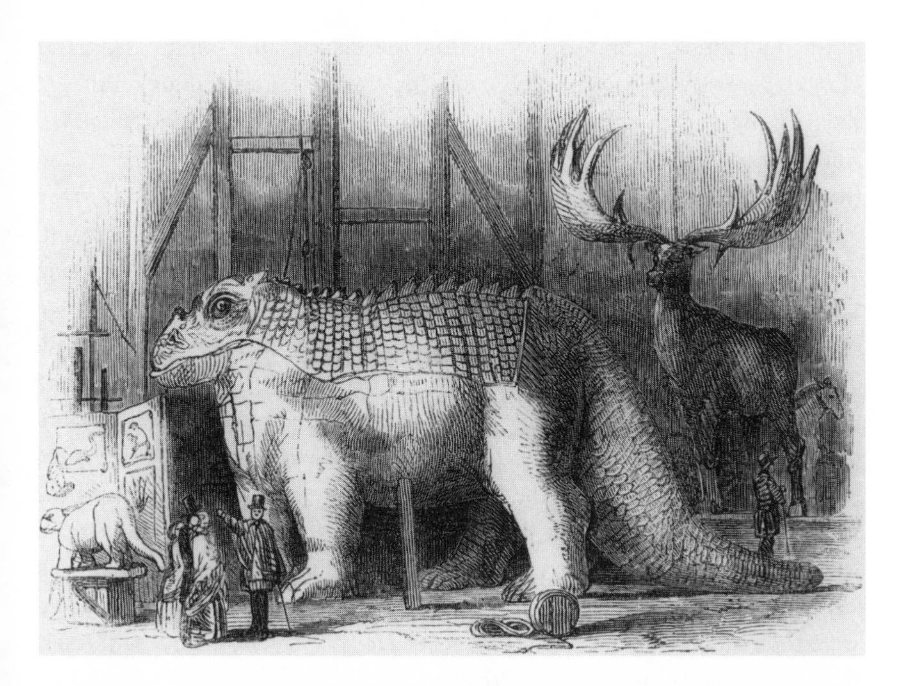

Figure 6.4 Visitors to the workshop were shown the different stages involved in making the sculptures, from drawings, to small-scale model, to full-scale model in clay. Illustrated here are the megaloceros and the iguanodon; the latter appears to be partly covered by the mould that would then be used to cast the final version. Source: [George Measom], *The Official Illustrated Guide to the Brighton and South Coast Railways and Their Branches* (London, 1855), 31, by permission of the Guildhall Library, Corporation of London.

Parameters for making the models were set by interpretations of the available fossil evidence. Palaeontologists had long made judgments about the life-habits and appearance of the creatures they studied, sometimes in considerable detail. This was done by comparing the fossil remains with those of their closest living analogues, according to the principles of functionalist anatomy. When remains were plentiful and the analogy seemed close, such comparisons could be made with confidence. This was the situation with most of the large extinct mammals planned for the Crystal Palace. The mammoth had been found frozen in the ice, so that even the colour and texture of its hair were known to a high degree of certainty. Some of the ancient sea-dwelling reptiles had left remains that made their restoration relatively uncontroversial: the British Museum possessed superb specimens of the dolphin-like ichthyosaurs and long-necked plesiosaurs. In other cases Hawkins was less constrained. The evidence for the land-dwelling reptiles of the Secondary era, such as iguanodon and megalosaurus, was fragmentary. The most complete specimens were crushed, distorted, and disaggregated, so that inferring even a basic body shape was controversial. Estimates as to the iguanodon's length, for example, ranged from 30 to 80 feet. In such circumstances, the artist had to choose from the range of conjectures in bringing the creature to life.

Hawkins fashioned his sculptures in a makeshift studio—an unheated, drafty, rat-infested shed, which Queen Victoria and other visitors reached through the muddy wastes of the construction site. The clay originals would have disintegrated in outdoor conditions, so they needed to be cast in more permanent materials. Once the clay dried, a mould was built around the sculpture in sections, which provided a template for casting the external features and shape of the creature. (It was in this mould, not the completed model, that Hawkins held his famous dinner.) The sections from the mould were then taken by sledge to the appropriate spot on the island, with tiles, bricks, and cement used to build the creature from the ground up (Fig. 6.5).

The largest creatures gained support from an internal iron frame. As was appropriate for Hawkins' aesthetic, this had the structure of a reinforced cube. Strength had to come from within: unlike many sculptors, Hawkins believed that the illusion of life would be destroyed if the creatures were buttressed by artificial trees, rocks, or foliage. To save weight and materials, the larger creatures were hollow. The final stages of the process were less like casting a sculpture (where marble or bronze were the usual materials) and more like constructing a series of small brick houses. Here Hawkins' belief that the cube was the "archetype" of all form proved immensely fruitful. Appropriately, the completed structures are today classed as buildings scheduled for preservation, and are signed inside by "B. Hawkins, Builder".[9]

BRAINS IN THE IGUANODON

To recoup the cost of producing the displays, the Crystal Palace Company would need to attract a huge number of visitors. The run-up to the official opening involved a massive publicity blitz, in which the most effective stunt was the dinner held inside the mould used to cast the iguanodon. On a wet and windy New Year's Eve in 1853, Hawkins and 21 guests—businessmen,

Figure 6.5 The geological islands under construction. Quantities of the appropriate strata were brought to the site as part of the display; temporary sheds for the construction of the larger animals are visible behind the completed iguanodons. Relatively small creatures were constructed in the workshop and dragged in on sledges. Source: Philip H. Delamotte, *Photographic Views of the Progress of the Crystal Palace Sydenham* (London, 1855), pl. 66; by permission of the British Library.

investors, artists, and naturalists—applauded the workshop's progress. The party was widely reported, most famously in the *Illustrated London News* (Fig. 6.6), whose editor, Herbert Ingram, had strategically been invited. Accounts of the dinner brought the claim for the scientific basis of the models to the emerging mass readership; 100,000 subscribed to Ingram's paper alone.[10]

The dinner was held in honour of the leading comparative anatomist, Richard Owen. He sat at the head of the table, so that this was literally an iguanodon with brains.[11] Most of the millions who were being encouraged to see the models had no experience of a progressive sequence of ancient life, let alone of the finer points of palaeontology. Geology, which notoriously challenged traditional interpretations of Creation and the Flood, had long been suspect. So the work done in constructing the models had to be

DINNER IN THE IGUANODON MODEL, AT THE CRYSTAL PALACE, SYDENHAM.

Sir W. Stewart, Bart.; with, in the middle ground, an old Scotch fir wood and forester's lodge.

Should the winter of 1854 prove to be as severe a one as has been predicted, and as it threatens to be, we should not be much surprised to see a number of Curling Clubs formed on this side of the Tweed.

THE CRYSTAL PALACE, AT SYDENHAM.

In our Number of last week we gave a whole-page Illustration of Mr. B. Waterhouse Hawkins's Model-room, or Studio, at the Crystal Palace, Sydenham, where he is constructing his gigantic restorations of the Extinct Inhabitants of the Ancient World. We then had only the opportunity briefly to allude to this novel and great undertaking; and repeated the speculations of enthusiastic discoverers of antediluvian remains which are now, with anatomical severity, being reconstructed and restored to a state of life-like nature by Mr. Waterhouse Hawkins; to whose talents and knowledge this department of the great educational scheme of the Directors of the Crystal Palace Company is now confided; and with how much credit to their judgment was most agreeably exemplified on Saturday evening last (the last day of the year 1853), when Mr. W. Hawkins, with the concurrence of the Directors, invited a number of his scientific friends and supporters to dine with him in the body of one of his largest models, called the Iguanodon, which occupies so conspicuous a place in our Illustration of last week. In the mould of this colossal work of art—for as such it must deservedly rank very high—Mr. Hawkins conceived the idea of bringing together those great names whose high position in the science of palæontology and geology would form the best guarantee for the severe truthfulness of his works; and, at the same time, show to the public the high tone of criticism and knowledge which the Directors of this truly national undertaking require those officers to sustain to whom they confide the carrying out of any important part of their plan which so particularly bears on the education of the people.

To carry out this extraordinary idea, cards were issued at the beginning of last week—and such cards! as startling as the invitation they bore: "Mr. B. Waterhouse Hawkins solicits the honour of Professor ——'s company at dinner, *in the Iguanodon*, on the 31st of December, 1853, at four p.m." The incredible request was written on the wing of a Pterodactyle, spread before a most graphic etching of the Iguanodon, with his socially-loaded stomach, so practicably and easily filled, as to tempt all to whom it was possible to accept, at such short notice, this singular invitation. Many have to regret the rapidity of executing this novel idea, at a season when almost all have a plurality of engagements. Nevertheless, Mr. Hawkins had one-and-twenty guests around him in the body of the Iguanodon on Saturday last; at the head of whom, most appropriately, and in the head of the gigantic animal, sat Professor Owen, supported by Professor E. Forbes; Mr. Prestwick, the geologist; Mr. Gould, the celebrated ornithologist; and the Directors and officers of the Company.

The dinner, which was luxurious and elegantly served, being ended, the usual routine of loyal toasts were duly given and responded to—allusion being gracefully made by Mr. Francis Fuller, Managing Director, to the great interest evinced and approbation expressed by H.M. the Queen and H.R.H. the Prince, on their recent visit to the extraordinary works by which the company were surrounded.

Professor Owen then took occasion to explain, in his lucid and powerful manner, the means and careful study by which Mr. Hawkins had prepared his models, and had attained his present truthful success; Professor Owen adding that it had been a source of great pleasure to him to aid so important an undertaking, by assisting with his instruction and direction a gentleman who possessed the rarely-united capabilities of an anatomist, a naturalist, and a practical artist, with a docility and eagerness for the truth which ensured Mr. Hawkins's careful restorations the highest point of knowledge which had been attained up to the present period. The learned Professor then briefly commented upon the course of reasoning by which Cuvier, and other comparative anatomists, were enabled to build up the various animals of which but small remains were at first presented to their anxious study; but which, when afterwards increased, served to develop [and confirm their confident conceptions—instancing the Megalosaurus, the Iguanodon, and Dinornis as striking examples.

Professor Forbes also bore testimony to the truthful care and study with which these great models were produced by Mr. Hawkins, and which would render them trustworthy lessons to the world at large in a branch of science which had hitherto been found too vast and abstruse to call in the aid of art to illustrate its wonderful truths.

After several appropriate toasts, this agreeable party of philosophers returned to London by rail, evidently well pleased with the modern hospitality of the Iguanodon, whose ancient sides there is no reason to suppose had ever before been shaken with philosophic mirth.

THE numbers attending the Museum of Ornamental Art at Marlborough House, during December, were—19,630 persons on the public days, free; 475 persons on the students' days, and admitted at 6d. each; besides the registered students of the classes and schools—an increase of 3,557 over last year.

GIGANTIC BIRD OF NEW ZEALAND.

IN the fine Museum of the Royal College of Surgeons, in Lincoln's-inn-fields, is a most interesting illustration of the pitch to which comparative anatomy has reached in this country; the result of an immense induction of particulars in this noble science. Such is the Skeleton of the *Dinornis* of New Zealand, which the visitor will immediately recognise on the left side of the old Museum, having the skeleton of O'Brien, the Irish giant, on its right. The means by which the Museum obtained this valuable acquisition is thus graphically described in Mr. Samuel Warren's truthful and eloquent lecture on "The Intellectual and Moral Development of the Present Age:"—

In the year 1839, Professor Owen was sitting alone in his study, when a shabbily-dressed man made his appearance, announcing that he had got a great curiosity which he had brought from New Zealand, and wished to dispose of it to him. Any one in London can now see the article in question, for it is deposited in the Museum of the College of Surgeons, in Lincoln's-inn-fields. It has the appearance of an old marrow-bone, about six inches in length, and rather more than two inches in thickness, *with both extremities broken off*; and Professor Owen considered that, to what-

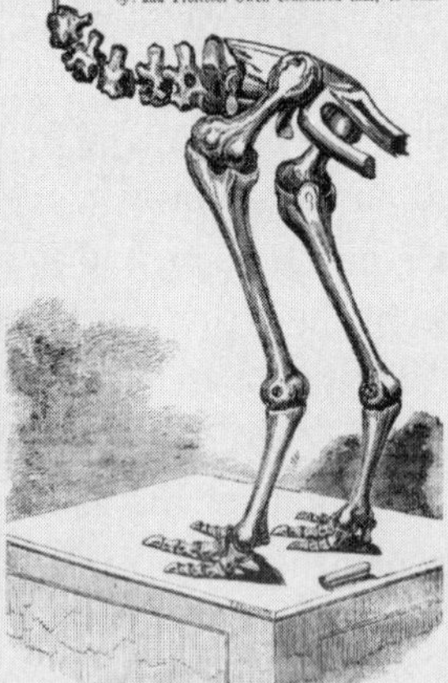

SKELETON OF THE DINORNIS, IN THE MUSEUM OF THE ROYAL COLLEGE OF SURGEONS.

ever animal it might have belonged, the fragment must have lain in the earth for centuries. At first he considered this same marrow bone to have belonged to an ox—at all events, to a quadruped; for the wall or rim of the bone was six times as thick as the bone of any bird, even the ostrich. He compared it with the bones in the skeleton of an ox, a horse, a camel, a tapir, and every quadruped apparently possessing a bone of that size and configuration; but it corresponded with none. On this he very narrowly examined the surface of the bony rim, and at length became satisfied that this monstrous fragment must have belonged to *a bird!*—to one at least *as* large as an ostrich, but of a totally different species; and consequently one never before heard of, as an ostrich was by far the biggest bird known. From the difference in the *strength* of the bone, the ostrich being unable to fly, so must have been unable this unknown bird: and so our anatomist came to the conclusion that this old shapeless bone indicated the former existence, in New Zealand, of some huge bird, at least as great as an ostrich, but of a far heavier and more sluggish kind. Prof. Owen was confident* of the validity of his conclusions, but could communicate that confidence to no one else; and, notwithstanding attempts to dissuade him from committing his views to the public, he printed his deductions in the "Transactions of the Zoological Society" for the year 1839, where, fortunately, they remain on record as conclusive evidence of the fact of his having then made this guess, so to speak, in the dark. He caused the bone, however, to be engraved; and having sent a hundred copies of the engraving to New Zealand, in the hopes of their being distributed and leading to interesting results, he patiently waited for three years, viz., till the year 1842, when he received intelligence from Dr. Buckland, at Oxford, that a great box, just arrived from New Zealand, consigned to himself, was on its way, unopened, to Professor Owen; who found it filled with bones, palpably of a bird, one of which was three feet in length, and much more than double the size of any bone in the ostrich! And out of the contents of this box the Professor was positively enabled to articulate almost the entire skeleton of a huge wingless bird, between ten and eleven feet in height, its bony structure in strict conformity with the fragment in question; and that skeleton may be at any time seen at the Museum of the College of Surgeons, towering over, and nearly twice the height of the skeleton of an ostrich; and at its feet is lying the old bone from which alone consummate anatomical science had deduced such an astounding reality: the existence of an enormous extinct creature of the bird kind, in an island where previously no bird had been known to exist larger than a pheasant or a common fowl!

* The paper on which he even sketched the outline of the unknown bird, is now in the hands of an accomplished naturalist in London—Mr. Broderip.

THE distinguished artist, Don Rafael Benjumea, has had the honour of a private audience of her Majesty the Queen of Spain, to deliver to her Majesty a splendid historical picture of the presentation of the Princess Royal in the Royal palace after the birth took place. This painting has been executed by the artist, by special command of her Majesty, and has taken the author two years to complete.

SOUTH SEA COMPANY.—On Thursday the half-yearly meeting of this company was held at the South Sea House, when a dividend of 1½ per cent for the half-year was declared. After some discussion, it was resolved that a special meeting should be called, to take into consideration the proposition for obtaining an Act of Parliament to continue the company as a trust company.

ASYLUM FOR FEMALE ORPHANS.—On Thursday the quarterly court of this charity was held at the Asylum, Westminster-road; W. Wild, Esq., in the chair; when it was reported that the school was now quite full, the number of children in the Asylum being 160, and their general health, notwithstanding the inclemency of the weather, very satisfactory. Since the last meeting, £500, being part of the proceeds of the triennial dinner, at which his Royal Highness the Duke of Cambridge presided, had been invested in the Funds. The report was adopted, and a special vote of thanks given to his Royal Highness for the great zeal evinced by him in the success of the institution.

ROYAL NATIONAL INSTITUTION FOR THE PRESERVATION OF LIFE FROM SHIPWRECK.—On Thursday last a meeting of the General Committee of this old and valuable institution, was held at the offices, John street, Adelphi; Mr. Alderman Thompson, chairman of the society, presiding. The gold medallion of the institution was voted to the gallant Captain Ludlow, of the American ship *Monmouth*, in admiration of his noble and humane conduct to the crew and passengers of the unfortunate emigrant ship *Meridian*. The silver medal of the Society was also presented to B. Harrington, and W. Waters, first and second coxswains of the *Southwold* life-boat, for their frequent services in saving life in the life boats of that place. The thanks of the committee, on vellum, were voted to the Rev. Mr. Williams, jun., and to Captain Hedenscht, of the Peninsular and Oriental Steam Navigation Company's Services, for their intrepidity in saving life. A reward of £4 10s was granted to the crew of the *Boulmer*, new life-boat, for saving the crew, consisting of seven men, of the brig *Robert Nicol*, of Perth. The boat was placed on this station by the Duke of Northumberland, K.G., and liberally presented by him, as president, to the institution. Other rewards having been voted, the committee decided that a public dinner should be held, early in March next, in aid of the funds of the society, and for the purpose of bringing the truly benevolent and national objects of the institution prominently before the public. It was reported that four new life-boats were nearly ready to be sent to the coast. Colonel Talloch, R.A., as successor to the late Colonel Colquhoun, R.A., having been elected a member of the committee, the proceedings closed.

A FIRE BROKE OUT IN DOCTOR'S-COMMONS at half past eight on Thursday morning, destroying the extensive manufactory of Messrs. Hodgkinson and Burnside, envelope makers, and the premises of Mr. Coombs, builder—these buildings lying between the Church of St. Benet and the College of Advocates, both of which narrowly escaped destruction; but the residence of Mr. Pritchard, adjoining the Hall, has met with some partial damage from the heat of the flames. Had the fire happened a few hours earlier, the library of the College (only separated a few feet from the burning premises), and the whole range of the buildings, including the Will-office, might have become a heap of ruins. The damage is estimated at from £3000 to £4000.

supplemented by the work of publicity; the illusion's success would depend on reinforcing a particular ideal of science. As Routledge's *Guide to the Crystal Palace* noted:

> Many of the spectators who for the first time gaze upon these uncouth forms, will be disposed to seek for the authority upon which these antediluvian creatures have been reconstructed, and to ask if it be possible from a few scattered fossil bones to construct an entire animal. A few years since, the answer to such a question would have involved something of doubt and uncertainty. Now, however, in matters connected with geology, as in many other sciences, brilliant hypotheses, and speculative ingenuity, have given place to well established facts, and clear, well-supported deductions from them (McDermott 1854, 206).

The extinct monsters had to seem as valid, as 'real', as the tons of brick and mortar of which they were composed. Hawkins' scientific friends served as the intellectual guarantors of the Crystal Palace Company's investment.

The authority of the displays could be enhanced most effectively by drawing on heroic images of the man of science. Publicity about the dinner never ceased to emphasise Owen's character as a scientific wizard who could restore lost creatures as if by magic. Owen featured in all the articles about the dinner; he gave the first toast; and he wrote the threepenny handbook to accompany the exhibits. The newspapers, relying on details from the Crystal Palace Company, carefully cultivated Owen's reputation. A parable illustrated his near-miraculous powers of prediction. This told how Owen, working from a single piece of broken femur bone from New Zealand, predicted the existence of a giant flightless bird, the extinct dinornis. This story was endlessly recycled in the publicity material about the Crystal Palace models, including the page of the *Illustrated London News* that invited readers to imagine themselves at the dinner (Fig. 6.6). The parable, especially when repeated in the context of such memorable events, became the keystone of the Company's strategy for associating the models with the authority of science.

←

Figure 6.6 The *Illustrated London News* reports the publicity dinner inside the iguanodon mould. Owen, at the 'head' of the table giving a toast to the memory of Mantell, is also featured in a separate article on the same page, which tells the story of his prediction of a huge flightless bird on the basis of a single bone. Source: *Illustrated London News*, 7 January 1854, 22, by permission of the British Library.

As the Routledge *Guide* explained, Owen's interpretative key was "as certain and as infallible as that with which a Champollion or a Rawlinson can unlock the mysteries of Egyptian or Assyrian inscriptions" (McDermott 1854, 206).

At the same time, Owen's work became the culmination of an ongoing history of interpretation. At the base of the pink and white drapery placed around the iguanodon, banners announced the pantheon of fossil reconstruction: William Buckland, Georges Cuvier, Richard Owen, Gideon Mantell. The first toast at the dinner emphasised the progressive nature of discovery, as Owen raised a glass to the recently-deceased Mantell, with whom he had quarrelled over everything from saurian anatomy to rights to describe new fossils. Only a few months before, Owen had written an appalling, anonymous obituary for the *Literary Gazette*, so the toast was an attempt to restore the façade of scientific solidarity. However, Owen could not resist a parting swipe at Mantell's tendency to maximise his beloved iguanodon's size.[12]

The Crystal Palace Company's publicity about Owen has proved remarkably successful. Historians associate the models with the great anatomist's archetype of the vertebrate skeleton, rather than Hawkins' archetypal cube. 'Designed' by Owen, the models become expressions of his opposition to evolution (Desmond 1979), his application of new taxonomic principles, or his wish to appear as an advocate of anatomical functionalism (Rupke 1994, 132–33). This is a remnant of a view of science as driven primarily by theory—a view that seems especially inappropriate in this case. Modelling did not simply involve transferring scientific ideas to stone, but rather a complex debate about how men of science related to the public use of their findings.

Most tellingly, Owen was not even the first choice as a consultant (Dean 1999, 260–61, Torrens 1997, 187–88). Because the Crystal Palace Company's directors knew that the sculptor would play the key role, it was only after Hawkins had presented his credentials that they approached men of science. Who this scientific advisor was mattered little; what was needed was an established public reputation, and in the first instance the directors approached Mantell. Famed as the iguanodon's discoverer, Mantell had been involved in the Great Exhibition, and was known to be favourable to reaching wide audiences through books like his *Wonders of Geology*, which had featured Martin's iguanodon scene. However, Mantell turned down the Company's request, on the grounds that the plan "was merely to have models of extinct animals". Presumably he wanted something that also showed actual fossil evidence—as in his museum at Brighton, which had Martin's painting of "The Country of the Iguanodon" above the cases of fossil bones (Fig. 6.3). This was a scheme he also advocated for the British Museum's palaeontological collections (Mantell 1851, 8). Moreover, the embittered Mantell did not share the projectors' optimism for the scheme. He had been dismayed by the "ignorant mobs" at the 1851 exhibition (Curwen 1940, 274), and felt a crowded show was scarcely a productive environment for the public to contemplate scientific truths.

After Mantell's refusal the directors turned to Owen, whose reputation came to underpin the credibility of the models. Hawkins himself set the tone for these eulogies, praising Owen's "mighty genius" at the Society of Arts. "As in the first instance it was by the light of his writings that I was enabled to interpret the fossils that I examined and compared", he explained, "so it was by his criticism that I found myself guided in and improved, by his profound learning being brought to bear upon my exertions to realise the truth" (Hawkins 1854, 445, 447). On seeing the proofs for the official handbook to the models, Hawkins wrote privately to thank Owen for giving his "sanction and approval", and signed himself "Your grateful servant and humble disciple".[13]

There can be no doubt that Hawkins did make extensive use of Owen's published descriptions of extinct reptiles and mammals. But he also depended upon works by other authors, as well as his own examination of museum collections. He based the final form of each creature on his estimate of the available evidence, infused by long experience in depicting the habits and structures of living animals. For him, works such as Martin's—with their careful merging of fact into living action—were at least as instructive as dry monographs and museums.

Owen appears to have had his only systematic input just before the small clay sculptures were scaled up to their final dimensions; and in many cases his advice was simply ignored. This can easily be demonstrated by the models themselves. The iguanodon, for example, had the overall body shape and short tail described by Owen, but the scales, grasping forearms, and long prehensile tongue were suggested by Mantell. Most notoriously, its nose sported a horn, a feature Owen dismissed but Mantell and most other geologists thought likely (Owen 1854, 17; Mantell 1851, 299). Another example is provided by the megalosaurus, which was displayed with a large hump (Fig. 6.7A). With his experience in modelling animals, Hawkins decided that

Figure 6.7 (A) Megalosaurus as reconstructed by Hawkins, with a large hump over the shoulder area (author's photograph); (B) the same animal as depicted in Richard Owen's official handbook (Owen, *Geology and the Inhabitants of the Ancient World* [London, 1854], 20). By using this picture, which revealed just how limited was the fossil evidence for reconstruction, Owen subtly distanced himself from the models.

the neck and humerus would have been supported by powerful muscles and tendons, attached to spiny projections over the shoulders. Contemporary newspaper reports—presumably leaked by Hawkins himself—indicate that Owen saw this feature only after the megalosaur was in place, but admitted that "'no one could say that the bump was not there'", as no evidence existed either for or against it.[14] Owen's handbook simply shows the paucity of available fossil remains, and the megalosaur without a hump (Fig. 6.7B). Such divergences indicate that Hawkins continued to believe that the true artist was far more than a passive instrument.

Behind their effusive mutual praise, Owen and Hawkins thus saw their respective contributions in different ways. In his official handbook, Owen implied that he had done all the preliminary sketches for the models. But this claim was implicitly rejected by Hawkins when he told the Society of Arts that the drawings (which still exist in his papers) were his own. Relations between the two men remained cordial, partly because of the disparity in social status, but Hawkins later dropped hints that Owen afforded relatively little assistance, at least after attending the feast.[15] Hawkins also began to tell parables that testified to his own skill in scientific prediction. One of these involved a newly-discovered specimen which supported his intuition about the iguanodon's skin texture, another concerned fresh fossil evidence for the megalosaur's conjectural hump. Such stories suggested that Hawkins was a wizard with powers of his own.

Conversely, Owen was quick to point out aspects of the display that he believed were conjectural or wrong. His 40-page handbook was sometimes critical, stating that the iguanodon's horn was "more than doubtful". He explicitly withheld his sanction from several of the creatures, especially when full bodies were conjectured only on the basis of skulls. Owen accordingly gave Hawkins full responsibility for modelling labyrinthodon as a giant frog and dicynodon as a monster turtle. In the case of the sea-dwelling mosasaur, the guidebook and the exhibit were at odds: Owen wrote that only the head had been reconstructed, as that was the only part which was known from fossils; but Hawkins had in fact modelled most of the body too. Owen's guidebook, far more detailed than those written for the interior courts at Sydenham, is simultaneously a claim to the leadership of comparative anatomy, and a distancing from inventive speculation (Owen 1854, 6, 17).

Owen's ambivalence reflected an uneasiness within the wider natural history community about public display. Mantell, as noted before, had refused to become involved because the project was not 'scientific' enough. The botanist John Lindley had serious doubts about depicting models of extinct animals in a garden full of modern vegetation. As he commented in the *Athenaeum*, "Martin's vision ... is present in all the wealth of life and beauty" in the palace itself, but the surrounding gardens were an anachronistic absurdity ([Lindley] 1854). A few cement cycads were a poor substitute for the vegetation in which the creatures would have originally lived. The most scathing criticisms were made by John Edward Gray, curator of the natural history collections at the British Museum. In a private letter, he denounced the models as speculative claptrap, akin to the spectacle mongering of the American showman P. T. Barnum:

> I feel assured if you were as well acquainted with Paleontologers on the study of fossil animals as with fine arts you could not have praised that "Crowning humbug" as a celebrated geologist called it, out Barnuming Barnum himself[.] The restored fossil animals are a gross delusion, Cuvier himself never attempted to give the external surface of the fossil animal, he merely gave an individual outline of what might be its general form. The models are not what they profess to be but merely enormously magnified representations of the present existing animals presumed to be the most nearly allied to the fossils—and often on very slender and sometimes on what is now known to be erroneous grounds.[16]

Behind such criticisms were worries that reconstructions on this scale were attempts to appeal to the lowest common denominator of sensual pleasure. They were akin to displays of Tom Thumb (or, more appropriately, of the Irish Giant). Even as spectacle, though, Gray condemned the models as failures: neither sufficiently accurate for knowledgeable visitors, nor interesting enough to attract, as he put it, "the curiosity of the less informed".

What then did the millions who saw the monsters actually make of them? The evidence we have is largely from reports that considered the remarks of exhibition-goers in the context of long-standing debates about public education. Secular reformers, such as Harriet Martineau, argued that even those who thought the creatures were relics of animals drowned in the Flood had at least begun to understand the possibility of a world before our own. Those even less educated, she suggested, tended to suppose that the models were constructed from genuine fossils, or that they reproduced creatures which still lived ([Martineau] 1854, 540). Later observers reported similar comments. They overheard holidaymakers complaining that they were being fobbed off with inferior imitations of animals living in the Zoological Gardens, where the admission charge was much higher. Others thought the models were "creations of some eccentric person's imagination" (Owen 1894, 1: 398–99).

The reporting of such remarks was intended to confound expectations that a visual display could work an educational revolution. In line with the Crystal Palace Company's philosophy of education, the creatures had no accompanying text. Opposing this view, critics claimed that the public would begin to interpret the models in an appropriate manner only if labels were provided (Sotheby 1855, 33). From their perspective, the entire Crystal Palace display was an encyclopaedia of nature whose connecting links could only be words. The reviewer in the *Mechanics' Magazine* insisted that "the Palace is an illustrated volume, and those who inspect its pages will need letter-press explanations". Without an appreciation of the associations of the places depicted, or an educated sense of the past, the visitor's understanding would be impoverished.[17] The Pestalozzian notion that the models alone would communicate understanding was rejected as fanciful. Soon afterwards, even those most involved in creating the displays began to acknowledge that the 'lessons of the eye' required printed supplements. The first edition of Samuel Phillips' official guidebook to the Crystal Palace, published before the arrangement of the "antediluvian department" was completed, had no key, but, like most other guidebooks to the Palace produced during the next few years, all the later ones did. It also appears that a card with the same information could be purchased on the spot (Fig. 6.2). Such expedients were never fully successful. Few visitors bought a guide; many could not read.

UTOPIA IN DECAY

Even before the monsters received their final coats of paint, the financial underpinnings of the project collapsed. Already in April 1854, as a crowd of 40,000 watched Queen Victoria open the Palace, critics began to hint that the Crystal Palace Company had been mismanaged, and that the projected visitor numbers were wildly optimistic. During the first summer, attempts were made to increase revenue from the refreshment stalls. A liquor licence was obtained, a move that created huge controversy but little additional income. Equally unsuccessful were efforts to enhance the commercial possibilities of the Palace as a place for displaying samples of goods for sale. These changes had to be handled carefully, lest the palace look too much like a shop—a similarity that critics were not slow to point out. In the early days, as the *Crystal Palace Herald* commented, "hopes were high and shares were high, and speculators became poetical and imaginative; now all who have observed it have awoke from the dream, now they see it in a true light ... a purely commercial speculation, that is expected to pay the proprietors a dividend."[18]

By the autumn of 1855, little more than a year after the opening, the Company faced catastrophe, and the monster models became the battleground for competing factions among the shareholders. As one infuriated correspondent complained to the *Daily News*, the directors had squandered a fortune on the "detestable absurd 'extinct animals'". Altogether, he wrote, "the 'ologies' in and about the Crystal Palace have together cost the proprietors the wild sum" of £23,625. His investment had declined in value by two-thirds.[19] Many commentators, while praising the exhibits, thought they should have been built more slowly, with funds being devoted in the first instance to the indoor displays. Another author suggested that the stone for the strata sequence and mine could have been brought in as ballast in ships. The process of constructing the models would itself have afforded additional interest. Even the Palace's most enthusiastic advocates recognized the dangers of excessive expense. Too much, it was argued, had been attempted too soon ([Eastlake] 1855, 353). After a shareholders' meeting in September 1855, Hawkins was abruptly told to stop work. Even the mammoth, the one creature specified in his original brief, was left half finished.[20]

The Company struggled on, but the aims of the Palace were radically transformed—a point that is rarely recognised. The utopian hopes for rational education were abandoned, as the park became the key mid- and late Victorian site for mass spectacle, music, and sport. As the American urban planner Frederick Law Olmsted noted after talking with the manager, some nine-tenths of the visitors came to relax in an atmosphere of festive entertainment.[21] The cafés were enlarged, and the interior courts left to decay. The financial situation was so bad that when a fire destroyed most of the ethnological and natural history exhibits in the nave in 1863, no attempt was made to replace them.

The monsters decayed more gradually. Critics lamented that Hawkins' models were falling apart, with toes, tails, and teeth crumbling away. A slip had to be inserted in Owen's guidebook asking readers not to vandalise displays set out for their benefit (visitors, thinking these were the real things, ripped the teeth from the sculptures as souvenirs). Recalling his first sight of the creatures at the age of ten, the playwright Thomas F. Plowman remembered how the "terrible antediluvian monsters" in their "pristine freshness"

> gave me a much better understanding of the stream of life upon the globe than ever I had had before, whilst the ethnological groups in the palace itself showed me what strange peoples were still inhabiting our planet. It was a veritable fairy-land in its variety and charm, and helped me to see visions and dream dreams for years afterwards. Ever since that, to me, eventful Sydenham day, I have made a pilgrimage to the old shrine whenever I could—and that has been many times—happy to renew such cherished associations. Since my first visit, the old palace has passed through many vicissitudes, and I have marked with melancholy its gradual decadence (Plowman 1918, 19).

Even the water levels in the lake were low, leaving (as an article in the *Garden* noted) exposed piping "like sea-serpents stranded in muddy waters". The ruin was "appalling and apparently beyond repair" (Fuller 1874, 13).

But the immediate lesson that the Crystal Palace experience taught exhibition-makers was not to build anything like this again. No display on this scale was completed anywhere in the world for the rest of the nineteenth century. In the wake of the exhibition, Barnum considered asking Hawkins to make a series of models for his American museum, but decided against it, apparently because too few North American species were as yet known (Mitchell 1998, 134–35). In 1867 Hawkins did start to build an exhibition of extinct life for New York's Central Park. But the expense of the work (and corrupt politics) led to the half-completed models being smashed with sledgehammers by thugs hired by the notorious Tweed gang (Colbert and Beneker 1959; Desmond 1974).

There were isolated exceptions. In Stuttgart, the taxidermist and animal sculptor Phillip Leopold Martin created an indoor museum with painted backdrops and reduced-scale papier-mâché models of prehistoric creatures. This opened to the public for a couple of years in the early 1870s, but was a commercial flop.[22] Other attempts to emulate the Crystal Palace show were more limited, such as a plaster hadrosaur built for the 1876 Centennial Exhibition at the Smithsonian in Washington. The Crystal Palace models thus remained for 50 years as the only large-scale display of extinct life in 3-D. The concept was revived only in the very different conditions of the early twentieth century. In Europe, the earliest large display appears to be that of the animal sculptor Joseph Pallenberg for Carl Hagenbeck's commercial zoo in Stellingen, Hamburg. By 1911 this included eighteen prehistoric creatures (Reichenbach 1980, 579–80). The revival was due in part to further developments in the use of reinforced concrete, which made building plesiosaurs and other long-necked creatures easier and cheaper.

While the models at Sydenham gently decayed, reproductions in other media led them to continue for decades as the leading representations of extinct animals. In the late 1850s, Hawkins produced six educational wall charts depicting views of ancient life. Most of these were based on the Crystal Palace displays, although their character as 2-D scenes situated the animals in characteristic environments rather than as elements in a progressive history. Reduced to 2-D, Hawkins' work became a template for nearly half a century of representations in elementary books. Illustrators relied upon them until the Rev. H. N. Hutchinson's *Extinct Monsters* in 1892 and E. Ray Lankester's *Extinct Animals* in 1905. Hawkins had created a sequence of relatively stable images of ancient life, and enhanced public interest in the antediluvian world, but the defeat of the hopes for Sydenham discouraged further experiments in this kind of visual education.

Instead, the ideals associated with the Crystal Palace Company were transferred to the campaign for universal secular education. This is apparent from the wall charts, which were issued as classroom aids by the Department of Science and Art, and also in the set of five small sculptures of the animals that Hawkins constructed for sale to museums and schools throughout Britain (Fig. 6.8).[23] The series could also be bought from Henry A. Ward of Rochester in New York, which made them available to educational institutions and museums throughout the United States (Ward 1866, 80; Kohlstedt 1980). Unlike the small plastic models available in the twentieth century, these models were not toys for individual children (as has often been assumed), but educational aids sold to public institutions. The idea for producing them was first broached in the discussion after Hawkins' Society of Arts lecture, and taken up by the London mineral dealer and lecturer, James Tennant. Initially Tennant had announced that these reduced-scale models would show skeletal anatomy on one side and the full restoration on the other; this form was common in zoological and anatomical modelling, but unprecedented in palaeontology.[24] As produced, though, the creatures were miniature replicas of those on display.

80 VERTEBRATA.

slender tail which left a groove-like impression. The stride from toe to toe measures thirteen inches; and the feet are about three and a half inches long. The hind-feet stepped upon nearly the same spot as the fore-feet, causing some obliteration of the first impression. The original slab is in the private Cabinet of Isaac Lea, Esq., of Philadelphia, by which gentleman it was first discovered and described. Size, 26 x 17. Price, $4.50.

No. 302–308. **Restorations of Fossil Reptiles.**

Pterodactyle, Megalosaurus, Iguanodon, Labyrinthodon, Ichthyosaurus, Plesiosaurus dolichodeirus and *P. macrocephalus*. They are reduced (one inch to the foot) from the gigantic models in the Crystal Palace, London; constructed to

No. 302. PTERODACTYLE No. 303. MEGALOSAURUS.

scale by B. Waterhouse Hawkins, F. G. S., F. L. S., from the form and proportions of the fossil remains, and in strict accordance with the scientific deductions of

No. 304. IGUANODON. No. 305. LABYRINTHODON.

Professor Owen. Preliminary drawings, with careful measurements of the originals in the Royal College of Surgeons, British Museum and Geological Society,

Nos. 306–308. ICHTHYOSAURUS WITH PLESIOAURI.

Figure 6.8 Soon after the display was completed, Hawkins designed small-scale models of seven of the creatures for sale to schools and local museums. The models were also sold in the United States, as this page from an American catalogue indicates. Source: Henry A. Ward, *Catalogue of Casts of Fossils* (Rochester, N. Y., 1866), 80, courtesy of the History of Science Collections, University of Oklahoma.

There is no evidence that the three years of model-building at the Crystal Palace produced any significant palaeontological discoveries or theoretical innovations. If anything, practitioners seem to have become even more cautious about engaging in speculative reconstructions of any kind. Paradoxically, the next half-century—one of the richest periods in the history of vertebrate palaeontology—produced almost no full reconstructions, either in two or three dimensions. Thomas Henry Huxley respected Hawkins, and they collaborated on an atlas of the vertebrates (Huxley 1864), but he never put his own radical vision of saurian anatomy in the form of a printed illustration, let alone a model in a public park (Ashworth 1996, 15). Huxley presumably saw the Crystal Palace display as one more sign of the way that commercial imperatives had tainted Owen's work. The boundary of the mould for the iguanodon, once enough to protect the reputation of science, was not enough to shield expert debates in the age of the mass media. As the great American fossil hunter Othniel Marsh wrote in 1895, the creatures "have suffered much from both their enemies and their friends. Many of them were destroyed and dismembered long ago by their natural enemies, but, more recently, their friends have done them further injustice in putting together their scattered remains, and restoring them to supposed life-like forms. So far as I can judge, there is nothing like unto them in the heavens, or on earth, or in the waters under the earth" (quoted in McCarthy 1994, 85).

CONCLUSION

The relations between spectacle, commerce, and expertise remain at the heart of current controversies about public science. Much of the research for this chapter was done in the Natural History Museum in London, where I passed stalls selling plastic dinosaur models, long lines of eager children, and displays of roaring latex-covered robots. Then I went into the archives. The extraordinary wealth of material there, ranging from Hawkins' rough watercolour sketches to copies of the wall charts, makes the Crystal Palace a tempting starting point for modern dinomania—the fascination with dinosaurs associated with *Jurassic Park* and *Walking with Dinosaurs* (Mitchell 1998).

Yet differences from our own concerns are striking. Those who originally made and saw the models, for example, called them "monsters", "extinct animals", or "antediluvian animals"; they were, in short, associated with older forms of display found in zoos, circuses, and churches—not those of department stores, cinemas, and spectator sports. Only a handful of the creatures represented (four out of thirty-one) were technically dinosaurs, but more significantly the word "dinosaur" itself did not become familiar to wider audiences until the very end of the nineteenth century.[25] This is a sign of the wider disparities between the context in which these models were originally seen and that in which they are viewed today. The invention of the dinosaur as a popular icon—what W. J. T. Mitchell has called the "totem animal of modernity"—took place decades later, through the emergence of institutions for secular mass entertainment and professional science. More generally, the "commodity spectacle" that Thomas Richards attributes to the early 1850s (Richards 1991) actually emerged only through the larger economic transformations that other historians have dated to the 1870s and after.

The models were thus not the first stirrings of modern dinomania, but survivals of the hope that science could be seamlessly united with capitalist commerce and moral improvement. In this vision, financial success would go hand in hand with rational edification and probity of character (Alborn 1995). The great aristocratic gardens had had their pagan gods and mythical heroes, but the gardens of the People's Palace would display the findings of science. In the ideal future, the body of shareholders in the Crystal Palace Company would be precisely co-extensive with those who went to see the displays. It was a moment, as Hawkins discovered to his cost, that was soon to pass.

Because the models were constructed to last indefinitely, they survived nearly a century of neglect. Their restoration from the 1950s marked a revival of interest in the Victorian era, and, more recently, the emergence of dinomania. Today the models can still be viewed on their original islands, and are a popular attraction of the Crystal Palace Park, easily reached by train from Waterloo Station in London. (Ironically, at the time of my research it was the Natural History Museum that charged for admission; the Crystal Palace Park was free.) In seeing the creatures today, however, it is vital to look beyond our own commercial spectacles. Just as the Great Exhibition of 1851 was not the forerunner of the modern shopping mall, so Hawkins' "antediluvian animals" are implausible precursors of Spielberg's rampaging velociraptors. They are not dinosaurs in our sense of the word at all.

When Hawkins and his guests dined in the belly of the iguanodon mould, they celebrated a remarkable experiment in visual education. The models were constructed by a sculptor naturalist, underwritten by men of science, and supported by the enlightened patronage of a joint-stock company. It had been a sign of the "progress of the masses", the *Crystal Palace Herald* reflected, "that capitalists should suppose that they could profit" by such an enterprise.[26] Within two years, decades before the Crystal Palace itself burned to the ground, this utopian union between commerce, education, and reason was in ruins. The models survived, but the circumstances that led to their making had vanished.

NOTES

1. J. E. Gray to J. L. Sotheby, 28 December 1854, Sotheby Scrapbook, vol. 1, Central Library, Bromley, Kent (hereafter Sotheby Scrapbook).

2. "The Crystal Palace at Sydenham", *Observer*, 7 January 1855.

3. "The Crystal Palace", *London Quarterly Review* 3 (October 1854): 232–79, at p. 239.

4. "Progress of the Crystal Palace", *Art-Journal*, 1 October 1853, 265–66, at p. 266.

5. A Sojourner, "A party at the people's palace", *Illustrated Crystal Palace Gazette* 1 (October 1853): 3.

6. Copies of the letters are in the Alexander Turnbull Library, Wellington, New Zealand, Gideon Mantell Papers, ms 83, folder 32.

7. A Sojourner, "A party" (see note 5).

8. "Dinner to Professor Owen in the iguanodon", *Morning Chronicle*, 2 January 1854, 3.

9. The principal explanation of the construction process is Hawkins 1854, 447. This can be clarified in important respects by Read's discussion of the making of Victorian sculpture (Read 1982), and by accounts from visitors to the workshop; see especially "The geology of the Crystal Palace", *Hogg's Instructor*, 2nd series, 2 (April 1854): 279–86. For the inscription inside the models, see McCarthy 1994, 17.

10. "The Crystal Palace at Sydenham", *Illustrated London News*, 7 January 1854, 22.

11. "Dinner to Professor Owen in the iguanodon", *Morning Chronicle*, 2 January 1854, 3.

12. Ibid.; "The Crystal Palace at Sydenham", *Illustrated London News*, 7 January 1854, 22. On the obituary, see Desmond 1982, 24.

13. Benjamin Waterhouse Hawkins to Richard Owen, 19 May 1854, Cambridge University Library Add. ms 5354, f. 64.

14. "The Crystal Palace at Sydenham", *Observer*, 7 January 1855. The discrepancy about the hump was pointed out to me by Steve McCarthy.

15. When Hawkins asked for testimonials before emigrating to the United States in 1867, Owen prepared a formal protest, claiming credit for the entire scheme and for all but the physical labour of making the models; Richard Owen to the editor of *Nature*, n. d., Natural History Museum, Owen Correspondence, vol. 21, f. 25.

16. J. E. Gray to J. L. Sotheby, 28 December 1854, Sotheby Scrapbook.

17. "The Crystal Palace at Sydenham", *Mechanics' Magazine* 60 (27 May 1854): 485–89, at p. 488. For wider Victorian debates about vision, see Flint 2000.

18. "The people's palace", *Crystal Palace Herald* 3, no. 4 (April 1856): 135. On Sydenham's finances, see Atmore 2000.

19. A Crystal Palace shareholder, "Crystal Palace. To the editor of the *Daily News*", *Daily News*, 26 September 1855, Sotheby Scrapbook.

20. "The Crystal Palace, Sydenham.—The geological restorations", *Observer*, 24 September 1855, Sotheby Scrapbook.

21. F. L. Olmsted to Charles K. Hamilton, 22 July 1870, in Schuyler and Censer 1992, 382–85, at p. 382; thanks to Carla Yanni for this reference.

22. I am grateful to Lynn Nyhart for sharing the results of her ongoing research into this fascinating episode.

23. A promotional flyer of July 1860, "Key to a coloured lithographic plate of Waterhouse Hawkins's restorations of extinct animals" (Hawkins Papers, Natural History Museum), offers a series of seven models for £5.5s, but it appears that only five were issued.

24. This is clear from the version of Hawkins' Royal Society of Arts lecture that Tennant published as a separate leaflet; it is reproduced in McCarthy 1994, 89–92.

25. The constructed character of the category "dinosaur" was first pointed out in Desmond 1979, but the consequences, especially for its public reception, have yet to be worked out.

26. "The people's palace", *Crystal Palace Herald* 3, no. 4 (April 1856): 135, in a comment referring to the entire exhibition.

REFERENCES

Alborn, Timothy L. 1995. "The moral of the failed bank: Professional plots in the Victorian money market." *Victorian Studies* 38: 199–226.

Ashworth, William B. 1996. *Paper Dinosaurs, 1824-1969: An Exhibition of Original Publications from the Collections of the Linda Hall Library.* Kansas City, Mo.: Linda Hall Library.

Atmore, Henry Philip. 2000. "The great Victorian way: Materiality and memory in mid-nineteenth-century technological culture." Ph. D. dissertation, University of Cambridge.

Auerbach, Jeffrey. 1999. *The Great Exhibition of 1851: A Nation on Display*. New Haven: Yale University Press.

Buckland, Frank. 1858. *Curiosities of Natural History*. London: Richard Bentley.

Cockle, James. 1854. "The Crystal Palace." *The Illustrated London Magazine* 2: 33–37.
Colbert, Edwin N., and Katharine Beneker. 1959. "The Paleozoic Museum in Central Park, or the museum that never was." *Curator* 2: 137–50.
Curwen, E. Cecil, ed. 1940. *The Journal of Gideon Mantell: Surgeon and Geologist*. London: Oxford University Press.
Dean, Dennis R. 1999. *Gideon Mantell and the Discovery of Dinosaurs*. Cambridge: Cambridge University Press.
Desmond, Adrian. 1974. "Central Park's fragile dinosaurs." *Natural History* 83: 63–71.
———. 1979. "Designing the dinosaur: Richard Owen's response to Robert Edmond Grant." *Isis* 70: 224–34.
———. 1982. *Archetypes and Ancestors: Palaeontology in Victorian London, 1850–1875*. London: Blond & Briggs.
Doyle, Peter, and Eric Robinson. 1993. "The Victorian 'geological illustrations' of Crystal Palace Park." *Proceedings of the Geologists' Association* 104: 181–94.
[Eastlake, Elizabeth]. 1855. "The Crystal Palace." *Quarterly Review* 96: 303–54.
Flint, Kate. 2000. *The Victorians and the Visual Imagination*. Cambridge: Cambridge University Press.
Fuller, Francis. 1874. *A Letter: To the Shareholders of the Crystal Palace Company*. London: Alfred Boot.
Hawkins, Benjamin Waterhouse. 1843. *The Science of Drawing Simplified; or, the Elements of Form Demonstrated by Models*. London: Smith, Elder, and Co.
———. 1854. "On visual education as applied to geology." *Journal of the Society of Arts* 2: 444–49.
Huxley, Thomas Henry. 1864. *An Elementary Atlas of Comparative Osteology in Twelve Plates, the Objects Selected and Arranged by Professor Huxley, F.R.S. and Drawn on Stone by B. Waterhouse Hawkins, Esq*. London: Williams and Norgate.
Kohlstedt, Sally Gregory. 1980. "Henry A. Ward: The merchant naturalist and American museum development." *Journal of the Society for the Bibliography of Natural History* 9: 647–61.
[Lindley, John]. 1854. "The Crystal Palace garden." *Athenaeum*, 780.
McCarthy, Steve. 1994. *The Crystal Palace Dinosaurs: The Story of the World's First Prehistoric Sculptures*. London: Crystal Palace Foundation.
McDermott, Edward. 1854. *Routledge's Guide to the Crystal Palace and Park at Sydenham: With Illustrations, and Plans of the Building, Grounds, and Courts. The Sixteenth Thousand. With Alterations and Additions*. London: George Routledge.
Mantell, Gideon. 1838. *Wonders of Geology*. London: Relfe and Fletcher.
———. 1851. *Petrifactions and Their Teachings; or, a Handbook to the Gallery of Organic Remains of the British Museum*. London: Henry G. Bohn.

[Martineau, Harriet]. 1854. "The Crystal Palace." *Westminster Review* 62: 534–50.
Mitchell, W. J. T. 1998. *The Last Dinosaur Book: The Life and Times of a Cultural Icon*. Chicago: University of Chicago Press.
Orvell, Miles. 1989. *The Real Thing: Imitation and Authenticity in American Culture, 1880–1940*. Chapel Hill: University of North Carolina Press.
Owen, Richard. 1854. *Geology and the Inhabitants of the Ancient World*. London: Crystal Palace Library.
Owen, Rev. Richard. 1894. *The Life of Richard Owen*, 2 vols. London: John Murray.
Piggott, Jan. 2004. *Palace of the People: The Crystal Palace at Sydenham, 1854–1936*. London: C. Hurst.
Plowman, Thomas F. 1918. *In the Days of Victoria: Some Memories of Men and Things*. London: John Lane.
Read, Benedict. 1982. *Victorian Sculpture*. New Haven: Yale University Press.
Reichenbach, Herman. 1980. "Carl Hagenbeck's Tierpark and modern zoological gardens." *Journal of the Society for the Bibliography of Natural History* 9: 573–85.
Richards, Thomas. 1991. *The Commodity Culture of Victorian England: Advertising and Spectacle, 1851–1914*. London: Verso.
Rudwick, M. J. S. 1992. *Scenes from Deep Time: Early Pictorial Representations of the Prehistoric World*. Chicago: University of Chicago Press.
Rupke, Nicolaas A. 1994. *Richard Owen: Victorian Naturalist*. New Haven: Yale University Press.
Schuyler, David, and Jane Turner Censer, eds. 1992. *The Papers of Frederick Law Olmsted*, vol. 6. Baltimore: Johns Hopkins University Press.
Secord, James A. 2000. *Victorian Sensation: The Extraordinary Publication, Reception, and Secret Authorship of* Vestiges of the Natural History of Creation. Chicago: University of Chicago Press.
Sotheby, Samuel Leigh. 1855. *A Few Words by Way of a Letter Addressed to the Directors of the Crystal Palace Company*. London: John Russell Smith.
Torrens, Hugh. 1997. "Politics and paleontology: Richard Owen and the invention of dinosaurs." In *The Complete Dinosaur Book*, ed. James O. Farlow and M. K. Brett-Surman, 175–90. Bloomington: Indiana University Press.
Ward, Henry A. 1866. *Catalogue of Casts of Fossils, from the Principal Museums of Europe and America, with Short Descriptions and Illustrations*. Rochester, N. Y.: Benton & Andrews.
Wyatt, M. Digby. 1854. *Views of the Crystal Palace and Park Sydenham*. Crystal Palace: Day and Son.

CHAPTER SEVEN

Plastic Publishing in Embryology

Nick Hopwood

With the creation of research careers and the industrialisation of communications, by the mid-nineteenth century the sciences depended more on publication than ever before. Journals, textbooks, handbooks, and manuals were founded that continue to this day. This chapter is about how a pair of publishers, Adolf and his son Friedrich Ziegler, shaped embryology between 1850 and 1918. Through close collaboration with leading researchers, including painstaking correction of proofs, the Zieglers set new standards for the representation of embryos' complex forms. Teaching became unimaginable without their publications, and involvement in new microscopical methods ensured the press a near monopoly in active fields of research. Yet the most remarkable feature of the firm's history is that—the language of print notwithstanding—the Zieglers published, not books, but wax models. Investigating the making and uses of these extraordinary objects reveals activities that were central to creating the embryonic forms that surround us today. Comparing the production of models and books, and exploring their relations, expands our view of scientific publication during the great age of print.

In the nineteenth century embryology was a central science of life (Churchill 1994; Hopwood, forthcoming). Especially in and around the German universities, it joined with comparative adult anatomy in morphology, the study of form; served obstetrics, gynaecology, and legal medicine by defining the stages of pregnancy; and, particularly from the 1860s, was hailed as the most compelling evidence of the evolution of life on earth. Many women still interpreted a missed period as signalling the clotting of excess blood or the growth of an unborn child. But doctors, midwives, biologists, and teachers all learned embryology, and in turn instructed ever wider audiences to see the progressive development of an embryo as representing human origins, individually in pregnancy and collectively in evolution. During the twentieth century, as experimental biology dominated embryological research and embryos became controversial subjects and objects in the clinic (Hopwood, forthcoming), new media took embryological images to very wide audiences and marginalized wax. Yet, to understand where the embryonic forms came from that are today checked off on a screen, or within which patterns of molecular staining are lit up, we have to go back to the shapes that were first drawn—and modelled—between the French Revolution and World War I.

Creating images of embryos constituted objects to study, and representing series of progressively more advanced stages showed those forms develop (Jordanova 1985; Daston and Galison 1992; Duden 1999; Hopwood 2000; Duden, Schlumbohm, and Veit 2002). Historical study of embryological work recasts development, which is usually taken for granted, as an effect that had to be produced. The routines of producing development have varied between species and changed over time, but we can distinguish several general steps: the medical encounters, expeditions, and controlled breeding from which often extremely scarce and inaccessible materials were collected; the anatomical, microscopical, and artistic procedures that converted these specimens into magnified images of embryos; their arrangement in developmental order and the selection of representatives; publication or exhibition; and the responses of scientists, students, and laypeople.

These practices raise many questions. What happened when embryologist-collectors reinterpreted as embryos objects that pregnant women, hunters, fishermen, and other suppliers of specimens had seen in very different terms? Or take the problem of choosing representative images, most challenging for human embryology: when the overwhelming majority of embryos came from abortions, how was normal development to be defined? And then there are the heated disputes over the results: did the series march in parallel or diverge? This chapter highlights the work that turned puzzling and unique specimens into vivid and widely distributed icons of development. Among imaging techniques, it concentrates, not on the more familiar dissecting, drawing, or photography, but on modelling. And of the several modellers, it focuses on the Ziegler studio, which dominated the academic market.

The Zieglers' models today comprise the majority of embryological waxes preserved in institutes and museums around the world. As successors to the Florentine masterpieces and contemporaries of dermatological moulages (Mazzolini, Schnalke, this volume), they appear briefly in histories of anatomical ceroplastics and descriptions of collections (e.g., Pyke 1973, 162; Mulder 1977, 428–30; Boschung 1980, viii; Fröber 1996, 97–98; Pelagalli, Esposito, and Paesano 1998). Bringing the models into the historiography of embryology, which has ignored teaching and the anatomical research that produced most of the waxes, allows us to see the difference they made to images of embryos at large. This chapter extends an article (Hopwood 1999) about scientists' adoption, between the 1860s and the 1880s, of reconstructive modelling from serial sections as the main technique in such major fields of research as human embryology. Placing the modellers at centre stage, I show here how, during the 1850s and 1860s, embryology teaching had already come to depend on models, and then explore the consequences of the shift to plastic reconstruction for publishing in the science.[1]

More generally, studies of the art of anatomy and embryology (e.g., Kemp and Wallace 2000) can revise the dominant disciplinary histories most effectively if we see how other media related to journals and monographs. As historians of science develop insights from the new book history (Frasca-Spada and Jardine 2000), I suggest that by becoming more aware of the changing ways in which modelling shaped pages, and letterpress joined with diagrams to guide the viewing of models, we gain a richer view of both print and wax.

"THROUGH DRAWINGS THE OBJECTS CAN BE MADE CLEAR ONLY WITH DIFFICULTY"

From the mid-eighteenth century a vigorous craft of anatomical ceroplastics and a new science of embryology were sometimes brought together, but in 1850, when Adolf Ziegler qualified as a physician, modelling still generally played secondary roles. He took the lead in making models essential, initially mainly in teaching.

Model embryos occupied prominent positions in late-eighteenth-century collections, and there was much traffic between wax and print (Jordanova 1985; Lemire 1990; Schnalke 1995; Düring et al. 1999; Mazzolini, this volume). Yet models are marginal in accounts of the making of a newly developmental embryology (Lenoir 1989; Churchill 1994; Duden 1999; Hagner 1999). This is in part because innovations in specimen preparation, microscopical observation, printing, and especially drawing, were more important, but may also be an artefact of concentrating on print.

In the late 1790s, for example, the anatomist Samuel Thomas Soemmerring published illustrations that are famous as the first representations of early human development. Soemmerring's artist, Christian Koeck, avoided the distortions of perspective by drawing almost as isometrically as an architect would draw a building (Soemmerring 2000, 173). This method has been identified as crucial to making the modern embryo, because it artificially distanced objects from the observer (Duden 1999; Hagner 1999, 206–9), and as making modelling possible (Choulant 1852, 133). It was the Swiss modeller Josef Benedikt Kuriger, using his experience in the anatomical theatre at Paris to move from portraits and devotional objects into anatomy and obstetrics, who brought the engravings into wax relief. Copied in wax and burnt clay, they were sold in the German lands to collectors, researchers, medical men, and institutes (Suter 1986; Schnalke 1995, 53–58). These models came after the atlas, but for studies of the ear, Koeck, a former modeller in stucco, had received a commission to make a model from sketches and specimens first. The experience was a great help when it was time to draw the plates, because "the true form of certain parts is really revealed . . . only during the enlargement, in the course of the cutting up and putting together again that this kind of work [i.e., modelling] demands" (Soemmerring quoted in Geus 1985, 278).

Soemmerring stopped at embryos' external surfaces, but in the German universities after the Napoleonic occupation anatomists and physiologists revealed the inner relations of their parts. Dissecting needles, pencils, and engraving tools—and especially magnifying glasses and microscopes—were used to create new visual languages of the embryo. Not limited to scarce human material, researchers concentrated on the more accessible chick, frog, and domestic mammals, showing first how development began with the organisation of primitive sheets of tissue or "germ layers"; then, as a second fundamental unit of analysis, they added cells (Churchill 1994). Heated debates over patterns of development stimulated interest in embryology. By mid-century, institutes of anatomy and of zoology put on special lectures and practical classes, and the general public was queuing to see embryological exhibits.

We lack evidence that modelling was important in researching germ layers and cells (but see Buffa 1977). Some professors were doubtless reserved towards a medium that was associated with popular shows, in metropolitan wax museums and country fairs, more than academic culture (König and Ortenau 1962; Altick 1978, 338–42; Lemire 1990, 323–65; Oettermann 1992; Yarrington 1996). Yet embryological entrepreneurs inside as well as outside the universities lamented the impossibility of demonstrating the scarce and tiny specimens on which the new science rested, and the insufficiency of drawings to capture the complex movements within. Many relied on models. In 1847 the Zürich anatomy professor Hermann Meyer presented "schematic models to explain some more difficult parts of embryology" in wax and stearine, and offered copies for exchange or money (Meyer 1847, 49). "The great superiority" of Dr Joseph Kahn's "celebrated Anatomical Museum", which the middle classes in several British cities paid to see, was a series of models "exhibiting the progress of the human ovum in the uterus, from its first impregnation to its birth"; he published an atlas depicting them in colour (Kahn 1851, iii; 1852; Altick 1978, 340–41). Nevertheless, when Ziegler began to model around 1850, much of the science still proceeded without ceroplastic assistance.

Adolf (or Adolph) Ziegler was born in 1820 at Mannheim in the Grand Duchy of Baden, but in 1824 his father moved to take charge of the tax office in Freiburg im Breisgau on the edge of the Black Forest, and it is here that Ziegler spent most of his life. Said "since his youth" to have been "skilful in drawing and modelling", he described himself as an artistic "autodidact". We do not know how he learned to model, but he attended drawing classes at Freiburg's classical grammar school and there were books that introduced the popular pastime of modelling in wax and clay.[2] After two semesters of preparatory studies at the small local university, Ziegler trained between 1839 and 1846 to become an apothecary, but then carried straight on as a medical student, gaining the doctorate in 1850. Next he travelled for two years' further training to the huge medical school in Vienna, and soon went off again, to work for a half-year with the leading microscopist and physiological institute-builder, Jan Purkyně, in Prague. In 1854 he came back to Freiburg to take up the position of zootomical (and briefly also physiological) assistant, responsible for the collection and helping with classes, and began to practise a little medicine.[3] Having modelled various objects while in Vienna and Prague, Ziegler probably owed his job to a collaboration that by 1851 he had already begun with Alexander Ecker, Freiburg's new professor of physiology, zoology, and comparative anatomy, and a driving force in the medical sciences' dramatic rise (Neuland 1941; Foerster 1963; Seidler 1991).

During the 1850s Ecker was occupied with making a Soemmerring-style atlas (Ecker 1851–59). As a student Ecker had received no embryology teaching at all (Ecker 1886, 64); now he was introducing the subject into his own lectures and setting up hands-on classes in the use of the microscope, including on embryological material (Nauck 1952, 76–79; Tuchman 1993, 56–58, 76–77). It was a commonplace that embryology relied especially heavily on "illustration through art". "The stages [*Zustände*] pass rapidly by and very many objects cannot be preserved because of their delicacy" (Ecker 1859). So published drawings and large schematic charts were essential in lectures and "to serve the student as a guide to his own investigations" (Ecker 1851–59, legend to pl. XXIII). At first glance Ecker's ten embryological plates seem self-contained, and—reprinted by embryologists and copied by students—the figures would have a life of their own. Looking more closely, however, we find that modelling was an integral part of the project. "[E]ven [numerous figures] do not always suffice", he reflected, "and for some relations plastic representations are absolutely essential" (Ecker 1859).

Ecker collected and prepared the specimens, then drew and/or had his artist draw them, ready for the engraver. But certain features were difficult to visualise in three dimensions, and as he worked, initially and for the earliest stages on frog development, he made rough clay models as an aid to more accurate drawings. So he leapt at the chance to have "a young doctor . . . of great dexterity" craft a series of 25 finished models in hard wax to accompany the plate.[4] Ziegler will have modelled initially in clay or wax, working freehand to rough out shapes that he improved using spatulas and knives, and then finished, painted, and mounted. He went on to model whole human embryos and the development of the face, external genitalia, and heart (Fig. 7.1).

Dissecting, drawing, and modelling, Ziegler and Ecker learned together to see embryos' tiny, complicated, and transient forms. At first, the anatomist had more to teach. "Dr Ziegler has much mechan[ical] and artistic sense", Ecker confided to a colleague in 1852, but—not unusually for a physician at the time—"unfortunately understands nothing at all of embryology and cannot see through a microscope, so that I have to show him everything on paper or in clay."[5] Ziegler appears to have mastered microscopy, either in

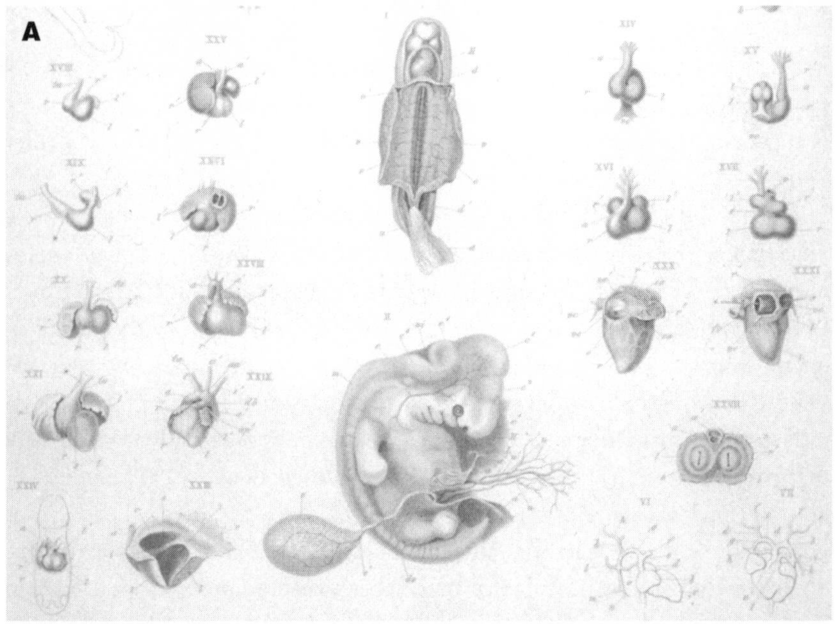

Figure 7.1 "Development of the heart in man". (A) Engraving from Alexander Ecker, *Icones physiologicae* (Leipzig, 1851–59), pl. XXX (detail). Balfour Library, University of Cambridge. (B) Adolf Ziegler's models that were sold to accompany Ecker's atlas (the first two are missing). Model 5 (third from left) is about 12.5 cm high (with stand), and corresponds to Ecker's fig. XX (third down on left-hand side), which is 1.5 cm across. So decontextualised are these organs, especially as the models, that it may take a moment's reflection to remember that each model represents a part of one body developing inside another. Institut für Anatomie I, Universität Jena.

Ecker's classes or in Prague, because by 1859 Ecker was vouching that the models of hearts and genitalia "were made entirely *from my specimens* and under my supervision at the same time as the drawings" for the atlas.[6] But if Ziegler did now model, "not from drawings, but *from nature*" (Ziegler 1867, 26), Ecker's illustrations must have begun to interpret the structures he did in wax. Conversely, as Ecker wrote of Ziegler's modelling, "one sees only, when one wants to represent something plastically, that all kinds of things one previously did not see at all".[7] As Ziegler later explained, the "objects [are] of such complicated form that through drawings they can be made clear only with difficulty" (Ziegler [circa 1886]).

Adding models to specimens that Ecker inherited and collected, Ziegler helped build up a separate embryological collection (Ecker 1857, 11), the need to house which was a strong argument for the new institute that Ecker would obtain in 1867. Ziegler contributed to institute-building in the rest of the university system by reproducing extra models to sell. Ecker helped him advertise at the annual meetings of German scientists and physicians (e.g., Ecker 1859), and wrote to colleagues, asking them to "support the young man" by buying models for their institutes.[8] The atlas lent the waxes authority through references in the text, as did a prominent review (Wagner 1859, 886): "Of all wax preparations that I have seen at home and abroad (even the excellent Florentine works not excepted), these of Ziegler show the most understanding. These preparations promote correct views, while other[s] . . . often only confuse them."

Several of Ziegler's medical contemporaries used university assistantships as stepping stones to chairs. He put ever more energy into modelling, initially after the drawings and specimens of Ecker and other local professors. Ziegler did not present himself as a modeller, though, and so risk being taken for a lower-status lay artist, but was a "Dr of medicine and surgery, assistant in the physiological institute of the university, and member of several learned societies" who made models.[9] He neglected his medical practice, resigned a position as an inspector of apothecaries' shops, and at the end of 1867 left his university job to "devote his whole time" to what had become "his life's work".[10]

This was an extraordinarily favourable time to model embryos. Numbers of medical and science students were increasing and ever more institutes maintained separate collections (Nyhart 1995, 65–102). Then in the 1860s Darwinism made embryology *the* most exciting science around. The

German evangelist of evolution, Jena zoologist Ernst Haeckel, taught that individual development repeats the evolutionary development of the species, or in his words, "ontogeny recapitulates phylogeny". Since fossils were often not available, evolutionists frequently took this short-cut (Gould 1977), and model embryos were in demand.

PUBLISHING IN WAX AND PRINT

Even once Ziegler set up on his own, he still does not appear to have called himself a modeller, but wrote, significantly, as a "plastic publisher".[11] Comparing plastic publishing to book publishing shows how models that we might expect to be, on the one hand, too bulky, fragile, and expensive (Latour 1990), and on the other, too low-status, could become accessible and credible enough to play major roles in embryology. Managing relations with scientists allowed Ziegler to label each series with the name of not just a species, but also a professor (Fig. 7.2). Advertising models "after Professor X" and listing the corresponding printed works made the waxes academically authoritative and relevant.[12]

First publisher and "author" (Ziegler [circa 1886]) had to agree to work together. Dr Ziegler attracted scientists with the claim that he "belongs to science himself" (Ziegler 1867, 25). But they were not just appraising him; before making the original models *he* had to be sure of the quality and commercial potential of *their* work. In 1875 he consulted Freiburg zoologist August Weismann about taking on Haeckel's "four main types of egg cleavage and gastrula formation". With controversy raging over the Darwinist prophet's 'gastraea', a speculative ancestor of the metazoa that Haeckel had deduced from what he called the 'gastrula' stage of ontogeny (Nyhart 1995, 181–93), Ziegler must have been excited about such a topical series, but worried lest it prove a costly flash in the pan or even damage his reputation. As Weismann reported to Haeckel, "He is a careful man, and first confirmed with me that your gastrula theory is truly sound and would not be overturned soon! Only when I reassured him about it, did he decide to make the models you requested!" (Uschmann and Hassenstein 1965, 38). Conveying his agreement, Ziegler explained how Haeckel, in turn, would approve his work. "I have read the 'Gastrula'", the offprint the zoologist had given him (Haeckel 1875), "and I hope to show through the models you will receive later that I have understood the matter correctly.... Without your proof

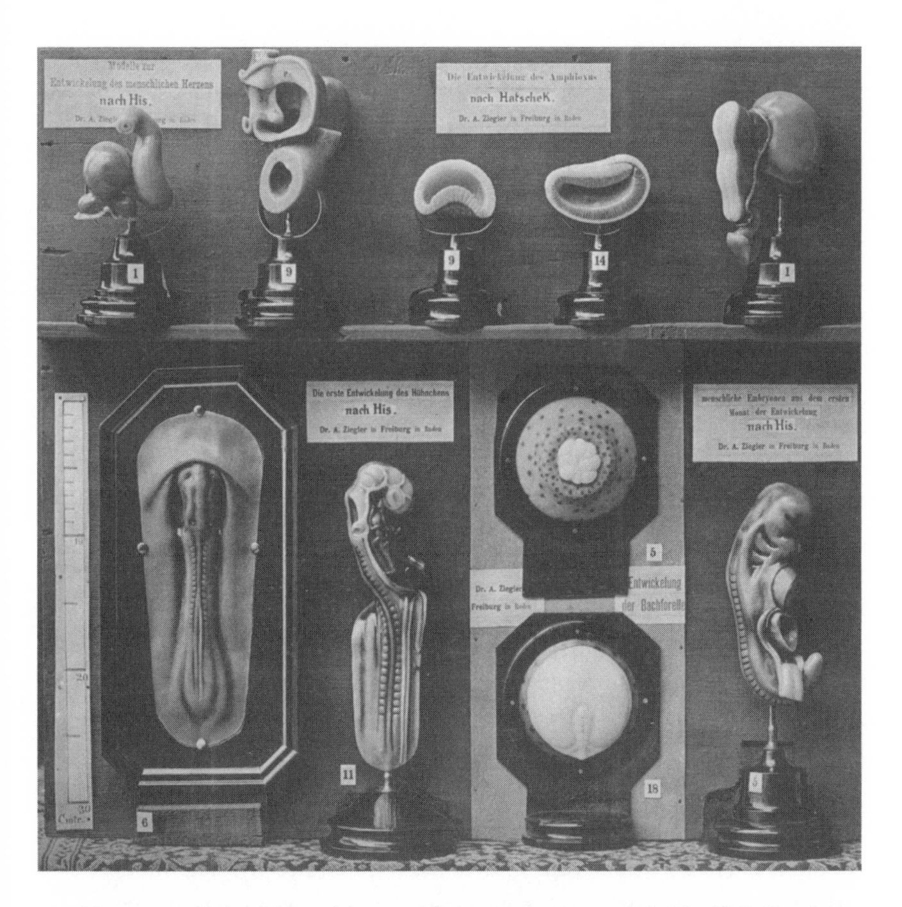

Figure 7.2 "Scientific teaching models in wax by Dr. med. A. Ziegler". Catalogue photograph of selected models from series (clockwise from top left) on the development of the human heart after the Leipzig anatomist Wilhelm His, the development of amphioxus after Haeckel's Austrian admirer Berthold Hatschek, human embryos of the first month after His, trout development after Innsbruck embryologist Josef Oellacher, and the early development of the chick after His. The His–Ziegler series are after the anatomist's original models, the others after drawings and specimens. The ruler at the left has a 30 cm scale. Photograph by Römmler und Jonas of Dresden, from *Wissenschaftliche Unterrichtsmodelle von Dr. med. Adolf Ziegler* (Freiburg in Baden [circa 1886]). Medizinhistorisches Institut der Universität Bern.

correction [*Correktur*] nothing will be finally finished."[13] Ziegler presumably did not expect authors to use a spatula, but to draw changes on paper.

The language of print—"author", "proof"—points up parallels between publishing models and publishing books, but contrasting processes of reproduction made the businesses very different. Far from having a run of hundreds or thousands printed on mechanised presses—or even, like Louis Thomas Jérôme Auzoux, owning a factory of nearly a hundred people mass-producing papier-mâché models (Davis 1977, 261)—Ziegler modelled by hand and at home. He presumably made the originals freehand in wax or clay, and reproduced them using plaster moulds, perhaps with the help of an assistant or two (Büll 1977, 1: 420–34; Schnalke 1995). To make many models from a single mould, it had to be preserved undamaged through the casting; for complex and undercutting structures this was achieved with 'piece moulds' of several parts, usually held together in a case. Or a model was cast in several pieces, which could then be made of different coloured waxes, and assembled later. The most time-consuming step was finishing the raw model by smoothing joints and imperfections, and painting. The products were fixed on wooden handles, put in cardboard cases, or—most commonly—mounted with wire on black-painted wooden stands, distinctively labelled with gold numbers on red paper (Figs 7.1B and 7.2).

Ziegler was much more independent than the underlabouring artists who drew and engraved the plates that accompanied scientific articles and monographs, but his own hands created the forms he sold. So the work contrasts with, more than it resembles, a book publisher's. Ziegler could not accumulate much stock, and often made copies in response to orders; not buffered by booksellers or dealers, he sold to customers directly. The models cost more than machine-printed books; depending on size and complexity, a series went for the price of from one or two to 15 big textbooks, or between one and ten volumes of a heavyweight journal (Engelmann 1911; Ziegler [circa 1912a]).

An unusually fraught exchange between Ziegler and Haeckel highlights these differences. Sending the zoologist his complimentary set of wax gastrulae in July 1876, Ziegler reported that he had three others ready, and asked if Haeckel would kindly place a notice in a learned journal. By November Ziegler had received only six orders, including a request for a second set from Haeckel himself. In early February 1877 he trusted that the zoologist had kept his promise to place "words of recommendation" in the new edition of the *Anthropogenie*, his latest semi-popular book (which would be translated as *The Evolution of Man*). Sending the models towards the end of the month, Ziegler reported only two more orders and some disappointment. "I have at last been able to obtain the 3rd edition of your *Anthropogenie* from the University Library, but not yet found anything referring to the preparations; perhaps I will find it when I am further on in the book." In May Haeckel confessed his oversight and promised to remedy the situation, but by October Ziegler was desperate.

> Since I have till today sold only eight copies, you will surely agree that at the low price of the series I cannot cover my costs as a plastic publisher [*plastischer Verleger*]. Now because I know what distribution your scientific works enjoy, I believe that it requires only a note of recommendation from your side . . . to direct the attention of interested parties also to my preparations, which make it so easy for anyone to convince themselves of the truth of your statements.[14]

Already the most talked about scientist in Germany, Haeckel's message hardly needed to be taken into wax. Was he even touched by the pathetic contrast that Ziegler implied between his own sales and those of Wilhelm Engelmann, the Leipzig publisher of the *Anthropogenie*? Complaining too, Engelmann was disappointed that of the 2,000 plus 2,500 copies of the first and second editions that he had optimistically produced in quick succession in autumn 1874, he would, as they prepared the third edition of 2,000 two years later, likely to be left with over 400.[15] The modeller ran less risk of unsold stock, but had to struggle harder for every order.

So Ziegler seized opportunities to advertise, especially at international exhibitions (e.g., Ziegler 1867). He also sent out catalogues containing price lists and high-quality photographs (Fig. 7.2), and issued prospectuses for individual series. These extracted the essential information from authors' publications and related the models to them. Professorial endorsements associated his products with the leading authorities in the field, and lists made the most of every university and museum supplied, medal awarded, and honour received (e.g., Ziegler [circa 1886]; Anderson, Burnett, and Gee 1990, ii–iv).

Advertising helped distribute the waxes, but could Ziegler really establish a general presence in embryology? In the late 1880s he claimed to have supplied institutions in at least 115 towns; by about 1912, the firm boasted some 180 homes on five continents of institutes, academies, and museums where its models were displayed, including the major public collections and all the main universities (Ziegler [circa 1886]; Ziegler [circa 1912b]). Less widely announced, the models found their way to public health museums and fairgrounds as well (Hopwood 2002, 72–75). Many institutes and museums bought so many waxes that separate registers were kept; Ziegler's dominate the catalogues in embryology.[16] Within a few decades of setting up as a plastic publisher, his products were a big budget item in starting new institutes,[17] and their export took embryology from its German powerhouse around the world.

"GRASPABLE FORM"

Because the dominant histories follow the progress of embryological research, Ziegler's "teaching models" might appear peripheral. But teaching was the backbone of a science practised mostly in the universities, and investigations, like Ecker's for his atlas, were organised around it. As embryology joined in a general mid-nineteenth-century move to instruction by seeing and doing, and hence to dependence on visual aids, Ziegler's models became central. What special qualities account for their adoption, and how did they work with other media?

Learning embryology was a process of disciplining the eye. Since schools taught hardly any biology, let alone topics associated with sex (Daum 1998, 43–83), most students probably entered university with little idea what embryos looked like. Embryology lectures and classes taught them to recognise and copy the initially bewildering forms produced with and observed through microscopes (Ecker 1886, 140; Gooday 1991; Jacyna 2001). Learning how to see depended on learning what to see, and so relied heavily on visual standards. Commerce and competition filled classrooms with illustrated books, and specimens, models, and charts, produced in-house and bought in (Lemire 1990, 324–65; Bucchi 1998). A photograph of embryological teaching in Freiburg shows off the range of visual aids that in one form or another were used in every university (Fig. 7.3).

With such a variety of teaching aids available, why buy expensive models? Natural preparations had an aura of authenticity, but also three major drawbacks: some objects were very scarce; most embryos had to be viewed under a microscope, an obstacle to group instruction; and even expert preparation did not display the structures of interest as clearly as a purpose-built device. Schematic drawings, charts, and models all magnified embryos for demonstration, and so focused attention on parts. For human and other

Figure 7.3 Visual aids for embryology, set up for a lecture in the Freiburg gynaecological clinic, 18 February 1893. The skeleton, blackboard drawings, and wall charts are arranged around a table that holds two trays of Ziegler models: on the left apparently Berthold Hatschek's amphioxus, and on the right Wilhelm His's two series of human embryonic anatomies. Photograph (detail), Universitätsarchiv Freiburg, D49/991.

mammalian specimens this created the illusion of development independent of the pregnant female body and, for organs, even of the rest of the embryo or fetus. All of these visual aids also mapped such abstractions as 'germ layers' on to embryonic bodies. But only models avoided reduction to 2-D. So though scale, texture, and colour worked together to convert delicate and shimmering but tiny and elusive forms into solid and opaque but huge and memorable shapes, wax models' most important advantage was their three-dimensionality (e.g., Peter 1911, 41).

What kind of engagement did the models invite? Different series used more or less depth and implied viewing angles of different width. At one extreme are the plaques on the development of the trout (Fig. 7.2), then come many models that were designed to be seen from one principal perspective, and at the other end of the spectrum are the heart models, which show "complicated cubic relationships" (Ziegler 1867, 26) in the round (Figs 7.1B and 7.2). Anatomists accepted Ziegler's claim that for difficult objects it was far better to offer the student developmental stages "in corporeal and, we might even say, graspable form" (Wiedersheim 1882).

It may seem obvious that avoiding the imposition of a single point of view, and bringing hands into play as well as eyes, would set struggling imaginations free. But how free were students, and to do what? Though only the nearest and keenest-eyed will have been able to appreciate the waxes as a lecturer held them up, some were handed round and there were demonstrations afterwards (Fischer 1957, 25–26). A photograph implies that Manchester students memorised amphioxus models before turning to their own microscope slides (Fig. 7.4). Even in the classrooms, encounters were probably

Figure 7.4 Learning to see microscopically. A class in a zoological laboratory at Owens College, Manchester, around 1900, with Hatschek–Ziegler amphioxus models on the table. Professor Arthur Milnes Marshall remounted and labelled the waxes, and published a guide that picked out what students should notice; the second edition warned of features that had gone out of date (Marshall 1902). Photograph (detail), John Rylands University Library, Manchester.

more visual than tactile; glass covers were sometimes added to keep off fingers and dust. Yet, however unrestricted the interaction, some students—especially those who were there, not for love of embryology, but to gain a professional qualification—will have experienced as a mixed blessing the freedom to learn more effectively the discipline's standard forms.

For all these unique properties, models never stood alone, and the company they kept set the frame within which they were viewed. On the one hand, any potentially inappropriate uses could be curbed. In a fairground collection of curiosities even these waxes might just have seemed titillating (*William Bonardo Collection* 2001, 5–11, 13, 17, 54), but with scholarly labels and charts on the same topic they looked academic. On the other hand, and more routinely, words and pictures bolstered the power of otherwise mysterious artefacts to convey the messages professors wanted to get across. Ziegler's prospectuses broke the colour codes, identified the parts, and referred to embryologists' printed works.

In the late 1880s, textbook authors created a new relationship between wax and print: they started to use figures, not of specimens, but of models. For example, instead of copying images from Ecker's plate of genital development, itself informed by modelling but not expressly depicting models, the new leading textbook carried a photomechanical reproduction of the Ecker–Ziegler waxes, and so invited students to compare pictures of models with the models themselves (Hertwig 1888, 299). If only after "rather long hesitation", authors wished to allow students to revise with the demonstration material in the classrooms (Bonnet 1907, v). Usually the books got away with a single picture, but occasionally acknowledged the advantages of models in the round by showing views from front and back.

If some prospective doctors and teachers meekly absorbed these powerfully interlocking images, and took them, perhaps into practice and the schools but at least into examinations, other students resisted the waxes' "rigid solidity" (Peter 1911, 41). The few who were inspired to do research produced new pictures, and sometimes new models, of their own. Meanwhile, teachers adapted the old Ziegler models to changing demands (Fig. 7.4). Yet, though it may sadden those fascinated by development, for every awe-inspired youngster ten probably found embryology the most difficult and tedious subject they ever encountered, and even Haeckel could not stop them switching off (Haeckel 1874, xv). When Ecker had "*mortui vivos docent*" chiselled in gold on the façade of his anatomy building, he meant that the dead should teach the living. Yawning students jokingly wished dead lecturers long lives (Fischer 1957, 21).

MODELS AS RESEARCH PUBLICATIONS

Already fixtures of embryology teaching, between the late 1860s and the late 1880s Ziegler models gained new significance in research. This shift accompanied a major innovation in microscopical practice, the routine conversion of specimens into serial sections. Though section-cutters, or microtomes, facilitated detailed and systematic exploration of organisms' interior structures, they would be criticised for alienating microscopists from whole forms. But the Basel (later Leipzig) anatomist Wilhelm His promoted one of the first efficient microtomes as allowing him to section a specimen completely, and so to build 3-D models that, far more than any mental construction, would "give body" to visual understanding again (His 1870, 231; Hopwood 1999).

At first His modelled chick embryos in leather and lead, but found these materials "rather unpliable as soon as one is dealing with more complicated shapes". So in about 1867 he invited Ziegler over the Swiss border to Basel to teach him to use the more malleable wax and clay. "[I]n joint work", His reported, "we manufactured a series of models from specimens and sections, in the preparation of which we strove for the greatest possible accuracy and fidelity" (His 1868, 182; see also His 1887, 384). His first formed the body freehand until it approximated the drawing of the whole that he had made before sectioning, and then measured with callipers on drawings of each section to judge how much wax to add or take away. From these originals Ziegler reproduced and marketed a series of models to complement His's plates of sections, adding colours to distinguish germ layers and organs (Fig. 7.2). The detour via the sections, and the work with hands as well as brain, gave unprecedentedly detailed, 3-D views of the internal development of the chick, which joined Ziegler's non-reconstructive series in every embryology classroom.

During the 1870s embryologists adopted the microtome, but few followed His into the new modelling. For the host of young Darwinists sectioning their way through the animal kingdom, the pickings must initially have seemed rich enough, especially from invertebrate embryos, and the pressure of competition sufficiently fierce, to make an extremely demanding technique unattractive. It did not help that His promoted plastic reconstruction in the same works as he opposed a mechanical embryology to Haeckel's phylogenetic approach, and accused Germany's leading evolutionist of forging pictures to suit his views. Haeckelians could use His's chick models, but modelling was initially picked up only by a comparative anatomist at Breslau (now Wrocław), Gustav Born, who for his studies of intricately developing head skeletons invented a simpler method (Born 1876, 578–80; Hopwood 1999, 487–91). He traced the relevant details of each section on to a wax plate as many times thicker than the section as the plane magnification, cut away the excess wax, and stacked the cut-outs up. But for several years this plate method was not adopted either.

In the early 1880s, His reformed human embryology, the field that Haeckel's controversial drawings had most destabilised, by putting large numbers of early human embryos under the microtome for the first time (Hopwood 2000). Railing against the mindlessness of displaying section after section, and the unreliability of reconstructing only in the head, His again demanded that those who wished to grasp complex anatomical structures must model, however laborious this was (His 1880, 1–13; 1885, 3–5).

Since Adolf Ziegler's health was failing, his second son Friedrich, a former teacher of drawing and modelling who had studied with a monumental sculptor in Vienna and attended morphological lectures in Freiburg (Bargmann 1936), finished His's human embryo models and reproduced them. Thirty years earlier Adolf had modelled from Ecker's specimens, drawings, and models; Friedrich now reproduced original models that His had built up from sections. A series of whole embryos superseded Ecker's models of the first month of development, correcting and showing far more detail; a second stripped down selected specimens to display His's syntheses of interior structure in schematic colours, and Friedrich Ziegler also reproduced isolated brains and hearts (Fig. 7.2).

Though now usually separated from what the antiquarian bookseller's bible of medical history lists as "3 p[ar]ts and atlas" (Morton 1983, 65), these models were once integral elements of His's monumental anatomy of human embryos. Thanking His for the gift of the work, his Basel successor Julius Kollmann showed that he appreciated its full complexity. The present consisted, he explained, of three "publication series" (*Publikationsreihen*): two folio atlases, three instalments of text plus related articles, and two "inseparable" sets of models (Kollmann 1890, 650, 659).

Insisting that published models were more than teaching aids, His campaigned against the persistent tendency to ignore them in print. This raised a question that went beyond Adolf Ziegler's self-presentation as a publisher: "*Should an author who has published* [veröffentlicht] *the results of his research in the form of plastic models claim for these models, just as for printed documents, the rights of scientific deeds* [Urkunden]?" His was in no doubt: "Just as I regard complicated spatial relationships as only really understood when they are available as plastic representations", wrote the editor of the leading anatomical journal, "so too I consider the model, even more than the written word, the decisive record [*Urkunde*] of the understanding of form of the researcher concerned" (His 1895, 359). A model recalled an object immediately into consciousness, and reproduced details that were difficult to describe in words. The effect was far more direct.

By the mid-1880s, His was already beginning to win younger anatomists over to reconstructive methods and the revaluation of models that accompanied them. Haeckel's evolutionary programme, though a huge popular success, was in serious trouble in the universities, and embryology in ferment (Gould 1977, 167–206; Nyhart 1995, 243–305). Historians of science have traditionally selected those new approaches that seem to lead most directly to twentieth-century biology, especially an experimental tradition that followed His's call to investigate the physiological causes of form (Maienschein 1991; Mocek 1998). These studies of relatively simple and accessible embryos had so little use for wax modelling that it was even written out of His's work on the chick (Hopwood 1999). Yet many vertebrate embryologists responded to the crisis not by abandoning comparative science, nor even by seeking simplicity, but with a thorough re-investigation of the relations of ontogeny and phylogeny in a greater variety of animals, and—inspired by His's anatomy of human embryos—especially the most difficult to study.

Born's plate method made this research possible. As a new generation of anatomists competed to improve his more mechanical technique, it superseded His's 'free' modelling, because it was seen to be easier, quicker, and more objective, while still involving anatomists in the mental exercise that some valued as an aid to understanding form (Hopwood 1999, 487–91). It effectively reduced the labour of each step to 2-D, and much of the work could be assigned to technicians. By the 1890s embryologists describing complex microscopic morphologies, especially human embryos and developing vertebrate skulls, found plate modelling indispensable.

Here is the standard method of reconstruction (Peter 1906). To aid correct mutual orientation of the stacked sections later, guide-lines or planes were first scratched into the block containing the embedded specimen. Then came the tedious and finicky work of cutting at known thickness without losing a section or introducing tears or folds, and distributing the sections on microscope slides (Fig. 7.5A). To an appropriate magnification (parts of) the sections were drawn from projections thrown on a board (this modelling was mostly drawing). Born had originally prepared 1 mm thick plates by pouring wax over hot water, but modellers went over to rolling them between brass strips on a lithographer's stone (Fig. 7.5B). This had the advantage that instead of tracing on to previously prepared plates, the sectional drawings could be soaked in turpentine and incorporated into the plates as these were rolled. In cutting out with a scalpel on glass, the guide-planes were left temporarily to aid alignment, as were 'bridges' to keep non-contiguous structures together (Fig. 7.5C). The plates were stacked (Figs 7.5D and E) and the reconstruction finished by removing bridges and guides, and smoothing the edges with a spatula (Fig. 7.5F). Painting eliminated the residual appearance of stripes; colours identified layers or regions.

Modelling allowed anatomists, whose working lives revolved around the dissecting room, to visualise microscopic structures on a scale similar to adult cadavers and to treat them in the same way. Having built up a model, an anatomist could "cut up, open, 'dissect' it like the object itself and easily gains insight into the most complicated relationships". Born's colleagues cut sections through the wax with warm wires or visiting cards soaked in turpentine, and let windows into interior cavities.[18]

To a limited degree, anatomists lent, copied, and displayed these original models at meetings, but to reach a wider audience, they had to publish them. So a researcher would make models, and write a research paper describing and depicting them. He, or now occasionally she, would send the manuscript to an editor, and the original models to a modeller for finishing and reproduction. Articles might even show the smoother forms and instructive colours of the reproductions. This completed a shift towards models as the primary representations of form, and the unit of publication as model series plus the viewing instructions that staked the author's claims. (Where

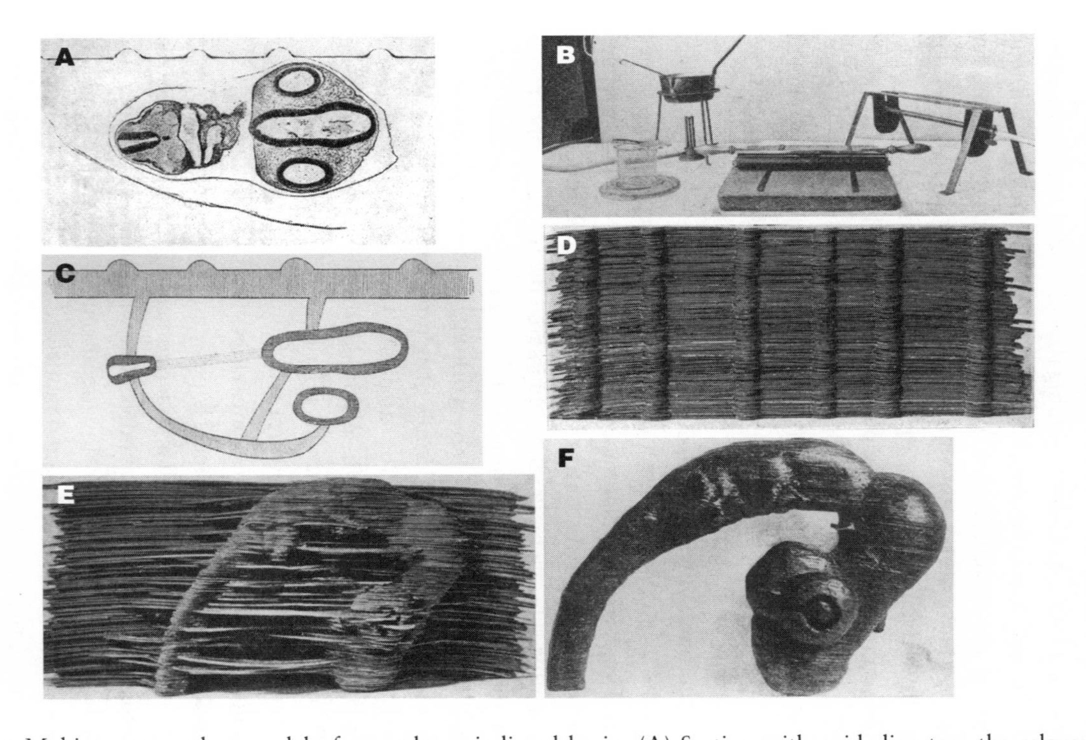

Figure 7.5 Making a wax-plate model of an embryonic lizard brain. (A) Section with guide-line top; the relevant parts were drawn. (B) Instrumentarium for rolling the plates, including stone and rolling pin; each sectional drawing was incorporated into a plate, and excess wax removed. (C) Diagram of cut-out plate, with 'bridges'. (D) Stacked cut-out plates seen from the guide-plane side. (E) Stacked plates from the object side. (F) Finished model, before painting. From Karl Peter, *Die Methoden der Rekonstruktion* (Jena, 1906), figs 12, 30, 33, 34, 35, and 36. Whipple Library, University of Cambridge.

no reconstruction was required, as often for invertebrate embryos, things went on as before.)

Adolf Ziegler had no cause, then, to regret having taught His to model. The "Studio for Scientific Modelling" that Friedrich Ziegler took over in 1886 might make fewer originals, but was in an unrivalled position to meet demand for published plate models. The most impressive single image from the heyday of evolutionary embryology may be a photograph of his display at the World's Columbian Exposition in Chicago in 1893 (Fig. 7.6). By the eve of World War I the "press" offered over 300 different models (Ziegler [circa 1912b]).

This success owed much to close relations with scientists. Friedrich Ziegler's brother, the zoologist Heinrich Ernst Ziegler, helped in many ways, but the modeller's key professional relationships were with the Freiburg anatomists. International production of embryological waxes for higher education centred on his studio and their institute.

"WORTH AT LEAST AS MUCH AS WHEN A RICH PUBLISHER GIVES A FEW THOUSAND TO PUBLISH BOOKS"

Fifty years after Ecker and Adolf Ziegler arrived, Freiburg's anatomy institute was an international centre of comparative vertebrate anatomy and embryology, and a hive of modelling activity, its staff busy as bees around the wax. Any teacher of embryology in a university around 1900 used Ziegler models; here *research* was unusually dependent on modelling, and hence on publishing with the firm.[19] The staunchly Darwinian chief, Ecker's successor Robert Wiedersheim, authored only one series. But the long-term prosectors, Franz Keibel and Ernst Gaupp, who did most of the work while they waited into late middle age for full professorships of their own, modelled with a passion bordering on obsession, and Americans flocked to work with them.

During the 1890s Keibel launched a major re-investigation of the relations between ontogeny and phylogeny, especially in the previously relatively little-studied mammals (Keibel 1898; Nauck 1937; Gould 1977, 174–75). He built a career as the world's leading descriptive vertebrate embryologist on modelling one complex structure after another. The visiting Johns Hopkins anatomist Florence R. Sabin was aghast that "Professor Keibel does nothing but make models—and has all of his students making models".[20] At Keibel's

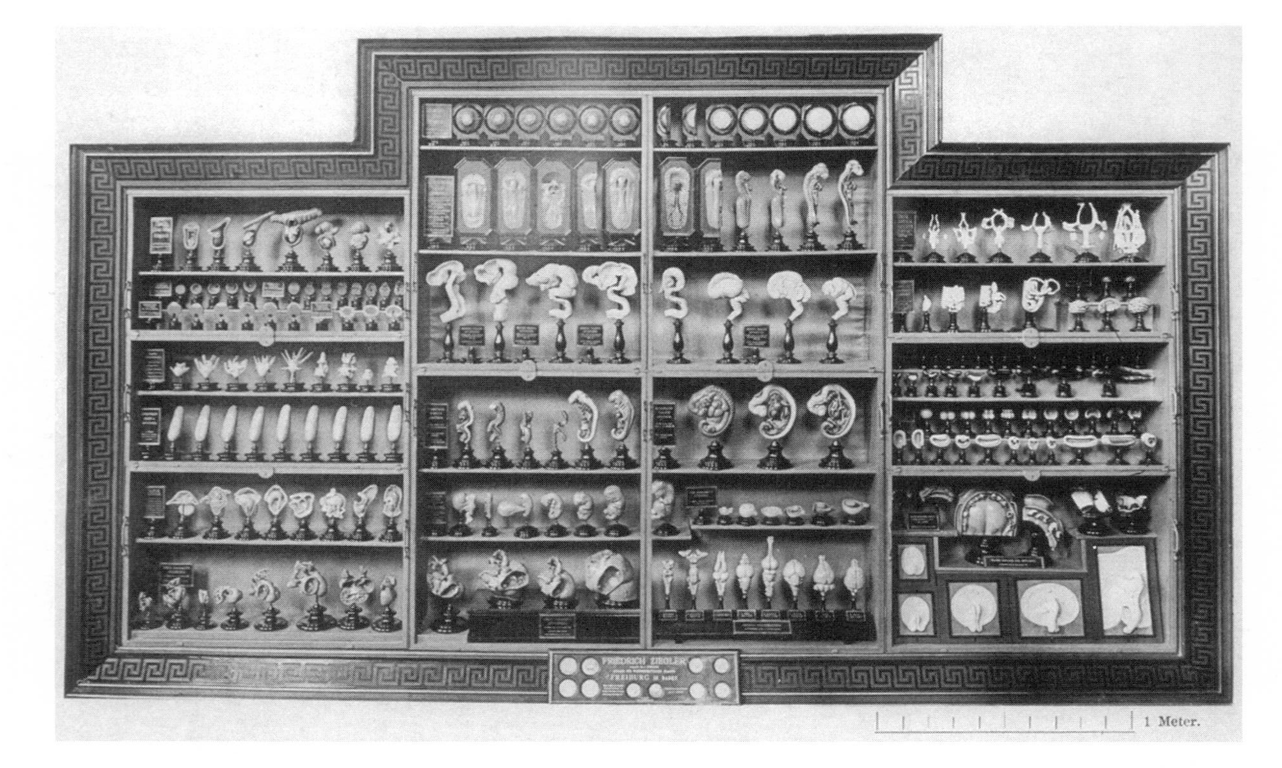

Figure 7.6 Friedrich Ziegler's prize-winning stand at the World's Columbian Exposition in Chicago, 1893. Photograph, with metre scale, from *Prospectus über die zu Unterrichtszwecken hergestellten Embryologischen Wachsmodelle von Friedrich Ziegler* (Freiburg in Baden, 1893). Division of Rare and Manuscript Collections, Cornell University Library.

farewell party in July 1914 the company bizarrely finished their meal with a human embryo modelled in marzipan (Wiedersheim 1919, 148).

Born had inspired Gaupp to study the evolution of the vertebrate skull (Fischer 1917). The significance of the head made its skeleton a favourite topic in comparative anatomy, but it was also the most complex. Thanks to Gaupp, one of the most productive and innovative comparative anatomists of his generation, the field came to depend as absolutely on modelling as did human embryology: of 83 figures in his 300-page handbook entry (Gaupp 1905), just over half show plate models (some the same model in different views), three-quarters of them made in Freiburg (Fig. 7.7).

So what happened when a raw model moved from what was effectively Ziegler's department of research and development to the institute's production facility? He did not just finish the originals—by smoothing away the steps between the plates—and make and distribute reproductions, but also put energy and initiative into converting stacks of wax plates into effective visual aids. Ziegler's later models are much larger than those made even for His, let alone Ecker—classes were getting bigger—and much more fragmented. In order selectively to bring out the greater interior detail that plate modelling revealed, Ziegler painted sectional images onto cut surfaces, carved windows, and mounted movable parts in iron frames.

Sabin's letters home from Freiburg vividly record her seven-week collaboration with Ziegler to publish an exceptionally complex model of "the medulla, pons, and midbrain of a newborn babe" (Sabin 1900; Andriole 1959). Ziegler first insisted that to show both sides of the brain (the original showed just one), they must also model them; he could not visualise the structure sufficiently clearly to model from Hopkins artist Max Broedel's perspectival drawings. Then Ziegler reproduced all parts of the model, checking each one with Sabin:

> I believe [she wrote] that there is no word in the English language that I know so well as [the German word] "stimm"—Every time Ziegler makes a new piece, it looks so odd to him that he is sure it won't "stimm" so he always greets me with—"Es stimmt nicht" ["It's not right"].—I have always gotten everything to stimm in every dimension but he never has any faith for the next time. It has gotten to be a great joke.[21]

It was probably Ziegler's decision not to make a single model, "[s]ince experience has taught that [wax] models to take apart and put together again

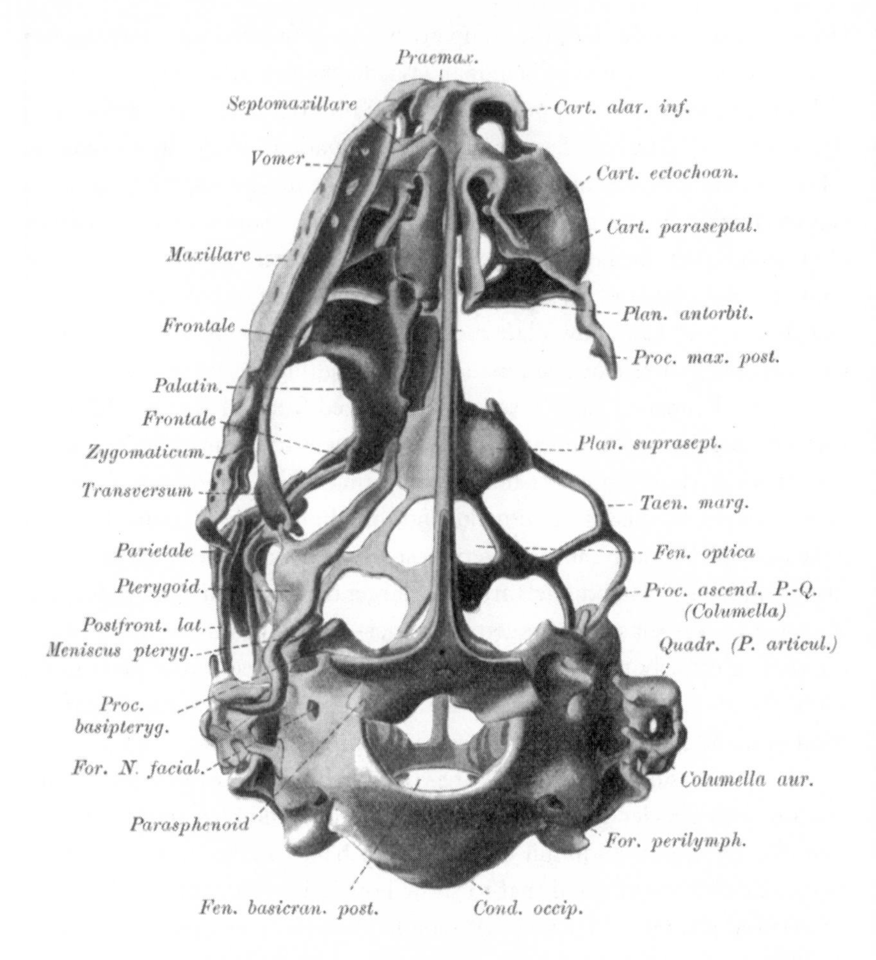

Figure 7.7 Reproduction of a drawing of the skull of a lizard embryo, after a plate model by Ernst Gaupp as copied by Friedrich Ziegler. This view is from the ventral side, and shows the secondary (investing) bones on our right side only. Colours distinguished cartilage (blue), and primary (white) and secondary (yellow) ossifications, as in Ziegler's model. From Gaupp's authoritative survey, "Die Entwickelung des Kopfskelettes", in *Handbuch der vergleichenden und experimentellen Entwickelungslehre der Wirbeltiere*, ed. Oskar Hertwig, vol. 3, part 2 (Jena, 1905), fig. 383. Whipple Library, University of Cambridge.

easily suffer damage when much used", but four separate models.[22] Three were swung in frames, Sabin reported. "Through each model passes a rod, one end of which is a pivot while at the other is a cog wheel, so that the model can be turned to any degree. . . . At present we are studying to see that everything is clear, putting in wires to represent tracts like the pyramids and painting the nuclei."[23] They made another model by cutting up the one Ziegler duplicated first and drawing diagrammatic sections onto what ended up as a series of metal slices, "to be used to relate the other models to transverse sections" (Sabin 1902, 282). These models dismantle the original model in different ways, and invite users actively to try out various viewing angles.

A catalogue photograph (Fig. 7.8) shows Ziegler and three assistants—"common workmen" but "excellent", wrote Sabin[24]—engaged in the time-consuming hand-finishing that was so much more than the routine smoothing of joints. Ziegler sits in the left foreground, leaning forward over a table

Figure 7.8 Catalogue photograph of Friedrich Ziegler's Freiburg workshop, first published circa 1912. Potential purchasers were shown artisanal work, which perhaps justified the high prices. Emphasising the commitment to 3-D form, even the wall decorations appear to be reliefs. From Friedrich Ziegler, *Embryologische Wachsmodelle* (Freiburg im Breisgau [circa 1924]). Medizinhistorisches Institut der Universität Bern.

to complete a model of a human fetus by Keibel; the man sitting near him appears to be adding labels to the last of Gaupp's models on the development of the primordial skull in the frog; and the assistant standing in the middle of the room is painting one of Sabin's models. Everything, she wrote, "is going to be far better than I ever imagined".[25] This had been no mere reproduction but a creative process in which the anatomist herself learned something new.

Relations between modeller and anatomists were not entirely smooth, however. Ziegler won high praise, but without his father's medical qualification, he felt the lack of status. When professors asked for discounts, he reflected that they not only grudged a fair price but also failed to recognise his scientific contribution.

> He tells himself [Gaupp explained to the Heidelberg anatomist Max Fürbringer in 1908] that he has for decades been giving medical science a large number of models, without which . . . embryological teaching is today hardly conceivable, and which also in scientific work we can hardly manage without. The importance of these aids in learning, researching, and teaching can be doubted by no one. Their manufacture demands that the maker knows the object exactly, reads all works relevant to it (which Z. actually does), and again and again works critically through every copy.

So it rankled with Ziegler that he had not received the kind of recognition from scientific circles that for a publisher of medical books was almost routine. What he did was surely

> worth at least as much as when a rich publisher gives a few thousand to publish books, in the making of which he otherwise has no involvement, and from which in the end he gets thousands back again. And when this service is valued by the representatives of science so highly that [Gustav] Fischer, [J. F.] Bergmann, [Ferdinand] Enke, [Paul] Siebe[c]k et al. become honorary doctors, then the maker of models [*Modell-Verfertiger*] would long since have had a claim to the same honour.[26]

In 1916 Wiedersheim finally arranged for the Freiburg medical faculty to acknowledge Ziegler's contribution in this way. And, like his father and the publisher of the German Anatomical Society's journal, he received an obituary in pages otherwise reserved for scientists (Wiedersheim 1889; Bardeleben 1910; Bargmann 1936). The discipline thus recognised that, for generations of students, teachers, and researchers, embryology and Ziegler's models had been inseparably linked.

EPILOGUE AND CONCLUSION

After World War I Friedrich Ziegler slowly lost his innovative role, and through the rest of the twentieth century his models gradually went out of use. Economic crises and increased competition from other firms reduced the number of new commissions, but two general shifts in embryology were decisive: experimenters who rarely modelled took over, and fields that still depended on methods of reconstruction ceased to rely on commercial modellers (Hopwood 2002, 77–82). Human embryologists, for example, needed so many models that they had in-house modellers produce them, not commercially, but for central reference collections (Blechschmidt 1968; O'Rahilly 1988). The general anatomical modeller Marcus Sommer took over the studio on Ziegler's death in 1936, but could not maintain the back list after he was expropriated by the German Democratic Republic in 1952 (Schnalke 1991).

Yet large numbers of the models remain in institutes and museums, and even in the 1990s a few were still taken from cabinets for annual demonstrations to students. Historical engagement with these, the oldest artefacts that developmental biologists regularly see at work, is timely. Three-dimensional morphology is on the embryological agenda again, with the results often published simultaneously in journals and on the World Wide Web, and increased traffic between laboratories, clinics, and the media is showing much wider audiences the results. Many of these forms were first visualised during the nineteenth and early twentieth centuries—in wax as well as on paper.

This chapter has explored the changing roles of wax models in an academic science during the heyday of print. By 1850 embryology had found various uses for models, but generally seems to have managed without. During the next decades it came to depend on the Zieglers to represent a wide range of developing forms, at first largely in teaching and then increasingly in research as well. Their models became so significant that they were used to produce book and journal illustrations, and joined print publications to make what we should appreciate were composite works.

NOTES

I thank Urs Boschung, Lynn Morgan, Lynn Nyhart, Michael Hagner, Alison Kraft, Willem Mulder, John Le Grand, Rosemarie Fröber, Jonathan Crisp, Hans Sommer, and the staffs of institutions holding manuscripts, catalogues, books, and models, for valuable advice and help in obtaining materials, Adrian Newman for photography, participants at the Wellcome symposium in November 1998, and seminar audiences in medical history at Manchester, Freiburg, Heidelberg, and Geneva for useful feedback, and Silvia De Renzi, Marina Frasca-Spada, Nick Jardine, Jim Secord, Jonathan Topham, the Press's referees, and especially Soraya de Chadarevian for helpful comments on drafts. This research was supported in part by the Wellcome Trust.

1. For a fuller and highly-illustrated history of the Ziegler studio and guide to its models, also limited by the absence of a major cache of Ziegler company or personal papers, see Hopwood (2002). All translations are my own, and any emphasis is in the originals.

2. Wiedersheim 1889. On Ziegler's father, see Generallandesarchiv Karlsruhe (hereafter GLAK): 76/8908; for youthful skills, [Rudolf] Dietz to Grossherzogliches Hof-Secretariat, 8 May 1868, GLAK: 56/1559; for description as an autodidact, Ziegler 1867, 27; for school attendance, the *Gymnasialprogramme*, Stadtarchiv Freiburg: Dwh 1135; and on modelling methods, Büll 1977, 1: 420–34; and Schnalke 1995.

3. Biographical information from Ziegler to Baden's sanitary commission, 17 October 1857, GLAK: 236/15540; to Friedrich Miescher-His, 25 February 1852, Öffentliche Bibliothek der Universität Basel: Miescher-His Papers; and to Jan Purkyně, 3 August 1854, Literary Archive, Museum of Czech Literature, Prague: Purkyně Papers. For dates of study and assistantships, see Adolph 1991, 520; and on Purkyně, Coleman 1988; and Kremer 1992.

4. Ecker 1851–59, legend to pl. XXIII; Ecker 1859. Quote from Ecker to Rudolph Wagner, 14 April 1852, Niedersächsische Staats- und Universitätsbibliothek Göttingen: Wagner Papers.

5. Ecker to Wagner, 3 June 1852, Wagner Papers.

6. Ecker to Wagner, 23 February 1859, ibid.

7. Ecker to Wagner, 16 April 1859, ibid.

8. Ecker to Wagner, 14 April 1852, ibid.

9. Prospectus on *The Development of the Frog* (Rana temporaria) *Represented in 29 Wax Preparations*, sent with a letter from Ziegler to Karl Ernst von Baer, 29 January 1858, Universitätsbibliothek Giessen: von Baer Papers. Ziegler had temporarily expanded the series from 25 to 29 models.

10. Quotes from [Rudolf] Dietz to Hof-Secretariat, 8 May 1868, GLAK: 56/1559. For the resignations, see Ziegler to Obermedicinalrath, 24 April 1866, GLAK: 236/15540; and Ecker to Medicinische Facultät, 31 December 1867, Universitätsarchiv Freiburg: B37/248.

11. For "*plastischer Verleger*", see Ziegler to Haeckel, 23 October 1877, Ernst-Haeckel-Haus, Universität Jena: Best. A, Abt. 1 (hereafter EHH). Long before the nineteenth century, '*Verleger*', once a general term for someone who took over the cost of making something, came to be used in the specific sense of the English 'publisher', a person who bore the cost and arranged the distribution of books

(*Deutsches Wörterbuch* 1984, 756–64). Ziegler seems to be attempting to extend its range. A more neutral term was "*Verfertiger*", or producer (Ziegler 1867, 27).

12. For Leopold and Rudolf Blaschka, modellers of glass animals for museums, institutes, and private collections, Henri Reiling has told a story of progress from copying scientists' illustrations, through collaborating with zoologists who provided specimens and plates, to working from their own living originals and so being recognised as sole authors, as of the renowned glass flowers at Harvard (Reiling 1998). By contrast, Ziegler's career displays the benefits of sharing authorship.

13. Ziegler to Haeckel, 30 November 1875, EHH.

14. Ziegler to Haeckel, 2 July 1876, 7 November 1876, 6 and 25 February 1877, and 23 October 1877, ibid.

15. Engelmann to Haeckel, 20 November 1876, ibid. On Engelmann's role in the zoological community, see Nyhart 1991.

16. For the anatomical collection in Jena, see Fröber 1996, 97–98, 128; and Anatomisches Institut I, Universität Jena: Sammlungsverzeichnis Nr. XX, Entwicklungsgeschichte und Missbildungen, VI. Modelle, circa 1859–1952.

17. When Oscar Hertwig set up a second anatomical institute in Berlin (Weindling 1991, 205–30, 344–46), Ziegler models accounted for 1,005 of 13,000 marks he requested for teaching instruments and aids (Hertwig to the Kultusminister, Gustav von Gossler, 26 April 1888, Geheimes Staatsarchiv Preußischer Kulturbesitz: I. HA Rep. 76 Kultusministerium V^{a} Sekt. 2 Tit. 10 Nr. 131 Bd. 1, Bl. 50–55).

18. Peter 1906, 96, 122–23. Compare the plea from Goethe's *Wilhelm Meister* that anatomists should model *rather than* dissect (Mazzolini, this volume).

19. For parallels to the importance of moulages in dermatology, see Schnalke 1995, and this volume.

20. Sabin to Franklin and Mabel Mall, 10 June 1902 (quote from the continuation of 14 June), Alan Mason Chesney Medical Archives, The Johns Hopkins Medical Institutions: Carnegie Institution of Washington Department of Embryology Papers, record group 1, series 1, box 15, folder 21 (hereafter Carnegie Department Papers).

21. Sabin to the Malls, 6 July 1902, ibid.

22. *Prospectus über die Modelle zur Erläuterung des Faserverlaufes und der Kerne des Mittelhirnes und des verlängerten Markes eines Neugeborenen, nach dem Plattenmodell von Florence R. Sabin M.D.* (Freiburg in Baden: Friedrich Ziegler, Atelier für wissenschaftliche Plastik). Held in Anatomical Museum, University of Groningen. Hence models in wood and papier mâché (Mazzolini, this volume; Davis 1977).

23. Sabin to the Malls, 24 August 1902, Carnegie Department Papers.

24. Sabin to the Malls, 10 June 1902, ibid.

25. Sabin to the Malls, 6 July 1902, ibid.

26. Gaupp to Max Fürbringer, 23 November 1908, Senckenbergische Bibliothek der Universität, Frankfurt am Main: Fürbringer Papers.

REFERENCES

Adolph, Thomas. 1991. *Die Matrikel der Universität Freiburg im Breisgau von 1806–1870, Teil 1 (1806–1849)*. Freiburg im Breisgau: Albert-Ludwigs-Universität.

Altick, Richard D. 1978. *The Shows of London*. Cambridge, Mass.: Harvard University Press, Belknap Press.

Anderson, R. G. W., J. Burnett, and B. Gee. 1990. *Handlist of Scientific Instrument-Makers' Trade Catalogues, 1600–1914*. Edinburgh: National Museums of Scotland.

Andriole, Vincent T. 1959. "Florence Rena Sabin: Teacher, scientist, citizen." *Journal of the History of Medicine and Allied Sciences* 14: 320–50.

Bardeleben, Karl von. 1910. "Dr Gustav Fischer †." *Anatomischer Anzeiger* 37, no. 4/5.

Bargmann, W. 1936. "Friedrich Ziegler †." *Anatomischer Anzeiger* 83: 156–58.

Blechschmidt, Erich. 1968. *Vom Ei zum Embryo. Die Gestaltungskraft des menschlichen Keims*. Stuttgart: Deutsche Verlags-Anstalt.

Bonnet, Robert. 1907. *Lehrbuch der Entwicklungsgeschichte*. Berlin: Parey.

Born, G. 1876. "Ueber die Nasenhöhlen und den Thränennasengang der Amphibien." *Morphologisches Jahrbuch* 2: 577–646.

Boschung, Urs. 1980. "Medizinische Lehrmodelle. Geschichte, Techniken, Modelleure." *Medita* 10, no. 5: ii–xv.

Bucchi, Massimiano. 1998. "Images of science in the classroom: Wallcharts and science education, 1850–1920." *The British Journal for the History of Science* 31: 161–84.

Buffa, Paolo. 1977. "On the accuracy of the wax models of biological microscopical preparations of Giovan Battista Amici (1786–1863)." In *La ceroplastica nella scienza e nell'arte. Atti del I Congresso Internazionale, Firenze, 3–7 giugno 1975*, vol. 1: 217–44. Florence: Olschki.

Büll, Reinhard. 1977. *Das große Buch vom Wachs. Geschichte, Kultur, Technik*, 2 vols. Munich: Callwey.

Choulant, Ludwig. 1852. *Geschichte und Bibliographie der anatomischen Abbildung nach ihrer Beziehung auf anatomische Wissenschaft und bildende Kunst*. Leipzig: Weigel.

Churchill, Frederick B. 1994 [1991]. "The rise of classical descriptive embryology." In *A Conceptual History of Modern Embryology*, ed. Scott F. Gilbert, 1–29. Baltimore: Johns Hopkins University Press.

Coleman, William. 1988. "Prussian pedagogy: Purkyně at Breslau, 1823–1839." In *The Investigative Enterprise: Experimental Physiology in Nineteenth-Century Medicine*, ed. William Coleman and Frederic L. Holmes, 15–64. Berkeley and Los Angeles: University of California Press.

Daston, Lorraine, and Peter Galison. 1992. "The image of objectivity." *Representations* 40: 81–128.

Daum, Andreas W. 1998. *Wissenschaftspopularisierung im 19. Jahrhundert. Bürgerliche Kultur, naturwissenschaftliche Bildung und die deutsche Öffentlichkeit, 1848–1914*. Munich: Oldenbourg.

Davis, Audrey B. 1977. "Louis Thomas Jerôme Auzoux and the papier mâché anatomical model." In *La ceroplastica nella scienza e nell'arte. Atti del I Congresso Internazionale, Firenze, 3–7 giugno 1975*, vol. 1: 257–79. Florence: Olschki.

Deutsches Wörterbuch von Jacob und Wilhelm Grimm. 1984 [1956]. Vol. 25, ed. E. Wülcker, R. Meiszner, M. Leopold, C. Wesle, and Arbeitsstelle des Deutschen Wörterbuches zu Berlin. Munich: Deutscher Taschenbuch Verlag.

Duden, Barbara. 1999. "The fetus on the 'farther shore': Toward a history of the unborn." In *Fetal Subjects, Feminist Positions*, ed. Lynn M. Morgan and Meredith W. Michaels, 13–25. Philadelphia: University of Pennsylvania Press.

Duden, Barbara, Jürgen Schlumbohm, and Patrice Veit, eds. 2002. *Geschichte des Ungeborenen. Zur Erfahrungs- und Wissenschaftsgeschichte der Schwangerschaft, 17.–20.* Jahrhundert. Göttingen: Vandenhoeck und Ruprecht.

Düring, Monika von, Georges Didi-Huberman, and Marta Poggesi, with photographs by Saulo Bambi. 1999. *Encyclopaedia Anatomica: A Complete Collection of Anatomical Waxes*. Cologne: Taschen.

Ecker, Alexander. 1851–59. *Icones physiologicae. Erläuterungstafeln zur Physiologie und Entwickelungsgeschichte*. Leipzig: Voss.

———. 1857. *Untersuchungen zur Ichthyologie angestellt in der physiologischen und vergleichend-anatomischen Anstalt der Universität Freiburg, nebst einer Geschichte und Beschreibung dieser Institute. Zur vierhundertjährigen Jubelfeier der Albert-Ludwigs-Universität*. Freiburg in Baden: Wagner.

———. 1859. "Ueber plastische Darstellungen aus der Entwicklungsgeschichte des Menschen." In *Amtlicher Bericht über die vier und dreissigste Versammlung Deutscher Naturforscher und Ärzte in Carlsruhe im September 1858*, ed. [Wilhelm] Eisenlohr and [Robert] Volz, 199. Karlsruhe: Müller'sche Hofbuchhandlung.

———. 1886. *Hundert Jahre einer Freiburger Professoren-Familie. Biographische Aufzeichnungen*. Freiburg in Baden: Mohr.

Engelmann, Verlag von Wilhelm. 1911. *1811–1911. Jubiläumskatalog der Verlagsbuchhandlung Wilhelm Engelmann in Leipzig*. Leipzig: [Engelmann].

Fischer, Eugen. 1917. "Ernst Gaupp †." *Anatomischer Anzeiger* 49: 584–91.

———. 1957. "Die Freiburger Anatomie um 1900. Ernste und heitere Geschichten." *Badische Heimat* 37: 20–27.

Foerster, Wolf-Dietrich. 1963. *Alexander Ecker. Sein Leben und Wirken*. Freiburg im Breisgau: Albert.

Frasca-Spada, Marina, and Nick Jardine, eds. 2000. *Books and the Sciences in History*. Cambridge: Cambridge University Press.

Fröber, Rosemarie, with Thomas Pester, Joachim Unger, Werner Linß, and Eberhard Burkhard. 1996. *Museum anatomicum Jenense. Die anatomische Sammlung in Jena und die Rolle Goethes bei ihrer Entstehung*. Jena: Jenzig.

Gaupp, E. 1905. "Die Entwickelung des Kopfskelettes." In *Handbuch der vergleichenden und experimentellen Entwickelungslehre der Wirbeltiere*, ed. Oskar Hertwig, vol. 3, part 2: 573–874. Jena: Fischer.

Geus, Armin. 1985. "Christian Koeck (1758–1818), der Illustrator Samuel Thomas Soemmerrings." In *Samuel Thomas Soemmerring und die Gelehrten der Goethezeit. Beiträge eines Symposions in Mainz vom 19. bis 21. Mai 1983*, ed. Gunter Mann and Franz Dumont, 263–78. Stuttgart: Fischer.

Gooday, Graeme. 1991. "'Nature' in the laboratory: Domestication and discipline with the microscope in Victorian life science." *The British Journal for the History of Science* 24: 307–41.

Gould, Stephen Jay. 1977. *Ontogeny and Phylogeny*. Cambridge, Mass.: Harvard University Press, Belknap Press.

Haeckel, Ernst. 1874. *Anthropogenie oder Entwickelungsgeschichte des Menschen. Gemeinverständliche wissenschaftliche Vorträge über die Grundzüge der menschlichen Keimes- und Stammes-Geschichte*. Leipzig: Engelmann.

———. 1875. "Die Gastrula und die Eifurchung der Thiere." *Jenaische Zeitschrift für Naturwissenschaft* 9: 402–508.

Hagner, Michael. 1999. "Enlightened monsters." In *The Sciences in Enlightened Europe*, ed. William Clark, Jan Golinski, and Simon Schaffer, 175–217. Chicago: University of Chicago Press.

Hertwig, Oscar. 1888. *Lehrbuch der Entwicklungsgeschichte des Menschen und der Wirbelthiere*. Jena: Fischer.

His, Wilhelm. 1868. *Untersuchungen über die erste Anlage des Wirbelthierleibes. Die erste Entwickelung des Hühnchens im Ei*. Leipzig: Vogel.

———. 1870. "Beschreibung eines Mikrotoms." *Archiv für mikroskopische Anatomie* 6: 229–32.

———. 1880. *Anatomie menschlicher Embryonen*, vol. 1: *Embryonen des ersten Monats*. Leipzig: Vogel.

———. 1885. *Anatomie menschlicher Embryonen*, vol. 3: *Zur Geschichte der Organe*. Leipzig: Vogel.

———. 1887. "Über die Methoden der plastischen Rekonstruktion und über deren Bedeutung für Anatomie u. Entwickelungsgeschichte." *Anatomischer Anzeiger* 2: 382–94.

———. 1895. "Ueber die wissenschaftliche Wertung veröffentlichter Modelle." *Anatomischer Anzeiger* 10: 358–60.

Hopwood, Nick. 1999. "'Giving body' to embryos: Modelling, mechanism, and the microtome in late nineteenth-century anatomy." *Isis* 90: 462–96.

———. 2000. "Producing development: The anatomy of human embryos and the norms of Wilhelm His." *Bulletin of the History of Medicine* 74: 29–79.

———. 2002. *Embryos in Wax: Models from the Ziegler Studio, with a Reprint of "Embryological Wax Models" by Friedrich Ziegler*. Cambridge: Whipple Museum of the History of Science; Bern: Institute of the History of Medicine.

———. Forthcoming. "Embryology." In *The Cambridge History of Science*, vol. 6: *The Modern Biological and Earth Sciences*, ed. Peter J. Bowler and John V. Pickstone. Cambridge: Cambridge University Press.

Jacyna, L. S. 2001. "'A host of experienced microscopists': The establishment of histology in nineteenth-century Edinburgh." *Bulletin of the History of Medicine* 75: 225–53.

Jordanova, L. J. 1985. "Gender, generation, and science: William Hunter's obstetrical atlas." In *William Hunter and the Eighteenth-Century Medical World*, ed. W. F. Bynum and Roy Porter, 385–412. Cambridge: Cambridge University Press.

Kahn, Joseph. 1851. *Catalogue of Dr. Kahn's Anatomical Museum, Now Exhibiting at 315, Oxford Street, near Regent Circus*. London: Golbourn. (Held in British Library.)

———. 1852. *Atlas of the Formation of the Human Body in the Earliest Stages of Its Development, Compiled from the Researches of the Late Professor, Dr. M. P. Erdl*. London: Churchill.

Keibel, Franz. 1898. "Das biogenetische Grundgesetz und die Cenogenese." *Ergebnisse der Anatomie und Entwickelungsgeschichte* 7: 722–92.

Kemp, Martin, and Marina Wallace. 2000. *Spectacular Bodies: The Art and Science of the Human Body from Leonardo to Now*. London: Hayward Gallery.

König, Hannes, and Erich Ortenau. 1962. *Panoptikum. Vom Zauberbild zum Gaukelspiel der Wachsfiguren*. Munich: Isartal.

Kollmann, J. 1890. "Die Anatomie menschlicher Embryonen von W. His in Leipzig." *Verhandlungen der Naturforschenden Gesellschaft in Basel* 8: 647–71.

Kremer, Richard L. 1992. "Building institutes for physiology in Prussia, 1836–1846: Contexts, interests, and rhetoric." In *The Laboratory Revolution in Medicine*, ed. Andrew Cunningham and Perry Williams, 72–109. Cambridge: Cambridge University Press.

Latour, Bruno. 1990. "Drawing things together." In *Representation in Scientific Practice*, ed. Michael Lynch and Steve Woolgar, 19–68. Cambridge, Mass.: MIT Press.

Lemire, Michel. 1990. *Artistes et mortels*. Paris: Chabaud.

Lenoir, Timothy. 1989 [1982]. *The Strategy of Life: Teleology and Mechanics in Nineteenth-Century German Biology*. Chicago: University of Chicago Press.

Maienschein, Jane. 1991. *Transforming Traditions in American Biology, 1880–1915*. Baltimore: Johns Hopkins University Press.

Marshall, A. Milnes. 1902. *Descriptive Catalogue of the Embryological Models*, 2nd edn, revised and extended by Sidney J. Hickson. London: Dulau; Manchester: Cornish.

Meyer, H. 1847. "Über Wachsmodelle zur Embryologie." *Mittheilungen der Naturforschenden Gesellschaft in Zürich* 1, no. 1: 49–52.
Mocek, Reinhard. 1998. *Die werdende Form. Eine Geschichte der Kausalen Morphologie*. Marburg: Basilisken-Presse.
Morton, Leslie T. 1983. *A Medical Bibliography (Garrison and Morton): An Annotated Check-List of Texts Illustrating the History of Medicine*, 4th edn. Aldershot: Gower.
Mulder, W. J. 1977. "Anatomic wax-models in the collection of Leiden." In *La ceroplastica nella scienza e nell'arte. Atti del I Congresso Internazionale, Firenze, 3–7 giugno 1975*, vol. 1: 427–32. Florence: Olschki.
Nauck, E[rnst] Th[eodor]. 1937. *Franz Keibel. Zugleich eine Untersuchung über das Problem des wissenschaftlichen Nachwuchses*. Jena: Fischer.
———. 1952. *Zur Geschichte des medizinischen Lehrplans und Unterrichts der Universität Freiburg i. Br*. Freiburg im Breisgau: Albert.
Neuland, Werner. 1941. *Geschichte des Anatomischen Instituts und des Anatomischen Unterrichts an der Universität Freiburg i. Br*. Freiburg im Breisgau: Schulz.
Nyhart, Lynn K. 1991. "Writing zoologically: *The Zeitschrift für wissenschaftliche* Zoologie and the zoological community in late nineteenth-century Germany." In *The Literary Structure of Scientific Argument: Historical Studies*, ed. Peter Dear, 43–71. Philadelphia: University of Pennsylvania Press.
———. 1995. *Biology Takes Form: Animal Morphology and the German Universities, 1800–1900*. Chicago: University of Chicago Press.
Oettermann, Stephan. 1992. "Alles-Schau: Wachsfigurenkabinette und Panoptiken." In *Viel Vergnügen: Öffentliche Lustbarkeiten im Ruhrgebiet der Jahrhundertwende*, ed. Lisa Kosok, 36–56, 294–302. Essen: Pomp.
O'Rahilly, Ronan. 1988. "One hundred years of human embryology." *Issues and Reviews in Teratology* 4: 81–128.
Pelagalli, Gaetano Vincenzo, Vincenzo Esposito, and Ciro Paesano. 1998. *I modelli in cera del Museo di Anatomia Veterinaria. Wax Models from the Museum of Veterinary Anatomy. Università degli Studi di Napoli Federico II. Facoltà di Medicina Veterinaria*. Naples: Arte Tipografica.
Peter, Karl. 1906. *Die Methoden der Rekonstruktion*. Jena: Fischer.
———. 1911. "Modelle zur Entwickelung des menschlichen Gesichtes." *Anatomischer Anzeiger* 39: 41–66.
Pyke, E. J. 1973. *A Biographical Dictionary of Wax Modellers*. Oxford: Clarendon.
Reiling, Henri. 1998. "The Blaschkas' glass animal models: Origins of design." *Journal of Glass Studies* 40: 105–26.
Sabin, Florence R. 1900. "A model of the medulla oblongata, pons, and midbrain of a new-born babe." *Johns Hopkins Hospital Reports* 9: 925–1023.
———. 1902. "A note concerning the model of the medulla, pons, and midbrain of a new-born babe as reproduced by Herr F. Ziegler." *Anatomischer Anzeiger* 22: 281–89.

Schnalke, Thomas. 1991. "Lernen am Modell. Die Geschichte eines Lehrmittelproduzenten im geteilten Land." *Ärztliches Reise & Kultur Journal* 15, no. 13: 95–97.

———. 1995. *Diseases in Wax: The History of the Medical Moulage*, translated by Kathy Spatschek. Berlin: Quintessence.

Seidler, Eduard. 1991. *Die Medizinische Fakultät der Albert-Ludwigs-Universität Freiburg im Breisgau. Grundlagen und Entwicklungen*. Berlin: Springer.

Soemmerring, Samuel Thomas. 2000. *Schriften zur Embryologie und Teratologie*, ed. Ulrike Enke. Basel: Schwabe.

Suter, Adrian Christoph. 1986. "Die anatomischen Reliefdarstellungen des Einsiedler Kleinkünstlers J. B. Kuriger (1754–1819)." Medical dissertation, Medizinhistorisches Institut der Universität Bern.

Tuchman, Arleen Marcia. 1993. *Science, Medicine, and the State in Germany: The Case of Baden, 1815–1871*. New York: Oxford University Press.

Uschmann, Georg, and Bernhard Hassenstein. 1965. "Der Briefwechsel zwischen Ernst Haeckel und August Weismann." In *Kleine Festgabe aus Anlaß der hundertjährigen Wiederkehr der Gründung des Zoologischen Institutes der Friedrich-Schiller-Universität Jena im Jahre 1865 durch Ernst Haeckel*, ed. Manfred Gersch, 7–68. Jena: Friedrich-Schiller-Universität.

Wagner, Rud[olph]. 1859. [Review of Alexander Ecker, *Icones physiologicae*.] *Göttingische gelehrte Anzeigen*, no. 89: 881–86.

Weindling, Paul Julian. 1991. *Darwinism and Social Darwinism in Imperial Germany: The Contribution of the Cell Biologist Oscar Hertwig (1849–1922)*. Stuttgart: Fischer.

Wiedersheim, Robert. 1882. "Notiz [of Adolf Ziegler's series after Philipp Stöhr's models on the development of the skull]." *Zoologischer Anzeiger* 5: 388.

———. 1889. "Adolf Ziegler †. Zum Gedächtnis." *Anatomischer Anzeiger* 4: 545–46.

———. 1919. *Lebenserinnerungen*. Tübingen: Mohr.

The William Bonardo Collection of Wax Anatomical Models. 2001. London: Christie's.

Yarrington, Alison. 1996. "Under the spell of Madame Tussaud: Aspects of 'high' and 'low' in 19th-century polychromed sculpture." In *The Colour of Sculpture, 1840–1910*, ed. Andreas Blühm and Penelope Curtis, 83–92. Amsterdam: Van Gogh Museum; Zwolle: Waanders.

Ziegler, Adolf. 1867. "Wissenschaftliche Wachspräparate als Unterrichtsmittel für vergleichende Entwickelungsgeschichte." In *Die Betheiligung des Großherzogthums Baden an der Universalausstellung zu Paris im Jahre 1867*, ed. Badische Ausstellungs-Commission, 24–27. Karlsruhe: Müller'sche Hofbuchdruckerei.

———. [circa 1886]. *Wissenschaftliche Unterrichtsmodelle von Dr. med. Adolf Ziegler, Ritter des Zähringer Löwenordens und des K. russisch. St.*

Annaordens, in Freiburg, Baden. Freiburg in Baden. (Held in Medizinhistorisches Institut der Universität Bern.)

Ziegler, Friedrich. [circa 1912a]. *Prospectus über die zu Unterrichtszwecken hergestellten Embryologischen Wachsmodelle von Friedrich Ziegler*. Freiburg in Baden: Atelier für wissenschaftliche Plastik. (Held in Biology Library, University of California, Berkeley.)

———. [circa 1912b]. *Verzeichnis der zu Unterrichtszwecken hergestellten Embryologischen Wachsmodelle von Friedrich Ziegler*. Freiburg im Breisgau: Atelier für wissenschaftliche Plastik. (Held in Division of Rare and Manuscript Collections, Cornell University Library [Adelmann QL 64 Z66 V5 1920].)

CHAPTER EIGHT

Casting Skin: Meanings for Doctors, Artists, and Patients

Thomas Schnalke

Moulages are pictures of diseases in wax. They are in colour. They are life-size. They are stunningly realistic. And above all, they are three-dimensional (Fig. 8.1). The particular realism of moulages results to a large extent from their production using a casting technique, in which three groups cooperated to make these clinical pictures-in-wax: physicians, the mouleurs or mouleuses, and patients. The setting in which these people interacted with moulages, and at times with each other, gave these objects meaning. This chapter is about both the distinctive imagery of moulages and their uses for the three groups of protagonists. To begin with, however, I shall examine the characteristics of the moulaging technique (Schnalke 1986; 1995; Parish et al. 1991; Hahn and Ambatielos 1994).

Wax is a special material. Since antiquity this natural product has served—among a wide range of other uses—as a pictorial medium. It proved to be particularly apt for representing organic life, especially the human body, for it was easy to melt, form, and colour. Not only could it be hardened, retouched, and durably fixed, but it was also possible to create a surface that mimics the internal and external surfaces of the body. These qualities of wax were used to generate two-dimensional pictures of the human body: the earliest remaining portraits are the 'encaustic' paintings of Egyptian mummies. At the same time, however, wax served to create 3-D representations of the body such as masks of the living and the dead, and bronze busts (Büll 1977).

Wax first became an effective medical medium in the late seventeenth century, when it was used to create life-like and life-sized models of human

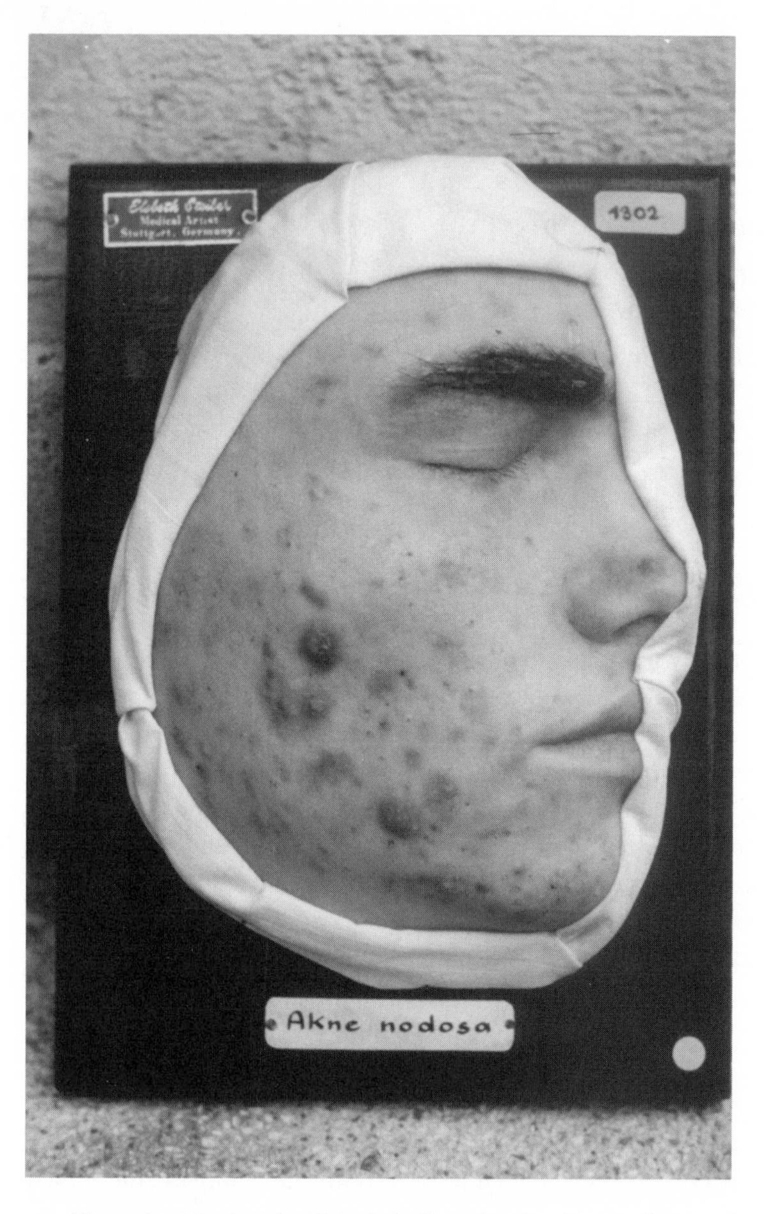

Figure 8.1 Moulage by Elsbeth Stoiber, showing Acne nodosa in the right half of the face, Zürich, 1956. Courtesy of the Museum of Medical Moulages, University Hospital and University of Zürich.

anatomy. The splendid collections in the northern Italian cities of Bologna and Florence, founded and developed in the mid- and late eighteenth century, marked the heyday of 'plastic anatomy' (Haviland and Parish 1970; Azzaroli Puccetti 1972; *Ceroplastica* 1977; Lanza et al. 1979; Boschung 1980; Lemire 1992; Schnalke 1995, 25–48; Düring et al. 1999; Skopec and Gröger 2002; Mazzolini, this volume). Despite this broad anatomical wax tradition, the medical moulage, which entered the stage in the first half of the nineteenth century, rarely developed in a direct line from these predecessors. Dressing up in the same material but taking on a new task—documenting diseases in three dimensions—it performed in a different arena. While the anatomical wax models in Bologna and Florence were shown to the public in natural history museums, at their debut moulages were presented exclusively to medical audiences in metropolitan hospitals (Schnalke 1995, 49–89; Brenner-Holländer 1996). They proved to be especially suited for representing afflictions that produced visible marks on human skin. For that reason it is no wonder that dermatologists and venerologists were from the outset the primary users of moulages.

Between 1890 and 1950 physicians in the West, Russia, and Japan used moulages widely, mainly as clinical collections for medical training (Parish et al. 1991). From the early twentieth century, moulages appeared in colour in dermatological atlases and handbooks. As shocking objects they were introduced into health museums and public health exhibitions, as well as into the lurid showrooms of stationary and mobile panopticons.[1] But from the late 1950s, this medical picture-in-wax lost its importance in every arena in which it had performed well for decades. Valuable stocks were broken up and the objects sometimes even melted down. In other places moulages were stored more or less professionally, so that a few collections can still be viewed, and continue to be used in medical training and public health education.[2] Finally, moulages have found their way into medical historical museums, where they perform completely new functions as representatives of historical patients.[3]

THE PRODUCTION OF MOULAGES

People formed waxen casts to create 3-D representations of parts of the human body long before moulaging began. Wax casts were made in antiquity to create facial masks, and in the late Middle Ages and Renaissance to shape votive offerings (Schlosser 1911; Brückner 1963; Kriss-Rettenbeck 1972; Angeletti 1980). In Bologna and Florence the anatomical wax modellers also used this method, though they primarily shaped their objects freehand, like a sculptor. The pioneers in medical moulaging—such as Franz Heinrich Martens in Jena, Johann Nepomuk Hoffmayr and Anton Elfinger in Vienna, Giuseppe Ricci in Florence, Joseph Towne in London, and Jules Pierre François Baretta in Paris—all followed similar programmes and aims. Nevertheless, they developed their techniques in relative isolation, so it is surprising that they all used the same casting method. Without common agreement they followed a scheme of production that has—despite a few local differences—remained relatively constant thoughout the era of moulage making.

In the classic procedure the mouleur or mouleuse starts by making a negative, typically in fine plaster, of the part of the patient's body to be depicted (Fig. 8.2). In the second step the mouleur or mouleuse casts a raw moulage from the negative using a wax mixture in several layers. This raw

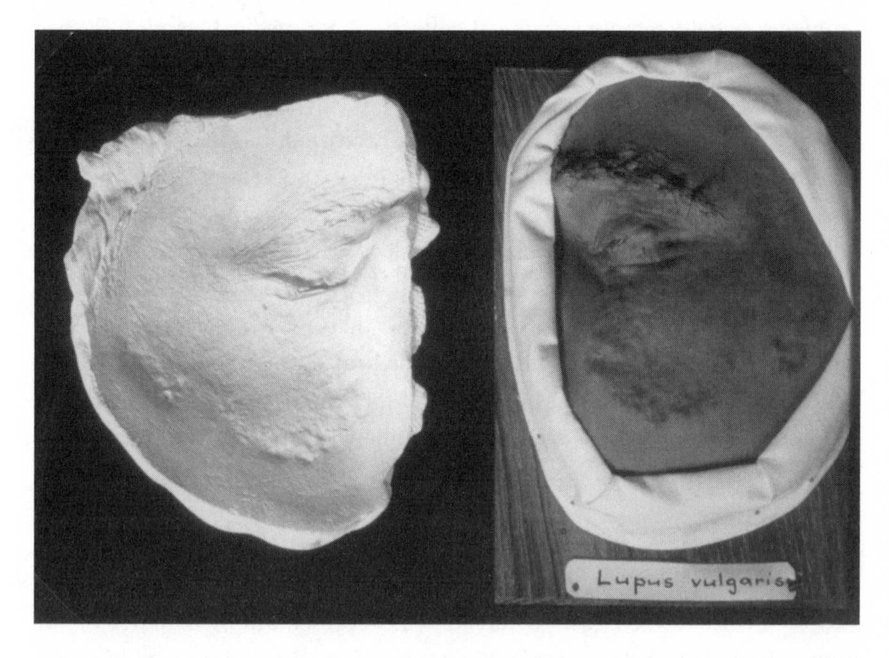

Figure 8.2 Plaster negative and finished moulage by Elsbeth Stoiber, showing Lupus vulgaris in the left half of the face, Zürich, 1964. Courtesy of the Museum of Medical Moulages, University Hospital and University of Zürich.

moulage is then refined in the presence of the patient in order to resemble the original body part as closely as possible. The mouleur or mouleuse colours the skin and the lesions either from the top or—like in a reverse painting on glass—from behind on the first very thin layer of wax which will eventually represent the external layer of skin. He or she models complex efflorescences such as irregular blisters, papules, or scales to a certain extent freehand, and then sometimes completes the realistic appearance of the moulage by implanting real hair in appropriate regions, by using artificial eyes in facial moulages, or by inserting real scales to imitate specific efflorescences. In the final step the moulage is attached to a wooden tablet and surrounded by linen. The mouleur or mouleuse adds a label with the diagnosis of the disease, and sometimes signs and dates the work (Schnalke 1986, 140–212; 1995).

IDEALISED ANATOMY AND THE SICK INDIVIDUAL

Looking at the principles of the moulaging technique it becomes clear that clinical pictures-in-wax reproduced only fragments of the living patients' skin. In order to understand the intention of their reduced and concentrated focus as well as their individualistic nature (Daston and Galison 1992, 94–98) more clearly it is helpful to contrast them with their predecessors, the anatomical wax models of the eighteenth century. Anatomical waxes always directed attention into the depths of the human body. Under the skin the viewer encountered an anonymous, largely ageless, and—apart from the few models showing male and female features (Jordanova 1985; 1989)—sexless 'body machine' (Fig. 8.3).[4] He or she could contemplate the perfectly reproduced details of a certain body structure, but also perceive larger anatomical systems, and so gain an idea of how the whole looks and functions. For these perspectives viewers were offered three basic types of anatomical models: detailed studies of single organs or discrete organic structures such as the heart or a bone; more complex presentations of functional units such as the respiratory or digestive tracts; or whole figures representing larger systems such as skeletons or muscles. Typically, several of these models were presented together in the careful arrangement of a museum collection. By 'walking through' the body, i.e., through the showroom of the model collection, where the waxes had been placed in their most telling positions, the visitor could narrow his or her perspective to the smallest detail, but also integrate it into a whole picture (Kleindienst 1989; 1990; Krüger-Fürhoff 1998; Dürbeck 2001; Wolkenhauer 2001).

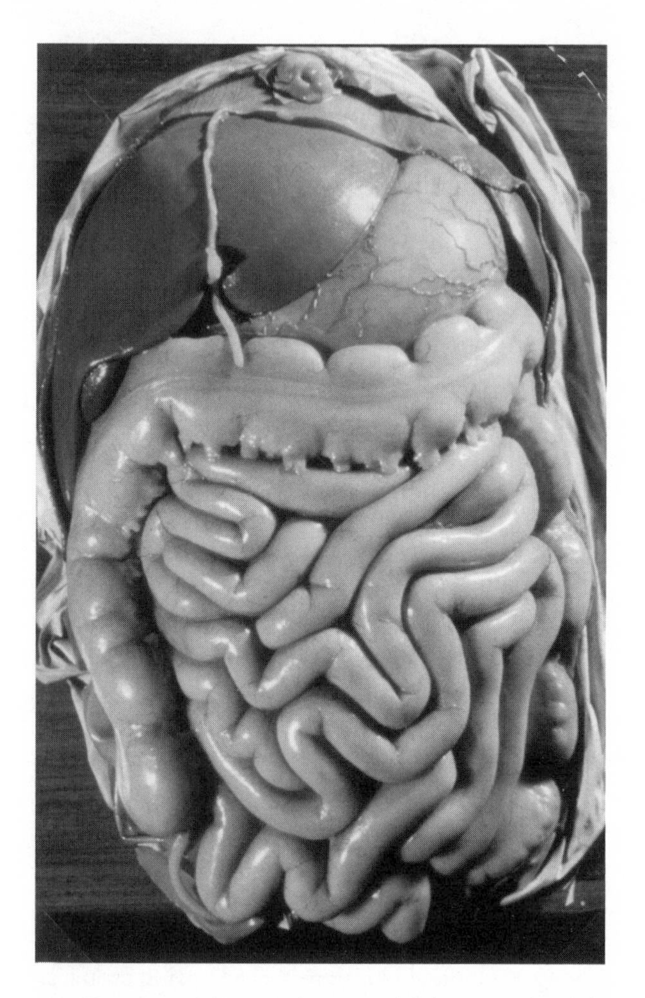

Figure 8.3 Anatomical wax model depicting the abdominal situs. The following organs are portrayed: oesophagus (cut off), liver, gall-bladder, spleen, stomach, and small and large intestines. Copy made by the studio of Felice Fontana, Florence, 1780–86, for the Vienna Academy of Military Surgery. Courtesy of the Institute for the History of Medicine, University of Vienna.

Anatomical models resulted from a complex process of synthesis (Wichelhausen 1798; Lanza et al. 1979; Schnalke 1995, 42–45; Düring et al. 1999; Skopec and Gröger 2002). The anatomists and the modellers usually decided in the first step on what the new model should show and how it should present specific bodily structures. The anatomists had to guarantee scientific standards which they usually found documented in the latest published anatomical illustrations—drawings, woodcuts, and/or copper-engravings (Wichelhausen 1798, 117; Jordanova 1985; 1989; Schnalke 1995, 43). According to the iconography they came across in these pictures, as well as the information gained from their own dissections, they planned how to shape the model. The modellers, who had artistic backgrounds, also based their work on fine-art representations of the body (Premuda 1972; Jordanova 1985; 1989, 45; Almhofer 2002). Once the didactic appearance of a new model had been decided, the anatomists supplied the modellers with corpses that showed the relevant pose as accurately as possible. For some models produced at the famous wax studio of the Florentine Natural History Museum ('La Specola') large numbers of corpses—up to 200—were dissected (Mazzolini, this volume).

On the basis of many pictures and preparations, the modellers finally started to create their models. They used three basic techniques. The body shapes were sculpted from a cold block of wax, and every now and then some structures—especially firmer tissues—were also cast, but most frequently the modellers worked freehand, shaping softened mixtures of wax including colour pigments and hardening ingredients. The finished model was mounted on a board, placed in a showcase, and presented in the collection. To enhance the didactic value, the Florentine waxes were all accompanied by a 2-D coloured drawing and a list of legends naming the various structures indicated in the pictures (Premuda 1972; Düring et al. 1999; Dürbeck 2001; Skopec and Gröger 2002).

As a product of synthesis, anatomical waxes presented an idealised picture of the human body (Fig. 8.4). Anatomists and modellers intended to show healthy organs in functional interaction, i.e., as normal living units of the 'body machine'. To achieve this goal, they had to ban every sign of disease, death, and decay. No damage is seen in these models, no dripping blood, or post mortem signs of decomposition. Moreover, 'normal' relations of a standard body were defined in the process of planning and creating a model. By capturing this 'normality' in wax and by colouring it realistically and sometimes even hyper-realistically, the model became a topos. It demonstrated how the normal human body ought ideally to be (Kleindienst 1989; Jordanova 1989; Krüger-Fürhoff 1998; Dürbeck 2001).[5]

At times the idealised nature of anatomical waxes was even used, according to baroque traditions of medical illustration, to convey theological and moralistic messages. The ideal body could be seen as the highest result of God's creation. It could therefore serve both as proof of the existence of

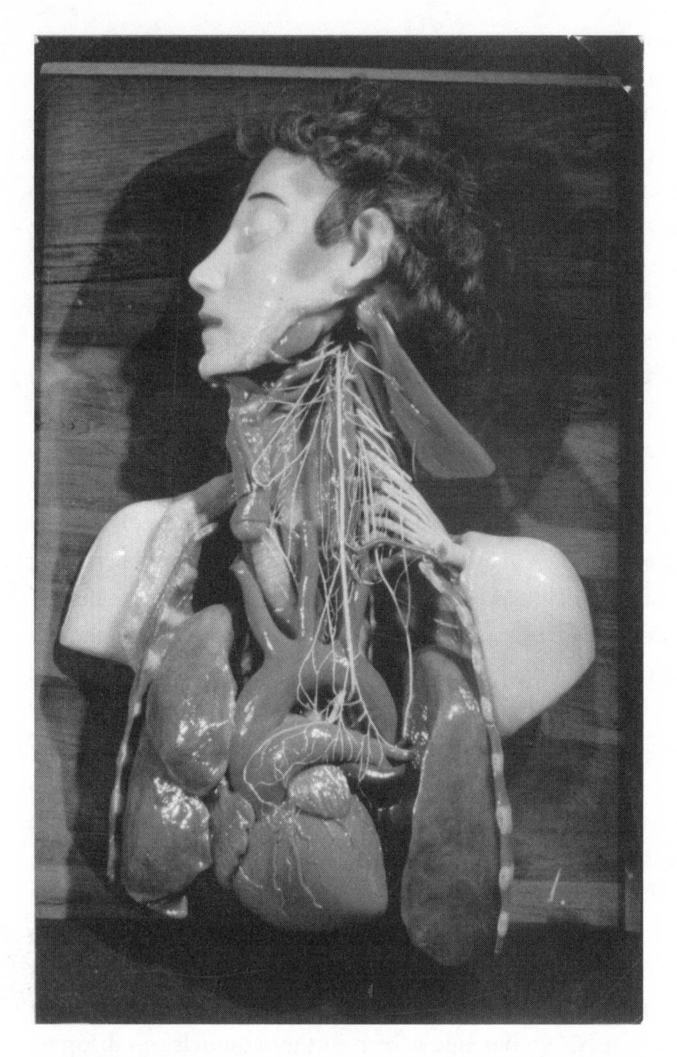

Figure 8.4 Anatomical wax model depicting the neck and chest situs. The following organs are in focus: course and branching of the left vagus nerve, the aorta with its upper arterial branches, heart, and lungs. Copy made by the studio of Felice Fontana, Florence, 1780–86, for the Vienna Academy of Military Surgery. Courtesy of the Institute for the History of Medicine, University of Vienna.

the Almighty and as a motivation for worship. At the same time, however, posing with a mournful gesture, showing external body surfaces that usually carry information pertaining to a distinct person such as whole faces, hair, eyes, noses, lips, ears, or fingers, but also 'wearing' jewelry and being draped on pillows in precious glass cases, some models seemed to point to the end of all existence. They served as anatomical memento mori (Lesky 1976; Kleindienst 1989; 1990; Dürbeck 2001; Wolkenhauer 2001).

Moulages were different. They performed by no means as synthetic and idealised images, but rather as documents of the characteristic signs of a disease in an individual (Schnalke 1995; 2001). Cast from living patients, they can be seen as single case studies that drew the attention of the viewer exclusively to the external façade of the body (Fig. 8.5). There the mouleurs or mouleuses aimed to depict particular lesions of a particular patient as accurately as possible. The reverse technique encouraged mouleurs and mouleuses to step back from the production process as much as possible in order to let

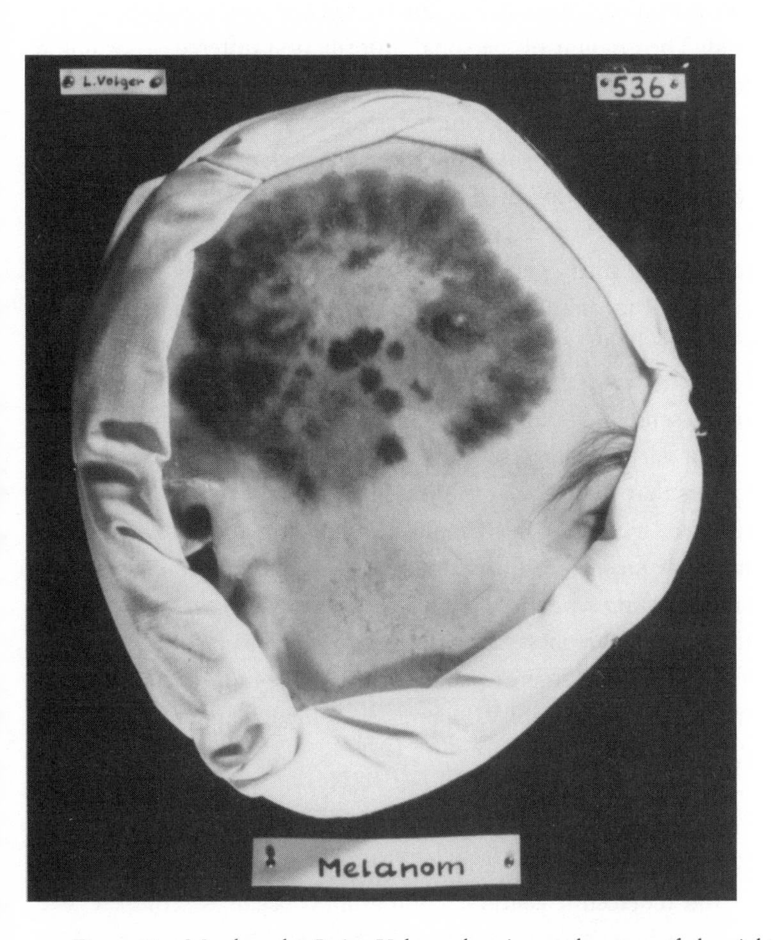

Figure 8.5 Moulage by Luise Volger, showing melanoma of the right temple region, Zürich, 1929. Courtesy of the Museum of Medical Moulages, University Hospital and University of Zürich.

nature imprint itself and to grasp as closely as possible what was characteristic (Schnalke 1986, 140–212).

However, the stigmata of a skin disease could never be separated from the surrounding healthy parts of the human integument. To give an idea of the nature of the illness, but also to help the viewer locate the signs of disease exactly, the unaffected areas included in a moulage had to be relatively large. Thus the patient never disappeared, but always lurked behind the clinical picture-in-wax (Schnalke 1993; 1999). Using the highly individualistic technique of replicating the skin textures in a cast, naturalistic colouring, and adding 'realistic' materials such as hair or artifical eyes to an extraordinary degree of perfection, moulages never failed to tell a second story, the story of an individual. The amount of 'personal' information differed according to how much a body region represented features of individual identity. A complete face with hair and open eyes sometimes gave a striking impression of the pain and emotional involvement of the patient. A rash or a group of papules on a part of the back incorporated only very little private information.

Like the anatomical waxes, moulages were not meant to be presented singly. While the former were supposed to give an idea of the whole 'body machine'—the anatomy of normal structures in their collective arrangement—the latter were put together in moulage collections to perform as 3-D textbooks presenting the 'anatomy of diseases'. Moulages usually made more sense in the company of other moulages in order to show the development or various manifestations of a disease, or to compare one disease with others that were similar or distinctively different. It was in clinical collections where moulages, with their characteristic and individual messages, found their final place and ultimate use. Here they brought together 'all' the patients to present the essence of disease, which was described in 1856 by the eminent Viennese dermatologist Ferdinand Hebra, one of the first renowned physicians to promote the art of moulaging, as "the eternally unchangeable of divine creation" (Hebra 1856, unpaginated introduction).

MOULAGES AND OTHER MEDIA

Despite the excellent quality of moulages, their heyday was short. Towards the end of the nineteenth century, photography began to compete with moulaging as a technique for presenting skin diseases (Ehring 1989, 214–33), and eventually superseded it. Initially, however, moulages kept the upper hand. Early photographs were not only two- instead of three-dimensional, but also black and white. Hand-coloured photographs did not generally achieve the same degree of accuracy in shading as moulages (Squire 1865; Hardy and de Montméja 1867/68; Fox 1881). Furthermore, photographs very often could not compete with moulages in reproducing organic textures.

A change in favour of photographs occurred as a result of developments in photography rather than any alterations in moulaging. The first advantage that emerged in photography's favour was its more mechanical production (Daston and Galison 1992, 98–117). Moulages had to be made one at a time, by hand, which required a large amount of both time and money. Although they had to sacrifice a certain degree of individuality through the loss of a dimension, many physicians came to prefer photographs because they were easily produced at any location and readily reproduced. The second reason photographs became more widely used was that the accuracy and availability of colour reproduction increased gradually, until by the middle of the twentieth century photographs of diseases became an acceptable alternative (Fox and Lawrence 1988).

Although by about 1950 the moulage finally lost the competition with photography, for a long time it had represented an excellent object for reproduction in skin atlases. In the first decades of the twentieth century, a large number of dermatologic atlases and handbooks presented photographic reproductions of moulages.[6] Eduard Jacobi, professor of dermatology at the Freiburg university hospital, justified in the preface to his *Atlas of Skin Diseases* (1903) why he had chosen pictures-in-wax as motifs for publication: "Although moulages do not exclude entirely the personal interpretation of the depicting artist, they can be produced today to such a degree of perfection that they are able to substitute completely for the living patient, at least to the eyes [of the viewer]" (Jacobi 1903, 3). For technical reasons, it was impossible for him to take photographs of the patients directly. With a newly developed photographic technique—"the method of Citochromy [*sic*] invented recently by Dr. Albert of Munich"[7]—he was, however, able to "produce colour pictures [from moulages] absolutely true to nature for a rather low price with the almost complete exclusion of a manipulating hand". Albert Neisser, then the leading authority in German dermatology, praised the plates: "Except for the work of the great Hebra there is no atlas which has yet come up with anything better, more true to nature, and which creates such a direct effect" (foreword to Jacobi 1903, 5).

At a time when the photograph of a whole patient or parts of a diseased body was the undisputed prime source of individualised images in other fields of medicine, the moulage still played an important role as an illustrative medium in dermatologic publications, for two main reasons. First, the art of moulaging had reached such a degree of perfection that physicians agreed these waxen objects could substitute for patients in pictorial reproductions. This was possible only because the mouleurs and mouleuses had learned to restrict themselves to reproducing the complete features—nothing more or less—of a particular patient's skin. Their products had convinced physicians that the factor of subjectivity was reduced to a minimum. Second, the moulage made it technically possible to add the decisive attribute of naturalistic colour not included in the printed photograph unless it had received a secondary manipulation at the hand of an artist-illustrator. The widely mechanised colour reproduction of moulages, however, presented a degree and quality of objectivity that led several dermatologists to rely on these images in their publications until the 1940s, when colour reproduction from photographs was refined to a sufficient level.[8]

Techniques alone do not tell the whole story of moulages. Their individual nature was also the result of the interaction of physicians, mouleurs or mouleuses, and patients involved in the process of moulage fabrication. Due to the scarcity of primary sources, this triad of cooperation has not yet been the focus of thorough investigation. The analysis that follows is mainly based on three detailed reports revealing the situation of moulage production in Berlin around 1900 (Photinos 1907) and the moulage-making traditions in Dresden (Walther 1986/87) and Zürich (Stoiber 1998) in the course of the last century.[9] These documents present the perspective of a physician who was trained as a mouleur himself, as well as the viewpoint of the last mouleuses to receive the classical training in the art of moulaging. To my knowledge there is no comparably extensive statement of patients' views. Therefore, I have had to reconstruct what the moulage might have meant for the patients through the eyes of physicians and mouleurs or mouleuses.

THE PHYSICIANS

Physicians who used moulages normally remained in the background of the actual fabrication, but were included primarily at two stages of the procedure that were of great importance to them. First, they determined which disease should be depicted. At the Zürich skin clinic, a list of desiderata was kept of interesting diseases that had not yet been captured in wax. This list was circulated to the dermatologists and the outpatient clinic so that the staff would be on the lookout for potential cases (Stoiber 1998). The physicians became responsible for judging whether their patients were able to stand the strains of having pictures-in-wax taken from their diseased bodies. Certain basic conditions in the patients—such as "heart disease, [great] pain, heat accumulation associated with local ischaemia, etc."—could be considered contraindications for casting a moulage (Walther 1986/87, 38).

A basic prerequisite for fruitful and successful moulage production was the trust physicians had in their mouleurs or mouleuses. Elfriede Walther, who worked as mouleuse at the German Hygiene Museum in Dresden from 1946 to 1980, as well as Elsbeth Stoiber, mouleuse at the Zürich University skin clinic from 1956 to 1998, each reported only one incident in their careers when the physician in charge supervised the whole procedure (Walther 1986/87; Stoiber 1998). Once the mouleurs or mouleuses had proved their abilities, physicians relied on their expertise. In Zürich, this confidence extended so far that—after it had been decided that a moulage ought to be made in a certain case—"the physician explained just a little bit to the patient about what would follow ... [The mouleuse] had to give the patient the detailed information" (Stoiber 1998).

The handling of information raises the question of whether and when in the history of making moulages physicians became sensitive to the issue of informed consent. Around 1900, for instance, an ethical debate arose concerning photographs taken of women for illustrations in obstetrical and gynaecological publications (Maehle 1993, 574–76). It was argued that women had been depicted without their consent while under anaesthesia and that most of the photographs unnecessarily presented the whole body, and so revealed identity. There was no similar discussion among physicians over moulages. Surely the medical picture-in-wax was always taken from conscious patients who—at least theoretically—had the chance to object. In addition, the moulage never presented the whole body, so the full identity of the patients became clearly visible only in those pieces that depicted faces. More important, however, was the fact that the moulage had not yet entered the stage of printed publication and popularisation. Until the beginning of the twentieth century, moulages were only accessible to a very restricted medical audience at a specific location, the clinical moulage collection. The Jacobi atlas, published in 1903, presented photographic reproductions of moulages, thus opening the medical picture-in-wax to a wider, yet still predominantly medical, circle (Jacobi 1903). Georgios Photinos, a Greek physician who had participated in a course in moulaging at Oskar Lassar's skin clinic in Berlin, did deal with the topic of informing the patients as early as 1907, but in a rather paternalistic way: "As we are dealing with a living person, we shall start [the procedure] by telling the patient what to do" (Photinos 1907, 141). In the following years sensitivity to patients' rights grew. Prospective patients received more thorough information aimed at gaining their consent, but also respecting their refusal. Although we lack clear evidence, this development seems to have been triggered by the presentation of moulages in public health exhibitions. At the First International Hygiene Exhibition in Dresden in 1911, medical pictures-in-wax were shown for the first time to an international audience in a public display. Later, the Dresden Hygiene Museum made use of the moulage in further public health exhibits and campaigns, and even ran a special workshop for producing, copying, and selling moulages (Schnalke 1995, 121–43). In addition to its restricted use as a medical teaching tool, the moulage had received a second identity, serving publicly as a means of health education. Its display on a public stage has only recently fuelled a debate on the ethical implications and limits of presenting medicine in museums. The core issue was the question of how to respect the wishes and values of patients who become visible in objects such as the moulages (Schnalke 1993; 1999).

The second point at which physicians became actively involved in moulage making was usually after the cast had been finished. They affirmed "the accuracy and likeness to nature" of the completed object and identified the disease shown in the moulage so that it could be labelled (Walther 1986/87, 41). With this step, physicians certified an object as a typical depiction of a common affliction or a pictorial document of some extraordinary finding which legitimised its inclusion in a moulage collection. Photinos listed some practical advantages to physicians of having a moulage collection at their disposal. For example, specialists would no longer forget rare diseases. And, when several moulages had been made in the course of one case, they could observe the development of a disease. They could also judge the consequences of a certain treatment, and use it as a document in conjunction with the case report and microscopic examination when a second opinion was necessary. Students could be offered depictions of a wide spectrum of different and even rare diseases. Finally, according to Photinos, physicians would be motivated to participate in health education (Photinos 1907, 135–36).

From the late nineteenth century, physicians increasingly attempted to use moulages and moulage collections to further not only their own professional status, but also the position of their clinic and even the reputation of their discipline. The discipline as a whole, namely dermatology and venerology, did not rise in esteem among the various medical specialties from having this impressive object at its disposal.[10] It was more within the field, at the level of individual physicians, the clinics where they built up moulage collections, and the national circles of dermatologists competing against each other, where the moulage was used to underline a certain status. In this context, the moulage—unlike some contemporary embryological wax models—was never used as an argument for the construction of basic medical theories or even nosological categories.[11] As a 'true' depiction of a disease, showing the pure morphology in a 'realistic' way, it followed the direction the Viennese dermatologist Ferdinand Hebra had laid out for colour lithographs in the preface and introduction to his skin atlas (Hebra 1856). Hebra was convinced he could give an "objective presentation" of the diseases depicted in his illustrations (Hebra 1856, preface), and so declined to arrange the plates according to a system. In a similar way, Hebra's successors who arranged moulage collections or published medical pictures-in-wax generally used these objects undogmatically as the closest surrogate for the patients.[12]

The meanings the moulage offered as a status symbol derived more from competition with other—mostly two-dimensional—media. In 1856, by the time the first instalment of his atlas appeared, Hebra already had an illustrator, Anton Elfinger, who was working for him not only on watercolours and lithographs, but also on the first moulages (Poch-Kalous 1966; Portele 1974; Skopec 1983; Schnalke 1995, 79–82). Hebra stuck to colour lithographs, indicating, however, that there were already three-dimensional teaching objects, such as the "preparations [*sic*] made by an artist's hand in wax, papier mâché, or plaster" that were largely able to substitute for the patients (Hebra 1856, preface). But, Hebra complained, there were too few of these items, and they cost too much. In the following 30 years, colour lithography remained an important technique in depicting skin afflictions, and photography enjoyed increasing use in this medical field (Ehring 1989). During the same period, French dermatologists carefully prepared the international debut of the moulage, promoting the work of the mouleur Jules

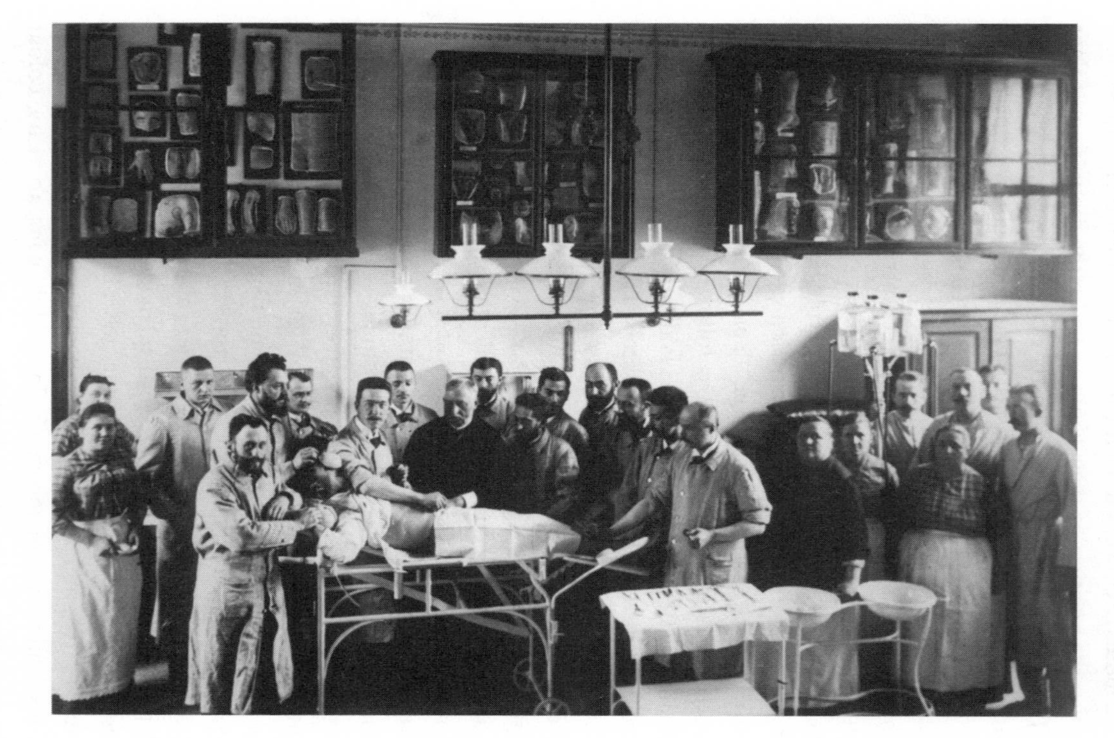

Figure 8.6 Isidor Neumann, professor of venerology, performing an operation sometime before 1904, probably in the lecture hall of Vienna's Second University Skin Clinic (Clinic of Syphilology), surrounded by colleagues, students, and clinic personnel. In the background moulages, mostly made by the Viennese mouleur Carl Henning, are hanging in glass display cases. Photograph courtesy of the Institute for the History of Medicine, University of Vienna.

Pierre François Baretta at the Hôpital Saint Louis in Paris (Solente 1983; Schnalke 1995, 85–89; Brenner-Holländer 1996, 89–96).

In 1889, when French dermatologists organised the first international congress of dermatology and syphilology in Paris, they aimed at regaining their international reputation and taking their place alongside the leading British and German representatives (Kaposi 1890; Wallach and Tilles 1990; Schnalke 1995, 91–92; Brenner-Holländer 1996, 97–99). Over 200 colleagues from 29 countries participated in the event at the newly-built pathological museum. Moriz Kaposi, Hebra's successor in Vienna, was impressed by this "large hall, in which the wax specimens of Baretta, deserving of their fame, were displayed in glass cases along the walls" (Kaposi 1890, 190). In a report in the renowned *Archiv für Dermatologie und Syphilis*, Kaposi judged that "Truly, a more suitable meeting hall could not have been devised for this congress than this room, which is so comfortable in its spaciousness and is serious and scientifically stimulating because of its unique wall decoration" (Kaposi 1890, 190).

European dermatologists followed the example of the Hôpital Saint Louis (Schnalke 1995, 93–201). Competing with each other for size, diversity, and quality, the growing collections became sources of local pride. It was important for physicians to carry out a decisive, authoritative act with the determination of the diagnosis with which the moulage was labelled. By naming medical pictures-in-wax, physicians documented that they had already dealt practically and scientifically with a disease.[13]

From the practical side, every nosologically identified moulage could be read as proof that a certain physician, or one of his or her colleagues, had been therapeutically active in that individual case. Thus, the moulages in a clinical collection represented the range of patients who had been treated at that institution. The specimens were all assembled as impressive documents of local practice, experience, and competence. This status function can be seen in some photographs taken in the Vienna University skin clinics in the early twentieth century (Fig. 8.6). They show the leading expert, the professor, in the centre, giving a lecture or preparing for an operation on a patient, surrounded by his assistants, other employees of the clinic, and students.[14] Despite the fact that so many people were acting and posing in the foreground, moulages also found their place in these pictures. Indeed, they were the indicators that truly showed what was happening. They stood for the large number of patients and the wide range of diseases the physicians had identified and treated in the clinic. That various well-respected European dermatologists eagerly sent moulages to Eduard Jacobi in Freiburg for reproduction in his atlas (Jacobi 1903, 3–4) attests to this local pride.

Physicians used moulages also as a means to a scientific reputation. While they had entered the international arena as "scientifically stimulating" objects in Paris (Kaposi 1890, 190), they were used repeatedly by dermatologists to document and publish newly described afflictions at other places too (Neisser and Jacobi 1906; Rüdlinger and Stoiber 1988a; 1988b; Barlag 1992). The Zürich moulage collection gives a clear picture of the scientific activities of the first promoters of moulaging in Switzerland, Bruno Bloch and Guido Miescher (Rüdlinger and Stoiber 1988a; 1988b). Elsbeth Stoiber from Zürich reported that Miescher, the professor of dermatology for whom she had first worked between 1956 and 1958, used to take moulages as far as the United States to present them to colleagues at conferences, while she arranged moulages from the local stocks for display upon request from dermatological scientists who were performing research on certain diseases (Stoiber 1998). At the same place special research casts—for instance, animal moulages—were made to document the effects of certain experiments.[15] Although by the mid-twentieth century moulages had lost ground to colour photographs, physicians still valued their accuracy in depicting historical diseases as well as organic textures. Nevertheless, moulages in general became more interesting to medical collectors than to practising physicians.

THE MOULEURS AND MOULEUSES

The mouleurs and mouleuses were, in a way, the main characters in the triad of moulage production,[16] for it was in their hands that the wax picture became the realistic, 3-D representation of the disease and the diseased. Yet if we pose more precise questions about how they created these pictures in order to find out what their art meant to them, we are met with a great silence. Around 1890, the Viennese physician and illustrator Carl Henning (Portele 1977; Schnalke 1995, 93–102) found it extremely difficult to discover information about the technique of the Parisian mouleur Baretta. The French author L. Roger-Miles, who painted a very positive picture of Baretta in his novel *La cité de misère*, published in 1891, faced the same problem (Roger-Miles 1891, 159–67). In 1905, Frederic Griffith, fellow of the New York Academy of Medicine, reported his lack of success after a visit to Guy's Hospital in London, where he had gathered information about Joseph Towne and his moulaging technique (Griffith 1905). Two years later Photinos gave the standard interpretation of this taciturnity, repeated in all later publications about the subject: "Before the mouleurs reach perfection, they have to invest effort and time. They have to learn two things. First, they need to gain at least basic knowledge in dermatology and venerology. Second, they have to invent some kind of method to produce the casts. The consequence was that until recent years no [mouleur] was willing to share his technique with another for fear of no longer being able to make a living from his art, if someone else learned it from him" (Photinos 1907, 133). Even the statements and publications of the last mouleuses who were trained in the classical manner—Walther and Stoiber—reflect that secrecy: both of them indicated explicitly or implicitly their desire to break with this tradition (Walther 1986/87; 1994; Stoiber 1979; 1988; 1998).

On the surface, the taboo about talking can be explained by the desire to guard trade secrets. Mouleurs and mouleuses who had developed techniques and materials with great effort over many years secured their existence with silence. Looked at this way, the art of moulaging served only to feed the producers, who belonged to a profession, just like many others.

The mouleurs' and mouleuses' silence, however, had other dimensions which were rooted in the special setting they created with their patients to form good moulages. The first main feature of this cooperation was an atmosphere of openness, trust, and credibility. Of central importance was the first contact, when the mouleurs or mouleuses were left alone with the patients to inform them in greater detail about what would take place. In the late nineteenth and early twentieth centuries, mouleurs or mouleuses did not seem to pay great attention to this issue, since the patients were primarily required to take their orders from the physicians in charge (Photinos 1907, 142). Later in the twentieth century, mouleuses such as Elfriede Walther and Elsbeth Stoiber approached their patients with greater sensitivity. As Walther stated, "The mouleur or mouleuse should try to explain the aim and use of the procedure to the person to be depicted. It was my experience that most of the patients were not familiar with the term moulage ... If the patients understand why and how certain things are done, they will behave cooperatively" (Walther 1986/87, 34–35). Stoiber reported that she had to deal with patients who came with rather diverse attitudes: "Some were very reverential from the beginning. Others, on the contrary, showed great distrust" (Stoiber 1998). Stoiber took time to explain at length the process of making moulages. She especially pointed out how she would deal with the warmth being generated by the hardening plaster, whether patients with a certain affliction would have to endure some pain and how she would cope with it, and—in cases where she had to take a negative from the face—how she would protect the patient's hair and eyes, and how she would provide for their respiration. To motivate sick persons, Stoiber presented examples of her work: "Of course, I never led the patients through the collection. I got some moulages for them which were not at all repulsive. In addition, I showed them some books, the Jacobi atlas for instance, where the moulages were reproduced. And then I just said: 'This work is very important for physicians in training"' (Stoiber 1998). And she added: "I told the patients that it was interesting to produce a moulage to see what effect therapy had, since the picture-in-wax is completely objective, much more objective than any photograph" (Stoiber 1998). Most important for the cooperation with patients was the ability of mouleurs or mouleuses to establish credibility with patients. According to Stoiber, "People always believed what I said. Nobody ever—even in other things—distrusted me. Maybe this was due to the fact that [the patients] believed me capable to doing my job" (Stoiber 1998). The atmosphere of trust between mouleurs or mouleuses and patients may have contributed to the inclination of mouleurs and mouleuses not to talk very much about the process of making moulages.

Secondly, mouleurs or mouleuses usually strove to set up working conditions that were characterised by a high degree of privacy, even intimacy, in order to generate an atmosphere of concentration and creativity. Roger-Miles noted that Baretta arranged his work like an operation: His great eagerness and good naturedness "made him an accurate servant of science. He used his instruments without crudeness but with the sensitivity of a mother and a never-ending patience. . . . To bridge the time at certain stages, when the work needed to harden, . . . he showed his patients his own sketches and paintings that hung on the walls of his studio. Then he sat at the piano and comforted his clients with some old melodies" (Roger-Miles 1891, 159–65). Griffith cited a coworker of Towne at Guy's Hospital in London: "Mr. Towne never liked to have anybody around when he worked and none of us ever saw him. Being forward then, I sometimes used to try and find out how he did it. But he would never let me get close enough to see anything before he would stop work and come over to me and say, 'Well, do you want anything?'" (Griffith 1905, 44).

From the description provided by Stoiber, who actively broke the taboo of secrecy, we get a rather detailed picture of what moulage production was like at a university skin clinic later in the twentieth century (Stoiber 1998). After she had explained to a certain patient what he or she would have to face during the process, Stoiber made dates for four or five meetings that would each last up to four hours. On average it took a week to complete the work. If possible, Stoiber asked the patient to come to her studio near the skin clinic. If the patient could not leave his or her bed, the mouleuse had to meet him or her either on the ward or in his or her home.

In the first sitting the negative was made. The mouleuse had to come physically very near the patient who had bared the part of the body to be depicted, and then she had to touch the sick person while she applied the modelling material to the body. The following meetings, which served to paint (Fig. 8.7) and prepare the moulage in a realistic way, formed the real heart of the mouleuse's work. She arranged the external working conditions with particular care. If possible, she attempted to meet with the patient alone. In any case she did not allow anyone else to look over her shoulder. Stoiber stated, "I always worked in a very quiet room, since I could not stand anybody around [except for the patient], when I worked. It was important to me that there was never anybody who walked back and forth, who came and went, or who wished to watch. I simply could not have somebody around me to look at how this all was done and I never allowed it" (Stoiber 1998). Stoiber made sure there was optimal lighting, and shut out irritating noises.

In everything she did, Stoiber strove to pay full attention to the patient, to sink herself into the pores of the patient's skin. She wanted to reproduce the 'anatomy of disease' in all its nuances and as close to nature as possible. "In the moulage", she pointed out, "everything had to be exactly like the patient, be it the surrounding skin or the specific location [of the affliction]" (Stoiber 1998). With this statement, however, Stoiber made clear that she, like her colleagues, was concerned to embed the correct replica of isolated bodily structures in the context of a 'true portrait' of sick individuals, who presented their personality not only in their faces but also through every wrinkle of skin. So, intentionally or not, by bringing their scientific and aesthetic efforts together, the mouleurs and mouleuses constructed the moulage as a hybrid: it became both a medical document and an independent work of art.

Thus the moulage can be read, on the one hand, as a scientific statement by the mouleurs and mouleuses, since—with the help of the third dimension—it

Figure 8.7 Elfriede Walther, mouleuse at the German Hygiene Museum, Dresden, at work on a moulage in 1977. Photograph courtesy of the German Hygiene Museum, Dresden.

brought the efforts of natural philosophers of the seventeenth and eighteenth centuries towards an unemotional, morphologic–macroscopic 'observatio' to completion.[17] On the other hand, the mouleurs and mouleuses made an artistic statement. Focusing on a certain region of the body, they showed a facet of a certain human being, in his or her nakedness and in a fragile phase of life, beset with the signs of illness, behind which the danger of death always lurked. Despite their efforts to remain 'objective', mouleurs and mouleuses inevitably represented in their portraits of the sick a concentration of the motifs from the dance of death. Moulages can therefore also be read as medical memento mori of the nineteenth and twentieth centuries.

The silence of the mouleurs and mouleuses can be explained as the silence of artists who do not want to give out information about aspects of their artistic creativity. But the taciturnity also resulted from the place of action—the medical sphere. Although an explicit ethical debate about this issue cannot be traced, the available sources strongly suggest that mouleurs or mouleuses in the nineteenth as well as in the twentieth centuries were greatly aware that 'their' patients had a right to enjoy respect, to remain intact, and not least important, to preserve their anonymity as far as possible. The latter value probably provided mouleurs and mouleuses with the main basis for adhering to professional secrecy.

THE PATIENTS

At first glance, patients seem to have had only a very passive role in the process of moulage production. They were chosen by the physicians to have pictures-in-wax made by the mouleurs or mouleuses. In this they were more or less free to cooperate or object, but once they had consented, were guided through the whole fabrication process by the mouleurs or mouleuses.

Looking at the situation from a different perspective, however, we see patients who had to bear up under the stress of moulage manufacture. First, they had to expose themselves to the wax-worker. Second, they had to place their sick bodies at the disposal of the mouleurs or mouleuses for protracted manipulations that did not promise to bring them any closer to recovery. And third, they had to face the fact that their disease would be made public through the moulage.

It is striking that patients only rarely—usually when the face or venereal diseases were to be depicted—refused to have a moulage made.[18] Indeed, for several obvious reasons most were eager to make themselves available. Physicians and mouleurs or mouleuses never ceased to remind the patients that the moulage made from them provided an important service to medical teaching and research. It promised the benefit of showing the progress of therapy in the individual case. In addition, some of the sick hoped that their consent would bring them a kind of quid pro quo from the doctor in the form of more attention and effort.

The readiness of some patients can be explained in other ways as well. We have to imagine how they experienced the process of making a moulage. In-patients always had to be released from the clinical routine for several hours. In the working sessions with the mouleurs or mouleuses, patients experienced, possibly for the first time during their treatment, the full and undivided attention of a medical person. Embedded in the concentrated working atmosphere created by the mouleurs or mouleuses, the contact between the patients and the wax sculptors attained a degree of privacy and trust that loosened patients' tongues. Stoiber frequently experienced this situation: "Of course, this was the place where patients had a chance to talk, and where they knew that somebody was listening to them. I did not say a whole lot. I just listened, so they got the feeling that I paid attention to what they told me. Certainly, this was a great relief for them.... They informed me about how it all had happened, what their friends thought about it, and what had happened to somebody else with the same disease at a different clinic or [under the treatment] of another physician" (Stoiber 1998). Most patients began to talk in front of the mouleuse: about themselves, their illness, their problems, and fears that related to their disease. Unintentionally and clearly as a secondary effect, the work on the moulage took on a therapeutic dimension.

This therapeutic aspect of the primarily 'diagnostic' moulage becomes even more striking when we hear what value some patients attached to this object. Stoiber reported on several patients who expressed their conviction after completion of their moulages that they "were healthier now, because the disease was gone". She even recalled one case when a patient verbalised a similar sensation during the working sessions, saying to her: "When you work on this object, I get the feeling it [the disease] has left me now. It is banished there in the moulage" (Stoiber 1998).

Experiencing over a long period of time the creation of an object that looked like a piece of themselves, some patients projected their illness onto this object. That is, they believed that in the struggle to gain a realistic, 3-D replica of themselves, the disease was transferred to the product of this effort and that they were freed from their suffering. In Zürich they expressed this idea to the mouleuse as an actual experience they had while the moulage was being created, as well as in the form of a personal certainty, when the wax sculpture had been finished and the signs of their affliction were gone.

For some patients, the moulage even acquired a religious value. Stoiber presented the case of a patient with a malignant skin tumour: "This woman asked me whether I could make a second moulage for her own use. I told her that this might be difficult, because these things take time. She then replied that this object did not have to be coloured, a cast would be enough. Of course, I fulfilled her wish. I handed the moulage to her and she took it [as an offering of thanks] to a church dedicated to a healing saint" (Stoiber 1998). Dealing in this way with a representation of the diseased region of the body, patients such as this woman transferred the moulage into a religious context. By presenting 'their' moulage at a sacred place, they could ask a higher authority for healing, or thank a heavenly protector for a successful recovery from an illness. In addition, they associated with their wax image the hope that the particular body part, now under the influence of higher powers, would be protected in the future.

Thus, for many patients, simply the opportunity to talk that was offered by the process of making a moulage served an unintended therapeutic function. Furthermore, for some patients, the moulage made from their bodies represented an object with magical, animistic, or religious dimensions. They saw the moulage as a sort of magic doll in which one could capture the demon of the disease. Finally, some sick people even recognised the moulage as a votive offering, just like other votives in the form of body parts that have been presented at other sacred places over the ages from antiquity into our Christian-influenced present (Schlosser 1911, 207–21; Brückner 1963; Kriss-Rettenbeck 1972; Angeletti 1980).

To round out the picture from the viewpoint of the patients, it is necessary to ask what meaning moulages had for their originals, the sick persons, after the wax work had been finished and added to a collection (Fig. 8.8). Walther and Stoiber encountered diverse reactions. In Dresden Walther found that moulages often had an unpleasant, even shocking, effect on the original patients. According to her, "To see the diseased part of their body alone, without the healthy surrounding area, to see the negatively connotated region that was placed in the spotlight, was often shocking for the patients. It even went so far that the patients glanced at the moulage only briefly and shyly" (Walther 1986/87, 42). Other sick people, however, showed quite the opposite reaction. They examined the moulage "thoroughly and critically and discovered symptoms which they had not noticed even on their own body" (Walther 1986/87, 42). Elsbeth Stoiber in Zürich reported on several patients who came for a visit after their treatment was over. "They told me how they were, that their sickness had disappeared, or what aspect the disease

Figure 8.8 Moulage collection (detail) of the University Skin Clinic, Freiburg im Breisgau, in 1954. The specially-designed demonstration cabinets presented moulages in upper and lower display cases; the sloping surfaces between them showed large-scale photographs and illustrations of the histological sections corresponding to the clinical syndromes that the moulages reproduced. Courtesy of the Institute for the History of Medicine, University of Freiburg im Breisgau.

had taken on in the meantime. And they wanted to see 'their' moulage. Some of them even wished to see the collection alone in order to find 'themselves'. Sometimes I also showed them around in the other parts of the collection" (Stoiber 1998). Stoiber found that for most of the patients it was important to know that their moulages were on display. "It might even be that some of these sick people felt protected not by the sacred halls of a church but by the shelter of modern medicine" (Stoiber 1998). The reactions of patients noted here reflect the usual ambivalence, the play between withdrawal and attraction, of any viewer of the moulage. It appears by contrast that the majority of patients maintained a certain inner bond to 'their' moulage. At this point it might be interesting to go on with research, using the few hints to follow the tracks to the patients.[19]

CONCLUSION

Moulages stand alone as representational media of human body parts. They served neither as ideals nor as mere documents. Because of both their individuality and their realism, they could be used in a socio-professional context by physicians, and they evoked a personal, to some extent even emotional, relationship with mouleurs or mouleuses and patients. The reverse casting method could reproduce organic textures in a more lifelike manner than had ever before been achieved in three dimensions. The realistic colouring as well as the insertion of natural products, such as hair or scales, and artificial items such as glass eyes, underscored the main intention of the moulage only to give a picture of the living original. It was the central programme of the moulage to reveal the characteristic 'anatomy' of a specific disease, embedded in the individuality of a particular patient. The partially mechanical nature of its production, as well as the depiction of selected aspects of the patients' bodies, the self-restraint of the mouleurs and mouleuses in representing diseased areas of the patients, and the individual picture that the wax work generated, defined the specific position of the moulage in nineteenth-century medical imagery.

With its realism, the moulage gained specific functions and meanings for its main protagonists. For physicians it served as a teaching tool for medical audiences and the general public, as well as a research document within the scientific community and a source for illustrating publications. In addition, it became the object of local pride as a document of the scientific and therapeutic competence of a certain physician, a specific clinic, or even the national representatives of a certain medical discipline. Mouleurs and mouleuses used the moulage as the basis for making a living, and were challenged to fulfil scientific requirements as well as artistic goals. Moreover, the close collaboration with the patients put mouleurs and mouleuses in a quasi-therapeutic role. For patients, the process of moulage fabrication frequently had the effect that they felt respected in their need to talk and express feelings and fears connected with the disease. For some patients, the moulage took on a protective or even religious function by representing the specific part of the formerly diseased body. The sickness could be banished, or protection of some kind could be guaranteed, by placing the moulage in the sacred space of a church or the sober rooms of a moulage collection.

NOTES

I wish to thank the editors, Soraya de Chadarevian and Nick Hopwood, as well as the referees for their helpful comments in revising this paper. Mason Barnett earns my special thanks for her support in discussing and translating the text.

1. On the use of moulages in public health education, see Sauerteig 1992. For the role the moulage played in the German Hygiene Museum at Dresden, see Walther 1986–87; Schnalke 1995, 121–43; and Frenzel 1997. On the connection between the art of moulaging and the voyeuristic world of panopticons, see Röhrich and Plewig 1979; Schade 1985; and Py and Vidart 1986.

2. For example, moulages are still in use in medical training for example at the Gordon Museum, Guy's Hospital, London (Schnalke 1989; 1995, 63–67) and the Moulage Museum of Zürich University Hospital (Universitätsspital Zürich 1993; Schnalke 1995, 145–176).

3. See, for instance, the holdings and/or displays of the London Science Museum, the Medical Historical Museums in Zürich, Ingolstadt, and Bochum, as well as the German Hygiene Museum in Dresden.

4. See also the presentations and publications of anatomical plastinations in the "Body Worlds" exhibition (Hagens and Whalley 2000).

5. For the presentation of ideal anatomy in wax, see also the ambitious photographic documentations of the Florentine models in Lanza et al. 1979, 173–239; Düring et al. 1999; and Skopec and Gröger 2002.

6. Jacobi 1903; Lesser 1904; Jacobi 1906; Neisser and Jacobi 1906; Jessner 1913; Riecke 1914; Mulzer 1924; Frieboes 1928. See also Hopwood, this volume, on the involvement of models in embryological illustration.

7. The technique of "citochromy" for generating coloured photographic reproductions of moulages, mentioned by Jacobi in 1903, has not yet been the subject of detailed historical study.

8. For discussion of medical moulages in the context of 'scientific objectivity', see the definitions and delineations of Daston and Galison 1992, as well as the contributions to Mazzolini 1993 and Daston 1994a. On the impact of medical and scientific photography in the debates on 'scientific objectivity', see in addition Fox and Lawrence 1988; Jones 1988; Maehle 1993; Schlich 1995; and Geimer 2001.

9. Besides the unpublished typescript of Walther (1986–87) and the unpublished documentation of an interview given by Stoiber to the author on 1 July 1998 (Stoiber 1998), see also the published accounts by these two last living mouleuses: Walther 1994; and Stoiber 1979; 1988. On the traditions of medical moulaging in Berlin, Dresden, and Zürich, including further references, see Schnalke 1995, 115–76.

10. On the reputation of dermatology and venerology among medical disciplines around 1900, see Eulner 1970, 222–56; and Scholz 1999, 1–75.

11. On the epistemic functions of embryological wax models in the late nineteenth century, see Hopwood 1999.

12. While the largest moulage collection, at the Hôpital Saint Louis in Paris, is arranged etiologically (Schnalke 1995, 85–89; Brenner-Holländer 1996), Elsbeth Stoiber at the Zürich moulage museum created a mixed presentation for didactic reasons (Schnalke 1995, 157–64).

13. As an example of the value of a high-quality moulage collection for the Zürich dermatologists Bruno Bloch and Guido Miescher, see Rüdlinger and Stoiber 1988a; 1988b.

14. See also the photograph of Gustav Riehl, a renowned dermatologist from Vienna, giving a lecture to his students and some of his colleagues before 1926, reproduced as fig. 66 in Schnalke 1995, 102.

15. 'Experimental moulages' have not yet been a topic of extensive historical investigation, but for an example, see Schnalke 1995, 156, fig. 123.

16. Over all, there are still only very few biographical data available concerning the various mouleurs and mouleuses; for some basic information and references see Barlag 1992; Schnalke 1995; Brenner-Holländer 1996; and Frenzel 1997.

17. On the conceptualisation of 'observatio' in the early modern republic of letters, see Findlen 1994a; 1994b; and Daston 1994b.

18. The following observations concerning the patient's attitudes and reactions were mostly reported in Stoiber 1998.

19. Among the Zürich moulages there is a stock of surgical items which have a concise case description, typed and attached to the back of the mounting board. An inventory number refers to the patient's records, but their location is uncertain and they have not been studied (Schnalke 1995, 145–51). Elsbeth Stoiber referred to a list of patients that she kept (Stoiber 1998); several of her clients might still be alive.

REFERENCES

Almhofer, Edith. 2002. "Beredte Bilder des Leibes." In *Anatomie als Kunst. Anatomische Wachsmodelle des 18. Jahrhunderts im Josephinum in Wien*, ed. Manfred Skopec and Helmut Gröger, 21–29. Vienna: Brandstätter.

Angeletti, Charlotte. 1980. *Geformtes Wachs. Kerzen, Votive, Wachsfiguren.* Munich: Callwey.

Azzaroli Puccetti, Maria Luisa. 1972. "La Specola: The Zoological Museum of the University of Florence." *Curator* 15: 93–112.

Barlag, Goetz. 1992. *Die Moulagensammlung der Universitäts-Hautklinik Freiburg im Breisgau. Katalog und Beiträge zu ihrer Geschichte*. Frankfurt am Main: Team Kommunikation.

Boschung, Urs. 1980. "Medizinische Lehrmodelle. Geschichte, Techniken, Modelleure." *Medita* 10, no. 5: ii–xv.

Brenner-Holländer, Hana. 1996. "Das Moulagenmuseum am Hôpital Saint-Louis in Paris. Zeit und Bedeutung seiner Entstehung." Medical dissertation, University of Zürich.

Brückner, Wolfgang. 1963. "Volkstümliche Strukturen und hochschichtliches Weltbild im Votivwesen: Zur Forschungssituation und Theorie des bildlichen Opferkultes." *Schweizerisches Archiv für Volkskunde* 59: 186–203.

Büll, Reinhard. 1977. *Das große Buch vom Wachs*, 2 vols. Munich: Callwey.

La ceroplastica nella scienza e nell'arte. Atti del I Congresso Internazionale, Firenze, 3–7 giugno 1975. 1977. 2 vols. Florence: Olschki.

Daston, Lorraine. 1994a. *Wordless Objectivity*. Berlin: Max-Planck-Institut für Wissenschaftsgeschichte.

———. 1994b. "Neugierde als Empfindung und Epistemologie in der frühmodernen Wissenschaft." In *Macrocosmos in Microcosmo. Die Welt in der Stube. Zur Geschichte des Sammelns 1450 bis 1800*, ed. Andreas Grote, 35–59. Opladen: Leske und Budrich.

Daston, Lorraine, and Peter Galison. 1992. "The image of objectivity." *Representations* 40: 81–128.

Dürbeck, Gabriele. 2001. "Empirischer und ästhetischer Sinn. Strategien der Vermittlung von Wissen in der anatomischen Wachsplastik um 1780." In *Wahrnehmung der Natur—Natur der Wahrnehmung. Studien zur Geschichte visueller Kultur um 1800*, ed. Gabriele Dürbeck, Bettina Gockel, Susanne B. Keller, Monika Renneberg, Jutta Schickore, Gerhard Wiesenfeld, and Anja Wolkenhauer, 35–54. Dresden: Verlag der Kunst.

Düring, Monika von, Georges Didi-Huberman, and Marta Poggesi, with photographs by Saulo Bambi. 1999. *Encyclopaedia Anatomica: A Complete Collection of Anatomical Waxes*. Cologne: Taschen.

Ehring, Franz. 1989. *Skin Diseases: Five Centuries of Scientific Illustration.* Stuttgart: Fischer.

Eulner, Hans-Heinz. 1970. *Die Entwicklung der medizinischen Spezialfächer an den Universitäten des deutschen Sprachgebiets*. Stuttgart: Enke.

Findlen, Paula. 1994a. *Possessing Nature: Museums, Collecting, and Scientific Culture in Early Modern Italy.* Berkeley and Los Angeles: University of California Press.

———. 1994b. "Die Zeit vor dem Laboratorium: Die Museen und der Bereich der Wissenschaft 1550–1750." In *Macrocosmos in Microcosmo. Die Welt in der Stube. Zur Geschichte des Sammelns 1450 bis 1800*, ed. Andreas Grote, 191–207. Opladen: Leske und Budrich.

Fox, Daniel M., and Christopher Lawrence. 1988. *Photographing Medicine: Images and Power in Britain and America since 1840*. New York: Greenwood.

Fox, George H. 1881. *Photographic Illustrations of Skin Diseases*. New York: E. B. Treat.

Frenzel, Michael. 1997. "Die Entwicklung und Nutzung der Moulagen in Sachsen." Medical dissertation, University of Dresden.

Frieboes, Walter. 1928. *Atlas der Haut- und Geschlechtskrankheiten*. Leipzig: Vogel.

Geimer, Peter. 2001. "Photographie und was sie nicht gewesen ist: Photogenic drawings, 1834–1844." In *Wahrnehmung der Natur—Natur der Wahrnehmung. Studien zur Geschichte visueller Kultur um 1800*, ed. Gabriele Dürbeck, Bettina Gockel, Susanne B. Keller, Monika Renneberg, Jutta Schickore, Gerhard Wiesenfeld, and Anja Wolkenhauer, 135–49. Dresden: Verlag der Kunst.

Griffith, Frederic. 1905. "Joseph Towne, wax modeller of Guy's Hospital." *Medical Library and Historical Journal* 3: 41–45.

Hagens, Gunther von, and Angelina Whalley. 2000. *Körperwelten. Die Faszination des Echten*. Heidelberg: Institut für Plastination.

Hahn, Susanne, and Dimitros Ambatielos, eds. 1994. *Wachs—Moulagen und Modelle. Internationales Kolloquium 26. und 27. Februar 1993*. Dresden: Verlag des Deutschen Hygiene-Museums Dresden.

Hardy, Louis P. A., and A. de Montméja. 1867/68. *Clinique photographique de l'Hôpital Saint Louis*. Paris: Chamerot and Lauwereyns.

Haviland, Thomas N., and Lawrence C. Parish. 1970. "A brief account of the use of wax models in the study of medicine." *Journal of the History of Medicine and Allied Sciences* 25: 52–75.

Hebra, Ferdinand. 1856. *Atlas der Hautkrankheiten*, instalment 1. Vienna: Braumüller.

Hopwood, Nick. 1999. "'Giving body' to embryos: Modelling, mechanism, and the microtome in late nineteenth-century anatomy." *Isis* 90: 462–96.

Jacobi, Eduard. 1903. *Atlas der Hautkrankheiten. Mit Einschluss der wichtigsten venerischen Erkrankungen*. Berlin and Vienna: Urban & Schwarzenberg.

———. 1906. *Supplement zum Atlas der Hautkrankheiten. Mit Einschluss der wichtigsten venerischen Erkrankungen*. Berlin and Vienna: Urban & Schwarzenberg.

Jessner, Samuel. 1913. *Lehrbuch (früher Kompendium) der Haut- und Geschlechtskrankheiten einschliesslich der Kosmetik*, 4th edn. Würzburg: Curt Kabitzsch.

Jones, Anne H., ed. 1988. *Images of Nurses: Perspectives from History, Art, and Literature*. Philadelphia: University of Pennsylvania Press.

Jordanova, Ludmilla J. 1985. "Gender, generation, and science: William Hunter's obstetrical atlas." In *William Hunter and the Eighteenth-Century Medical World*, ed W. F. Bynum and Roy Porter, 385–412. Cambridge: Cambridge University Press.

———. 1989. *Sexual Visions: Images of Gender in Science and Medicine between the Eighteenth and Twentieth Centuries*. New York: Harvester Wheatsheaf.

Kaposi, Moriz. 1890. "Bericht über den I. internationalen Congress für Dermatologie und Syphilographie zu Paris." *Archiv für Dermatologie und Syphilis* 22: 190–204.

Kleindienst, Heike. 1989. "Ästhetisierte Anatomie aus Wachs. Ursprung—Genese—Interpretation", vol. 1. Philosophical dissertation, University of Marburg.

———. 1990. "Das Apologiemodell einer Anatomia aesthetica in cera im Zeitalter Papst Benedikts XIV (1740–1758)." *Archiv für Kulturgeschichte* 72: 367–80.

Kriss-Rettenbeck, Lenz. 1972. *Ex voto. Zeichen, Bild und Abbild im christlichen Volksbrauchtum*. Zürich and Freiburg im Breisgau: Atlantis.

Krüger-Fürhoff, Irmela M. 1998. "Der vervollständigte Torso und die verstümmelte Venus. Zur Rezeption antiker Plastik und plastischer Anatomie in Ästhetik und Reiseliteratur des 18. Jahrhunderts." *Zeitschrift für Germanistik* 8: 361–73.

Lanza, Benedetto, Maria Luisa Azzaroli Puccetti, Marta Poggesi, and Antonio Martelli, with photographs by Liberto Perugi. 1979. *Le cere anatomiche della Specola*. Florence: Arnaud.

Lemire, Michel. 1992. "Representation of the human body: The coloured wax anatomical models of the eighteenth and nineteenth centuries in the revival of medical instruction." *Surgical-Radiologic Anatomy* 14: 283–91.

Lesky, Erna. 1976. "Wiener Lehrsammlung von Wachspräparaten." *Gesnerus* 33: 8–20.

Lesser, Edmund. 1904. *Lehrbuch der Haut- und Geschlechtskrankheiten,* 11th edn. Leipzig: Vogel.

Maehle, Andreas-Holger. 1993. "The search for objective communication: Medical photography in the nineteenth century." In *Non-Verbal Communication in Science prior to 1900*, ed. Renato G. Mazzolini, 563–86. Florence: Olschki.

Mazzolini, Renato G., ed. 1993. *Non-Verbal Communication in Science prior to 1900*. Florence: Olschki.

Mulzer, Paul, ed. 1924. *Atlas der Hautkrankheiten*. Munich: Lehmann.

Neisser, Albert, and Eduard Jacobi. 1906. *Ikonographia dermatologica. Atlas seltener, neuer und diagnostisch unklarer Hautkrankheiten*. Berlin and Vienna: Urban & Schwarzenberg.

Parish, Lawrence, Gretchen Worden, Joseph A. Witkowski, Albrecht Scholz, and Daniel H. Parish. 1991. "Wax models in dermatology." *Transactions and Studies of the College of Physicians of Philadelphia*, ser. 5, 13: 29–74.

Photinos, Georgios T. 1907. "Die Herstellung und Bedeutung von Moulagen (farbige Wachsabdrücke)." *Dermatologische Zeitschrift* 14: 131–57.

Poch-Kalous, Margarete. 1966. *Cajetan—Das Leben des Wiener Mediziners und Karikaturisten Dr. Anton Elfinger*. Vienna: Wiener Bibliophilen-Gesellschaft.

Portele, Karl A. von 1974. "Dr. med. Anton Elfinger, ein vergessener medizinischer Modelleur." *Annalen des Naturhistorischen Museums Wien* 78: 95–102.

———. 1977. "Die Moulagensammlung des Pathologisch-anatomischen Bundesmuseums in Wien." *Mitteilungen des Pathologisch-anatomischen Bundesmuseums in Wien* 1: 5–13.

Premuda, Loris. 1972. "Wachsbildnerei und Medizin." *IMAGE Roche* 48: 17–24.

Py, Christine, and Cécile Vidart. 1986. "Die Anatomischen Museen auf den Jahrmärkten." *Freibeuter* 27: 66–77.

Riecke, Erhard. 1914. *Lehrbuch der Haut- und Geschlechtskrankheiten*, 3rd edn. Jena: Fischer.

Röhrich, Heinrich, and Gerd Plewig. 1979. "Pathologisch-anatomische Lehrmodelle in Moulagen. Herstellungstechnik und Geschichte." *Der Hautarzt* 30: 259–63.

Roger-Miles, L. 1891. *La cité de misère*. Paris: Librairie Marpon et Flammarion.

Rüdlinger, René, and Elsbeth Stoiber. 1988a. "Die Ära Bloch im Spiegel der Moulagensammlung der Dermatologischen Universitätsklinik Zürich 1917–1933." *Der Hautarzt* 39: 314–17.

———. 1988b. "Auf den Spuren von Guido Mieschers wissenschaftlichem Werk in der Moulagensammlung der Dermatologischen Universitätsklinik Zürich 1933–1958." *Der Hautarzt* 39: 457–60.

Sauerteig, Lutz. 1992. "Lust und Abschreckung. Moulagen in der Geschlechtskrankheitenaufklärung." *Medizin, Gesellschaft und Geschichte* 11: 89–105.

Schade, Doris. 1985. "'Durch die Kunst blüht das Gewerbe'. Fotografien aus dem Nachlaß der Wachsfiguren-Fabrik Gebrüder Weber Berlin." *Fotogeschichte—Beiträge zur Geschichte und Ästhetik der Fotografie* 5: 33–48.

Schlich, Thomas. 1995. "'Wichtiger als der Gegenstand selbst'—Die Bedeutung des fotografischen Bildes in der Begründung der bakteriologischen Krankheitsauffassung durch Robert Koch." In *Neue Wege in der Seuchengeschichte*, ed. Martin Dinges and Thomas Schlich, 143–74. Stuttgart: Steiner.

Schlosser, Julius v. 1911. "Geschichte der Portraitbildnerei in Wachs." *Jahrbuch der Kunsthistorischen Sammlungen des Allerhöchsten Kaiserhauses* 29: 171–258.

Schnalke, Thomas. 1986. "Moulagen in der Dermatologie. Geschichte und Technik." Medical dissertation, University of Marburg.

———. 1989. "Joseph Towne: British pioneer of wax dermatological modelling." *The American Journal of Dermatopathology* 11: 466–72.

———. 1993. "Der Patient dahinter—Gedanken zum Umgang mit Moulagen im Medizinhistorischen Museum." *Medizin im Museum. Jahrbuch der Medizinhistorischen Sammlung der Ruhr-Universität-Bochum* 1: 69–71.

———. 1995. *Diseases in Wax: The History of the Medical Moulage*, translated by Kathy Spatschek. Berlin: Quintessence.

———. 1999. "Veröffentlichte Körperwelten. Möglichkeiten und Grenzen einer Medizin im Museum." *Zeitschrift für medizinische Ethik* 45: 15–26.

———. 2001. "Vom Modell zur Moulage: Der neue Blick auf den menschlichen Körper am Beispiel des medizinischen Wachsbildes." In *Wahrnehmung der Natur—Natur der Wahrnehmung. Studien zur Geschichte visueller Kultur um 1800*, ed. Gabriele Dürbeck, Bettina Gockel, Susanne B. Keller, Monika Renneberg, Jutta Schickore, Gerhard Wiesenfeld, and Anja Wolkenhauer, 55–69. Dresden: Verlag der Kunst.

Scholz, Albrecht. 1999. *Geschichte der Dermatologie in Deutschland*. Berlin and Heidelberg: Springer.

Skopec, Manfred. 1983. "Anton Elfinger (1821–1864): A forgotten medical illustrator." *International Journal of Dermatology* 22: 256–59.

Skopec, Manfred, and Helmut Gröger, eds. 2002. *Anatomie als Kunst. Anatomische Wachsmodelle des 18. Jahrhunderts im Josephinum in Wien*. Vienna: Brandstätter.

Solente, G. 1983. "Le Musée de l'Hôpital Saint Louis." *The American Journal of Dermatopathology* 5: 483–89.

Squire, Alexander J. B. 1865. *Photographs (Coloured from Life) of the Diseases of the Skin*. London: Churchill.

Stoiber, Elsbeth. 1979. *Tonbild-Schau: Herstellung einer dermatologischen Wachsmoulage*. Zürich: Typescript.

———. 1988. *The Moulages Collection of the University of Zürich: Present Utilisation and Methods of Preservation*. Zürich: Typescript.

———. 1998. *Documentation of an Interview Given to Thomas Schnalke on 1 July 1998*. Erlangen: Typescript.

Universitätsspital Zürich, ed. 1993. *Moulagen-Sammlungen des Universitätsspitals Zürich*. Zürich: Universitätsspital.

Wallach, Daniel, and Gérard Tilles. 1990. "Le premier congrès international de dermatologie et de syphiligraphie, Paris, 5–10 août 1889." *Histoire des Sciences Médicales* 24: 99–104.

Walther, Elfriede. 1986/87. *Moulagen und Wachsmodelle 1945–1980 in Dresden*. Dresden: Typescript.

———. 1994 "Moulagen und Wachsmodelle am Deutschen Hygiene-Museum unter besonderer Berücksichtigung der Zeit von 1945–80." In *Wachs—Moulagen und Modelle. Internationales Kolloquium 26. und 27. Februar 1993*, ed. Susanne Hahn and Dimitros Ambatielos, 91–102. Dresden: Verlag des Deutschen Hygiene-Museums Dresden.

Wichelhausen, Engelbert. 1798. *Ideen über die beste Anwendung der Wachsbildnerei, nebst Nachrichten von den anatomischen Wachspräparaten in Florenz und deren Verfertigung*. Frankfurt am Main: Zeßler.

Wolkenhauer, Anja 2001. "'Grausenhaft wahr ist diese wächserne Geschichte.' Die Wachsfiguren von Don Gaetano Zumbo zwischen Kunst und medizinischer Anatomie." In *Wahrnehmung der Natur—Natur der Wahrnehmung. Studien zur Geschichte visueller Kultur um 1800*, ed. Gabriele Dürbeck, Bettina Gockel, Susanne B. Keller, Monika Renneberg, Jutta Schickore, Gerhard Wiesenfeld, and Anja Wolkenhauer, 71–85. Dresden: Verlag der Kunst.

CHAPTER NINE

Molecules and Croquet Balls

Christoph Meinel

For much of its history chemistry had an ambiguous attitude to visual representations. While a rich tradition depicted chemical laboratories and (al)chemists at work, the language of chemistry and its theoretical notions remained verbal and abstract (Crosland 1962; Knight 1993; 1997). Even John Dalton's highly figurative atomic theory of 1808 and his speculations about atoms as little spheres arranged in space were incorporated only in the non-figurative version of Berzelius' algebraic notation (Thackray 1970, 264–66; Rocke 1984).[1] During the 1860s, however, the dominant way chemists thought about matter changed from an abstract and verbal to a constructivist and pictorial approach. This gradual transition was closely related to the new interest in molecular constitution stimulated by the rise of organic chemistry during the first half of the nineteenth century, and intimately linked to the advent, from the late 1850s, of a new theory of chemical structure. As a result, molecules were considered to be composed of atoms, the relative positions of which were determined by their respective valency or binding force. Yet to what extent these structures represented the true arrangement of atoms within the molecule remained controversial. Most chemists preferred to use these formulae as mere aids to classification and were reluctant to take their spatial properties for physical reality.

As a consequence of a new vision of chemistry's future, proposed by a London-based group of chemists engaged in organic synthesis, this attitude changed and scientists began to realise that a 'chemistry in space' would not only allow them to relate chemical behaviour to physical, for example optical, properties, but could also be used as a blueprint for a new laboratory practice based upon the idea of a molecule as a truly spatial arrangement. This was indeed one of the great revolutions in nineteenth-century chemistry, as announced in Jacobus Henricus van't Hoff's manifesto, *The Arrangement of Atoms in Space*, first published in Dutch in 1874 (van't Hoff 1874; Ramsay 1975).

Historians of chemistry have treated the emergence of stereochemistry as a sequence of arguments and discoveries within the development of chemical theory. Molecular models have been seen as merely illustrating theoretical concepts such as atom, valency, or space (Ramsay 1974; 1981; Spronsen 1974). In this chapter I argue that the change that eventually resulted in a three-dimensional representation of molecules was led, not by theory, but by modelling—a kind of modelling invented, not primarily to express chemical theory, but rather as a new way of communicating a variety of messages. By manipulating tin boxes or tinkering with little spheres and toothpicks, chemists not only visualised their abstract theoretical notions but also impressively testified to the claim that they would build a new world out of new materials. For this purpose molecular models supplied the elements of a new symbolic and gestic language by which chemists conquered new spaces: material space in the form of new substances, notional space in the new stereochemistry, and social space by expressing professional claims to power. Though the use of these models remained epistemologically problematic, their social and cultural message was much more easily understood, and this predisposed younger chemists to accept their implicitly constructivist and three-dimensional approach. The active construction of space was not peculiar to chemistry, but part of a more comprehensive change in perceiving the world and making it one's own, an attitude present in cultural domains from pedagogy to architecture.

SYNTHESIS AND CHEMICAL STRUCTURE

The notion of synthesis and the notion of structure mark the beginning of a reorientation in mid-nineteenth-century chemistry. Both concepts originated with a group of London chemists who, in the mid-1840s, proposed the idea that chemistry's foremost task was no longer analysis and understanding, but rather the making of compounds. 'Chemical synthesis' was the new slogan, introduced by Hermann Kolbe in 1845 and subsequently turned into a research programme by Edward Frankland and Kolbe at the Royal School of Mines in London. At a meeting of the Chemical Society of London in April 1845, August Wilhelm Hofmann, then head of the Royal College of Chemistry, proclaimed that the old-fashioned analytical approach dating from Lavoisier would soon give way to a new era of synthetic chemistry (Muspratt and Hofmann 1845; Russell 1987; Rocke 1993b). And by the late 1850s many chemists were convinced that even complicated natural compounds could be prepared in the laboratory, provided their molecular constitution was established. Finally, Marcellin Berthelot's *Organic Chemistry Based upon Synthesis*, published in French in 1860 (Berthelot 1860), became the manifesto of this new approach, the counterparts of which were the advent of chemical industrialisation and the success of artificial dyestuffs prepared according to the principle of chemical synthesis.

The original basis of chemistry's new self-image was the so-called substitution theory of chemical combination. This theory treated the molecule as a unit in which individual atoms or groups could be replaced by other elements without fundamentally changing the general character of a substance. Compounds that could be formally reduced to a common scheme belonged to the same 'chemical type'. Hydrochloric acid, water, ammonia, and marsh gas (methane) were believed to represent the four basic patterns, or types, out of which the entirety of chemical compounds could be derived. In the late 1850s, the type theory was supplemented by the theory of chemical structure developed by Frankland and Hofmann in London, and by August Kekulé, who had been a member of this London group before he moved to Heidelberg and, in 1858, to Ghent in Belgium. The interpretation favoured by this group of young chemists took the chemical type in a 'mechanical' sense, indicating how atoms or groups are linked together depending on the number of valencies or binding units peculiar to each sort of element.

In the beginning, however, neither type nor structural formulae were meant to represent the true intramolecular arrangement of the atoms. Rather, the formulae were regarded as a mere aid to classifying reactivities and searching for analogies—a taxonomic model with no correlation to a reality that was assumed to be fundamentally unintelligible (Brooke 1976). Nevertheless, the idea of a 'mechanical type' offered the tremendous advantage of permitting, for the first time, predictions regarding possible and still unknown compounds. As a consequence, type formulae became the most powerful tools of the new synthetic chemistry.

TYPE MOULDS AND ATOMIC CUBES

Hofmann's model substance was ammonia (NH_3), the three hydrogen atoms of which could be replaced, one after another, by other groups. In this way a whole array of homologous primary, secondary, and tertiary amines could be obtained, and the number of possible combinations was almost unlimited. In this combinatorial game, the chemical type provided an aid to construction, a template in the spaces of which atoms or atomic groups could be inserted like bricks in a wall—very much in the same way that Berzelian formulae had been used as paper tools for modelling chemical reactions since the 1830s (Klein 1999). To visualise this way of chemical reasoning Hofmann prepared three-dimensional frames, consisting of two, three, or four wire cubes made to receive solid painted cubes representing atoms or atomic groups. These "type moulds", first presented in a publication of 1862, were "mechanical types", transformed into pedagogical tools that could be used in chemistry lectures (Hofmann 1862).

Anschaulichkeit, the ability to appeal to the mind's eye by transforming abstract notions into vivid mental images, was Hofmann's chief pedagogical method. Impressive demonstration rather than abstract reasoning was supposed to transmit scientific knowledge. In this way theoretical notions turned into mental images could be read as a language, "presented to the mind in neat pictures" (Hofmann 1871, 119). Hofmann wanted the student to acquire the kind of pictorial representation that would emerge "if we use these formulas as types, as blueprints for as many classes of chemical compounds. All members of each class are cast, so to speak, into the same mould, and thus they render the peculiarities of the models as in true imitation" (Hofmann 1871, 108).

The choice of the model and the language that came with it disclose a modeller's or an architect's approach, and it was indeed architecture where the notion of model originated. From Vitruvius through the eighteenth century a model meant a concrete exemplar or mould after which something else was prepared (Baker, this volume). Architectural templates teach us how to make something. Epistemology or ontology does not matter in this case;

Figure 9.1 Samuel M. Gaines' "chemical apparatus" as patented by U.S. patent no. 85299 of 29 December 1868. The box contains coloured wooden cubes in two sections marked "Metalloid" and "Metals". The cubes represent chemical weights from 1 to 39. Case size 225 × 225 × 85 mm. National Museums of Scotland, Edinburgh, acc. no. T.1985.112; copyright Trustees of the National Museums of Scotland.

moulds may prove useful in practice, but the notion of truth would be alien to them.

This was exactly the way Hofmann used his type moulds. Convinced as he was that symbolic notations in chemistry were purely formal tools that did not immediately correspond to reality, this approach explicitly avoided the question of truth. Consequently, Hofmann's type moulds and atomic cubes were not meant to represent the physical arrangement of the atoms.

They rather supplied a pattern according to which the chemical operations of elimination and substitution could be classified and analogies found.

Shortly afterwards Hofmannian model kits were manufactured commercially. The 1866 sales catalogue of John Joseph Griffin's London instruments company advertised a kit of painted atomic cubes made from white biscuit ware and sold "in a neat black wood cabinet" for 31s 6d (Griffin 1866, nos 194–95; Gee and Brock 1991).[2]

> Atomic Symbols for the illustration of Theoretical Subjects at Chemical Lectures: consisting of Coloured Cubes of Pottery, about two inches square, intended to represent chemical atoms or gaseous volumes. They can be easily grouped, so as to easily illustrate the atomic composition of compounds, the theory of combination in volumes, and the double decomposition of salts, and to illustrate various chemical doctrines by equations. The following series of sixty models is sufficient to explain the formulae of most frequent occurrence (Griffin 1866, no. 194).

No set of the Griffin models seems to have survived, but the National Museum of Scotland has an American derivative, a wooden box marked "Gaines' Chemical Apparatus" (Fig. 9.1). It contains numbered cubes of wood, coloured and of various sizes, representing atomic masses between 1 and 39. The item was patented for a certain Samuel M. Gaines of Glasgow, Kentucky, in 1868, and the patent claims a new "method of teaching the rudiments of chemistry by means of moveable material bodies" (U.S. Patent 85299 1868).[3]

CIRCLES AND LINES

The models discussed so far were used to illustrate chemical reactions by means of physical objects that could be manipulated according to certain rules. These objects represented relations rather than specific particulars of the natural world. Their meaning and iconography remained closely linked to the type-mould pattern and to the chemistry of elimination and substitution reactions. By marking the cubes with the respective atomic masses the model could even be used as a calculating device, as in Gaines' apparatus. However, it operated within the framework of the type theory; its units could be individual atoms or whole groups, such as phenyl, but it was not meant to convey the idea of structure or molecular constitution.

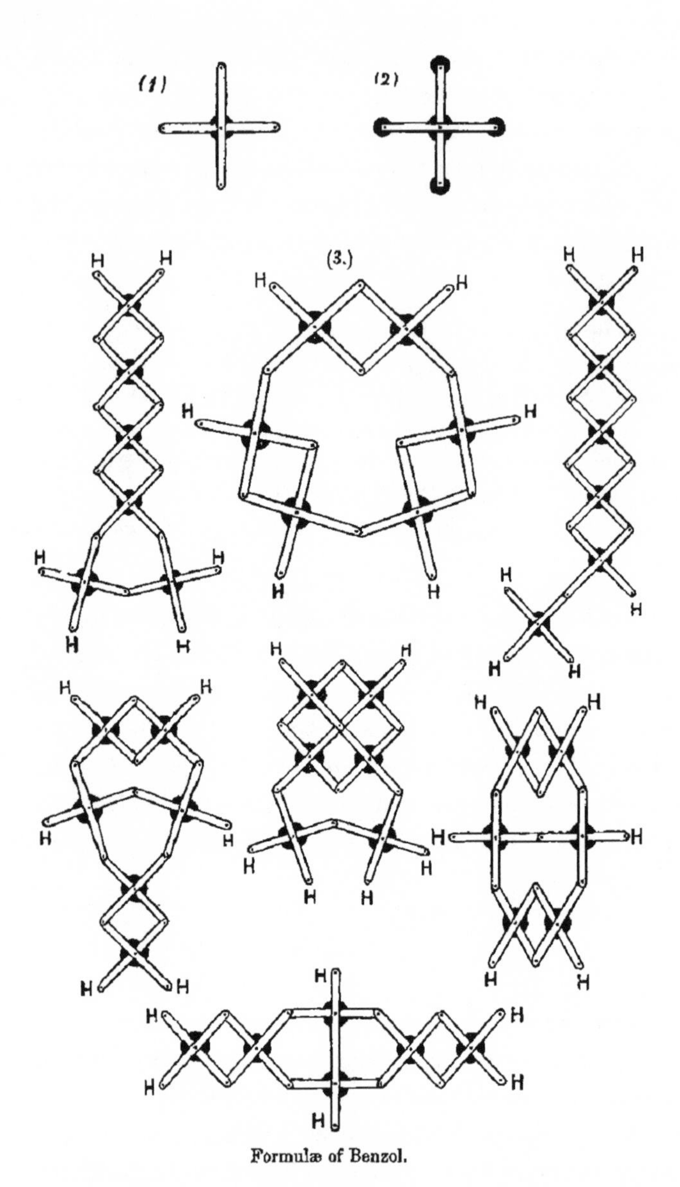

Figure 9.2 James Dewar's brass bar formulae of 1866, used to illustrate the tetravalent carbon (1, 2) and various C_6H_6 combinations, including Kekulé's benzene formula (3) and what is now called Dewar benzene (third row, right). Source: Dewar, "On the oxidation of phenyl alcohol", *Proceedings of the Royal Society of Edinburgh* 6 (1866–67): 85.

In 1861 a new way of representing chemical formulae was introduced by the young Edinburgh medical graduate Alexander Crum Brown to be used with the new structural theory. Adopting an earlier proposal, made by Archibald Scott Couper while working at the chemistry laboratory in Edinburgh, to draw dotted lines between the atomic symbols to denote the valencies, Crum Brown suggested drawing circles around the atomic letters, a move John Dalton had already proposed. This first attempt at a graphic representation of molecular constitution was published in 1864 (Crum Brown 1864; Crosland 1962, chap. 3; Larder 1967; Russell 1971, 100–4).

Although Crum Brown's atomic design did not add anything new in terms of chemical theory, his 'graphic formulae' appealed more vividly to the imagination than the alphanumeric formulae of Berzelius' atomic symbols used so far, even though they were by no means meant to be more realistic. In fact Crum Brown stated that "by [this notation], it is scarcely necessary to remark, I do not mean to indicate the physical, but merely the chemical position of the atoms", alluding to the then familiar distinction between physical (i.e., real) and chemical (i.e., functional) atoms. As to the use of this notation for teaching purposes, he added that "while it is no doubt liable, when not explained, to be mistaken for a representation of the physical position of the atoms, this misunderstanding can easily be prevented" (Crum Brown 1864, 708)—namely by reminding the students of the purely formal nature of these formulae. Correspondingly, they were intended to be read in the plane of the paper and not to imply any spatial relationship.

Inspired by Crum Brown's graphic formulae, James Dewar, another student of Lyon Playfair, professor of chemistry in Edinburgh, prepared a model kit in 1866 (Dewar 1866–67) (Fig. 9.2). A series of narrow thin bars of brass of equal length were taken and clamped together in pairs by a central nut to form an X-shaped unit, the arms of which could be adjusted to different angles. This combination represented a single carbon atom with its four places of attachment. "In order to make the combination look like an atom" [!], Dewar even recommended placing a thin disc of blackened brass under the central nut. Little holes at the ends of the arms allowed one carbon atom to be connected with another. Hydrogen atoms were represented by white discs equipped with a screw to be fixed at the free carbon valencies; and oxygen was a red disc in the middle of a brass bar long enough to fit exactly between the free ends of two carbon valencies.[4] In this way mechanical structures could be assembled and disassembled, but besides being an assembly kit Dewar's model did not add anything new to the graphic notation on paper.

CROQUET BALLS AND MOLECULAR SCAFFOLDING

In 1865 Crum Brown's paper tools were developed into a much more elaborate device by Hofmann in London (Hofmann 1865; Torracca 1991). The context is well documented. On 9 January, Hofmann lectured at the Royal College of Chemistry on the determination of chemical equivalents, and he may have used his type mould models described above. After the lecture, he went to Herbert McLeod, at that time his assistant, and told him "about a new mechanical dodge of his for showing the atomic constitution of bodies" (James 1987, 9 January 1865). The wording in McLeod's diary suggests that no new chemical theory had come to his master's mind, but rather a pedagogical trick to improve his teaching. Six weeks later McLeod recorded attempts at trying "some dodges for colouring the spheres" which Hofmann wanted to show in the next day's lecture (James 1987, 14 February 1865). The test must have been a success, for when Hofmann began to prepare for one of the famous Friday Evening Discourses at the Royal Institution several weeks later, we find him again thinking about mechanical models. On 4 April, McLeod was sent "to the painter to see how he is getting on with the balls and cubes. Told him how some lines are to be painted on the cubes" (James 1987, 4 April 1865).

The model Hofmann presented at his Friday Evening Discourse on 7 April 1865 to a distinguished audience, including the Prince of Wales, the Duke d'Aumale, and the Prince de Condi, was borrowed from one of the most popular games in Victorian England: table croquet (Fig. 9.3). Croquet balls were Hofmann's atoms, painted in white for hydrogen, green for chlorine, red for the 'fiery oxygen', blue for nitrogen, and black for carbon—colour codes still in use today which seem to originate from that remarkable evening lecture. To exhibit the 'combining powers', i.e., the valencies, metallic tubes and pins were screwed into the balls "to join the balls and to rear in this manner a kind of mechanical structures in imitation of the atomic edifices to be illustrated" (Hofmann 1865, 416). Hydrogen and chlorine croquet balls received one single arm each, oxygen two, nitrogen three, and carbon four. In this way chemical formulae could be realised in a three-dimensional,

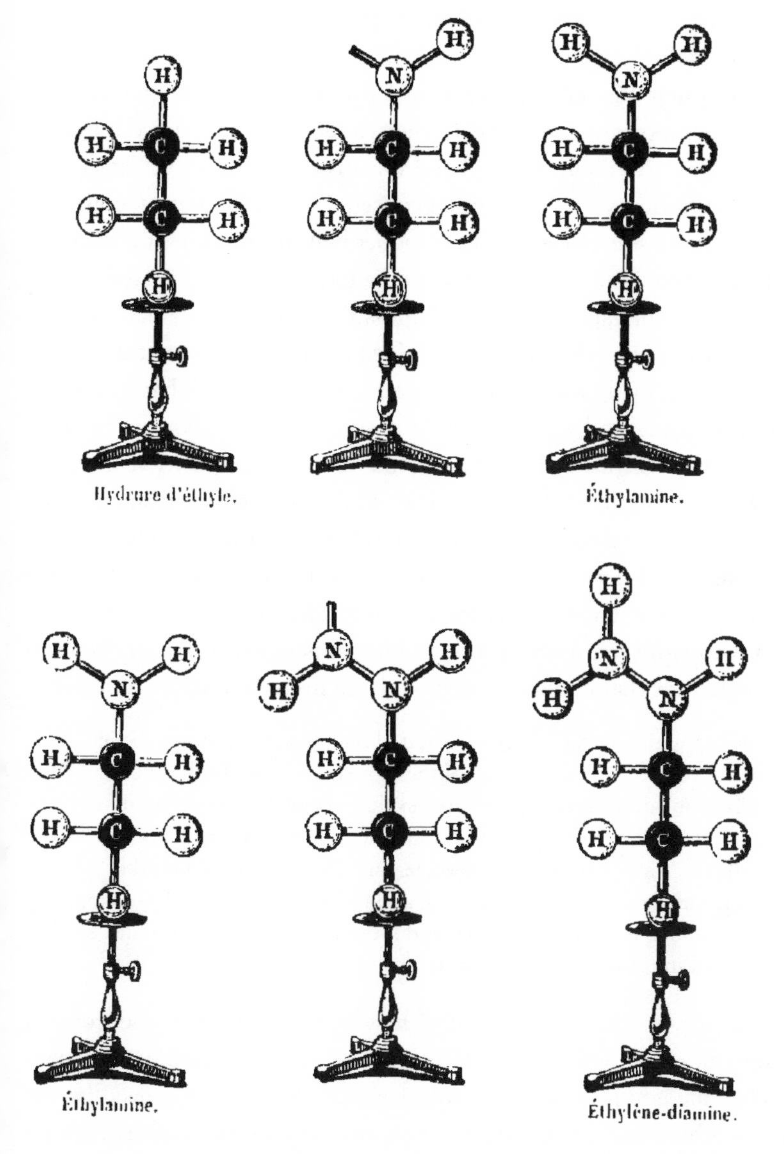

Figure 9.3 August Wilhelm Hofmann's glyptic formulae of 1865: amino derivatives of ethane. Source: Hofmann, "On the combining power of atoms", *Proceedings of the Royal Institution of Great Britain* 4 (1865): 421.

spatial way. They were therefore called "glyptic [i.e., 'sculptured'] formulae" ("Glyptic formulae" 1867, 78).

Hofmann's new model was based on Crum Brown's 1864 publication, but went one important step further. The croquet balls translated a flat arrangement into a three-dimensional and visually more attractive device. Yet its spatial properties were clearly not a consequence of theoretical considerations, but a mere side effect of using croquet balls to turn lines on paper or a blackboard into a mechanical device to be put on the table of the lecture theatre. Consequently, Hofmann's ball-and-stick model maintained two features of its forerunner: the valencies retained the planar symmetrical orientation, and it was by no means meant to represent the physical arrangement of the atoms.

But the chemical theory did not really matter in this Friday Evening Discourse. What Hofmann delivered in front of the powerful and the leisured was not meant as an introduction to organic chemistry. Instead, it was a most carefully composed performance primarily meant to convey the idea of the chemist as someone who knows how to manipulate matter according to his will, and who will eventually be able to build a new world out of chemical building materials that could be assembled and disassembled *ad libitum*.

> The facility with which our newly-acquired building material may be handled, enables us to construct even some of the more complicated substances. . . . We are thus enabled, by availing ourselves exclusively of oxygen as building material, to convert the two-storied molecule of hydrochloric acid successively into a three-, four-, five-storied molecule, and ultimately even into the six-storied molecule of perchloric acid; and there is no reason why a happy experimentalist, by using additional and more complicated scaffolding, should not succeed in raising still loftier structures (Hofmann 1865, 418–19).

The chemist as the architect of a new world: this was the core of Hofmann's message. In this context the ball-and-stick models created a symbolic space, the conquest and control of which the chemist proved by the skilful use of his hands. The lofty and colourful structures of hydrocarbons that covered the desk at the end of the lecture may have conveyed to the audience the imaginary skyline of a world within the reach of man's constructive power: "It is scarcely necessary to expand on these illustrations, and if I venture to raise up a few more of these mechanico-chemical edifices, it is because I want to show you that our building stones are available for many purposes" (Hofmann 1865, 424). Appropriately, the lecture ended by referring to the "sense of mastery and power" associated with the "great movement of modern chemistry" and by invoking "the grandeur belonging to the conception of a world created out of chaos" (Hofmann 1865, 430).

In Hofmann's hands the structural theory of organic chemistry was translated into rules for construction and put into practice. It was a maker's vision of the future of chemistry. Whether his models were true representations of reality mattered little as long as they supplied precepts for chemical syntheses. For Hofmann chemistry was a "magical tree, reaching out in every direction with its branches and twigs and ramifications" (Hofmann 1890, 41; Brock, Benfey, and Stark 1991; Brock 1992; Meinel 1992; 1995) and his research programme followed this same agenda. In the laboratory the assembly-kit principle worked surprisingly well, and the systematic charting of whole classes of substances soon became the standard approach to the chemistry of synthetic dyestuffs and thus to the first major success of a science-based industry.

"ANYTHING BUT ABSTRACT CHEMICAL TRUTH"

Hofmann's Friday Evening Discourse appeared with plenty of illustrations in the 1865 volume of the *Proceedings of the Royal Institution* and was reprinted in the widely read *Chemical News* of 6 October 1865 (Hofmann 1865). In May 1867 the journal *The Laboratory* inserted a brief editorial note advertising a set of Hofmannian "glyptic formulae" for "those teachers who think, with Dr. Frankland and Dr. Crum Brown, that the fundamental facts of chemical combination may be advantageously symbolised by balls and wires, and those practical students who require tangible demonstrations of such facts" ("Glyptic formulae" 1867, 78). Made by a certain Mr. Blakeman of Gray's Inn Road, London, and supplied in a box of 70 balls with brass rods, straight or bent, and some rubber bands, the set was praised for the striking constructions that could be assembled from it, "more likely to rivet the attention of students than chalk symbols on a blackboard" ("Glyptic formulae" 1867, 78).

It is difficult to tell how much such model kits were used. The scarcity of examples in museum collections argues for their being used up in classroom teaching and thrown away afterwards: transient objects that could be made out of cheap materials by even the least skilled laboratory assistant. Accordingly, most surviving molecular model sets have no manufacturer's name, although size, design, and colour codes are very similar. An early boxed set, now kept in Oxford (Fig. 9.4), contains coloured wooden balls

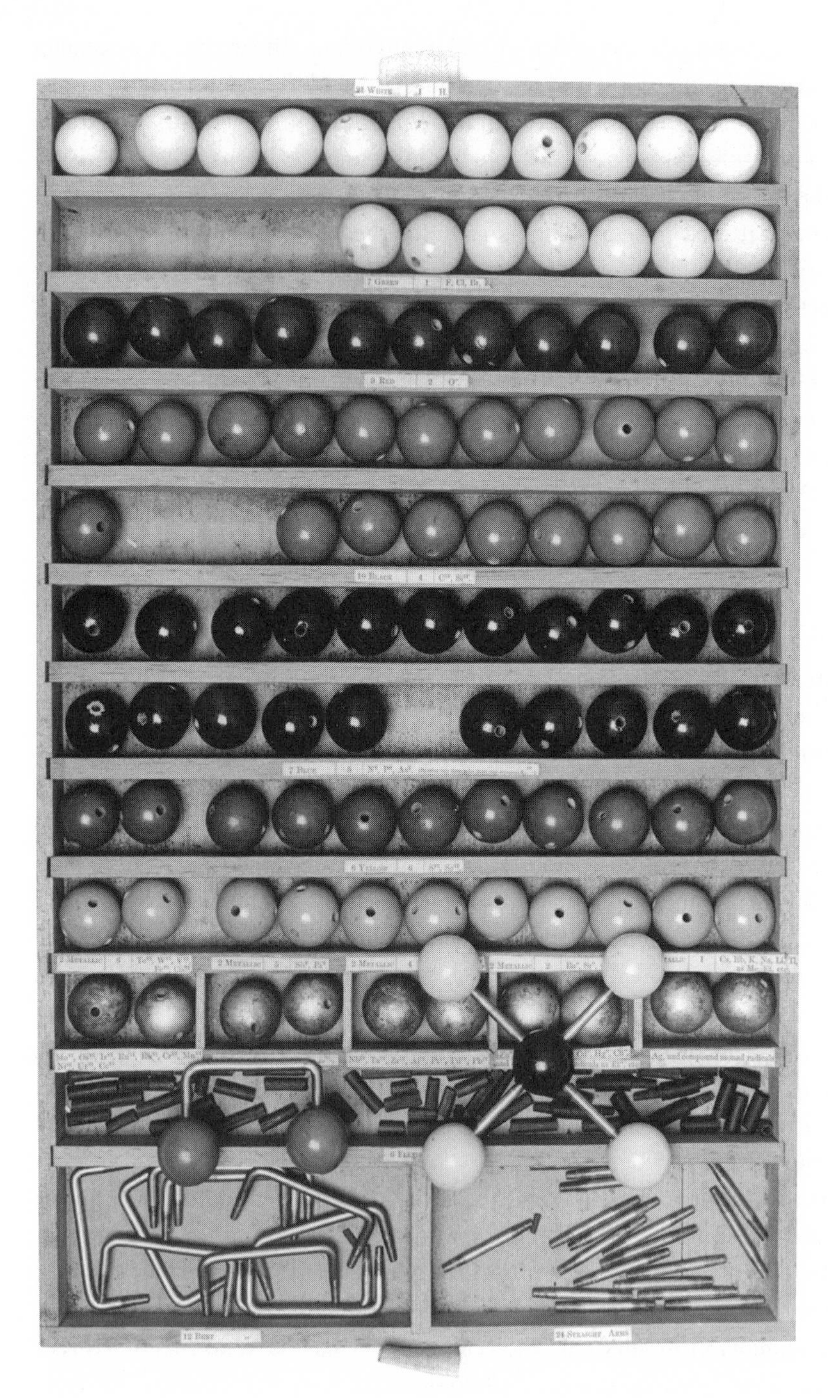

12 Bent
24 Straight Arms
Ag, and compound monad radicals

of 30 mm diameter to represent hydrogen (white), halogen (green), oxygen (red), carbon or silicium (black), nitrogen or arsenic (green), sulphur or selenium (yellow), and the metals (silver). Holes drilled into the atom balls mark the respective number of binding units to be joined mechanically by means of the straight, bent, or flexible arms contained in the same box.[5]

Contemporary debates about the use of such models in chemistry courses are almost non-existent, and in general very little is known about the methods of classroom teaching at that time. Molecular models may have been popular for disseminating scientific education, but it is likely that their adoption met with scepticism from the beginning. Even the very first advertisement—most unusually for this genre—contains two warnings. The first makes clear that it might be dangerous to mix up serious science and children's toys: "At first sight, the collection of bright-coloured and silvered balls suggests anything but abstract chemical truth, and a very young philosopher might excusably convert them to purposes of exclusively recreative science." The second warning is even more interesting, for the unknown author writes somewhat cryptically: "Whether they [these models] are calculated to induce erroneous conceptions is a question about which much might be said" ("Glyptic formulae" 1867, 78).

Physical models were indeed looked upon with suspicion by many a scientist. In a meeting of London's Chemical Society on 6 June 1867, Benjamin Brodie, the most ardent critic of atomism in Britain, ridiculed the ball-and-stick model as a materialistic bit of joiner's work. Referring to Mr. Blakeman's kit advertised in *The Laboratory*, he continued:

> [T]he promulgation of such ideas—even the partial reception of such views—indicates that the science must have got, somehow or another, upon a wrong track; that the science of chemistry must have got, in its modes of representation, altogether off the rules of philosophy, for it really could only be a long series of errors and of misconceptions which could have landed us in such a bathos as this (Brodie 1867, 296; Brock 1967).

←

Figure 9.4 Hofmann's glyptic formulae, circa 1870. The case (630 × 310 × 60 mm) contains 109 wooden atom balls (diameter 30 mm) plus 24 straight, 12 bent, and 6 flexible arms. Source: Museum of the History of Science, Oxford, inv. no. 42347 (for further pictures and descriptions, see Hill 1971, no. 395; Turner 1983, fig. xix; and Turner 1991, 295).

Two years later William Crookes, a former assistant of Hofmann and at the time well known as an independent chemist and experimenter, advised an unknown correspondent of *Chemical News*:

> As you are a student of chemistry, take our advice, and leave atoms and molecules alone for the present. Nobody knows how the atoms are arranged in elements of different atomicities. Graphic formulae, diagrams, etc., are only artificial aids to fix certain properties of bodies on the memory; but no one intends them to represent the architectural plan and elevation of the body. Avoid theory: stick to experiment (Crookes 1869).

Edward Frankland on the other hand, Hofmann's successor at the Royal College of Chemistry, was one of the early converts to using models in teaching. His *Lecture Notes for Chemical Students*, written on the basis of lectures given in the winter of 1865–66, was the first textbook to use Crum Brown's graphic formulae systematically—despite the problems they created for the printer (Russell 1996, chap. 10). Frankland is also known to have used Hofmann's ball-and-stick models in the classroom, as he believed that their pedagogical advantages would outweigh their epistemological deficiencies.

> I am aware that graphic and glyptic formulae may be objected to, on the ground that students, even when specially warned against such an interpretation, will be liable to regard them as representations of the actual physical position of the atoms of compounds. In practice I have not found this evil to arise; and even if it did occasionally occur, I should deprecate it less than ignorance of all notion of atomic constitution (Frankland 1866, v–vi).

For similar reasons the cautious use of models was recommended "for lecture illustrations" by Carl Schorlemmer, the first professor of organic chemistry at Owen's College in Manchester. In this context Schorlemmer explicitly referred to the use of globes in geography; there was no danger that any student "should acquire curious notions about the brazen meridian, or the wooden horizon". Nevertheless he recalled that, due to the naïve use of atomic models in chemistry courses, "it happened indeed that a dunce, when asked to explain the atomic theory, said: 'Atoms are square blocks of wood invented by Dr. Dalton'" (Schorlemmer 1894, 117).

As a rule, these early molecular models appeared in the context of teaching, not research, and they seem to have been fairly popular—at least in Britain (Russell 1996, 284–303). On the European continent the situation was different. In France the atomic theory was a minority view vigorously opposed by the powerful Paris-based schools of Jean-Baptiste Dumas and Marcellin Berthelot. There was an early French translation of Hofmann's 1865 Royal Institution Discourse, issued by the prolific Jesuit populariser and physicist Abbé François Moigno (Hofmann 1866), but this favourable reception remained an exception. Even Adolphe Wurtz, one of the few defenders of chemical atomism in France, did not adopt the new visual aids—with one notable exception: At the 1876 meeting of the French Association for the Advancement of Science Wurtz used them in front of a general audience; but after the lecture he admitted in a somewhat sceptical tone, "I have constructed this formula [rosaniline] from black, white, green balls, which represent the atoms of carbon, of hydrogen, of nitrogen. They understood that, or they believed they understood, for they clapped. I am almost proud of this success for the theory".[6]

In Germany, too, the prevailing attitude required that science should stick to facts and data produced in the laboratory, and refrain from speculation which was still tainted by association with Romanticism. The usual distinction between chemical and physical atoms provided a common denominator for those who did not want to engage in metaphysical debates about the existence of atoms, but rather sought to pursue chemistry pragmatically (Nye 1989; Görs 1999). For the same reason most publications continued to use the old empirical formulae, which gave only the quantitative composition, or a modified type-theory notation. It is interesting to note that even Hofmann, when he returned from England in late 1865, turned to a facts-oriented, atheoretical way of teaching, and there are no hints that he ever used glyptic formulae in his Berlin lectures.

However, the issue was never discussed publicly. In a private letter only Hermann Kolbe, professor of chemistry in Leipzig and one of the most ardent opponents of structural chemistry, having received Frankland's *Lecture Notes* from the author, replied by appealing to the most weighty metaphysical argument, the biblical "thou shalt not make unto thee any graven image":

> Frankly, I believe that all of these graphic representations are out-of-date and even dangerous: dangerous because they leave too much scope for the imagination, as for example happened with Kekulé: his imagination bolted with his understanding long ago. It is impossible, and will ever remain so, to arrive at a clear notion of the spatial arrangement of the atoms. We must

therefore take care not to think of it in a pictorial way, just as the Bible warns us against making a sensual image of the Godhead.[7]

BREAD ROLLS AND SAUSAGES

To what extent such protestations were but part of a widespread rhetoric without being necessarily characteristic of the way chemists thought and practised their science in private, is a difficult question to answer. There was clearly much pressure within the discipline to disclaim realism and to pay lip-service to a stick-to-the-facts attitude. The most notable exception was the group of young chemists who gathered in Kekulé's laboratory at the University of Ghent in Belgium.

By this time Kekulé had established the doctrine of constant valency and the tetravalent carbon atom; he had made major contributions to the theory of molecular structure and was struggling with the benzene problem (Gillis 1967; 1996; Russell 1971, 61–71, 100–7). For it was one of the challenges to the new structural chemistry that it account for the peculiarities of non-saturated and aromatic hydrocarbons. Kekulé's 'bread rolls' or 'sausage formulae', first introduced in his Heidelberg lectures of 1857/58, represent the number of affinity units of the individual atoms by the length of the sausage. Multiple bonds—and this was the advantage of this notation—could now be symbolised by the lateral contact of several valency units. Through Kekulé's *Textbook of Organic Chemistry*, which began to appear in German in 1859, these formulae reached a wider audience. They were adopted in a modified form by Wurtz (Wurtz 1864, 132–38), reappeared in 1865 in Kekulé's sausage formula of benzene (Kekulé 1865; 1866) and were used systematically the same year in Alfred Naquet's *Principles of Chemistry Based on the Modern Theories* as a kind of graphic algorithm or "algebraic scheme" to map the number of possible molecules from a given number of atoms (Naquet 1865).

In Kekulé's view aromatic compounds contain a nucleus of six carbon atoms that form a closed chain of alternating single and double bonds. In print this was rendered as a linear chain of a length corresponding to the number of valencies, with markers on the two unsaturated positions at each end to indicate that they link to make a closed ring. It goes almost without saying, however, that the odd sausage-shaped form of the carbon was not meant to represent the true form of the atom, as this would have clearly offended the traditional iconography of the atom as a little sphere (Lüthy 2000).

It is likely, therefore, that Kekulé favoured the more abstract, linear representation in order to avoid a too-realist reading of his sausage formulae (Kekulé 1861–66, 1: 159n).The original version, however, was more tangible. Already in the 1857 Heidelberg lectures Kekulé had used 3-D models assembled from well-turned wooden balls; he used the same type of device when he discussed the benzene hexagon in 1865 (*Tussen kunst* 1992, 93) (Fig. 9.5).

Thus between 1860 and 1865 a number of chemists were engaged in the search for a new visual language that would give mental images to the new structural chemistry. Paul Havrez's curious speculations about molecular symmetry, which took Kekulé's benzene model as their point of departure (Havrez 1865; Kekulé 1861–66, 2: 515n; Heilbronner and Jacques 1998a and b), and Joseph Loschmidt's "Constitutional formulae in graphic representation", designed after a quasi-cosmological idea of atomic interaction with different "spheres of action" (Loschmidt 1861; Wotiz 1993; Schiemenz 1994; Eliel 1997; Heilbronner and Hafner 1998), belong to the same context. Though clearly not meant to represent physical reality, these were attempts to overcome the merely stoichiometric and a-visual tradition of chemistry and tentatively to inscribe the new notions of valency and molecule into a form that would give them, in Loschmidt's words, "immediate figurative quality" (*unmittelbare Anschaulichkeit*) (Loschmidt 1861, 4).

In 1865 Kekulé adopted Crum Brown's graphic formulae and their didactic counterpart, namely Hofmann's ball-and-stick models. The latter may have been brought from London by Kekulé's assistant Wilhelm Körner, who went there to buy laboratory equipment for his master. From the practical point of view, however, Hofmann's model was deficient in one regard: multiple bonds, a key problem for the doctrine of valency in explaining the structure of non-saturated and aromatic compounds, were difficult to realise with the stiff metal joints. Commercial Hofmann/Blakeman model kits provided U-shaped bridges or rubber joints for this purpose, but the resulting structures were not easy to assemble. In addition, the use of two different sorts of bonds, namely the short straight line between the two atomic centres and the much longer curved bond that bridged them, added another question mark to an already questionable model. Furthermore, Hofmann's glyptic formulae did not really provide what Kekulé was looking for

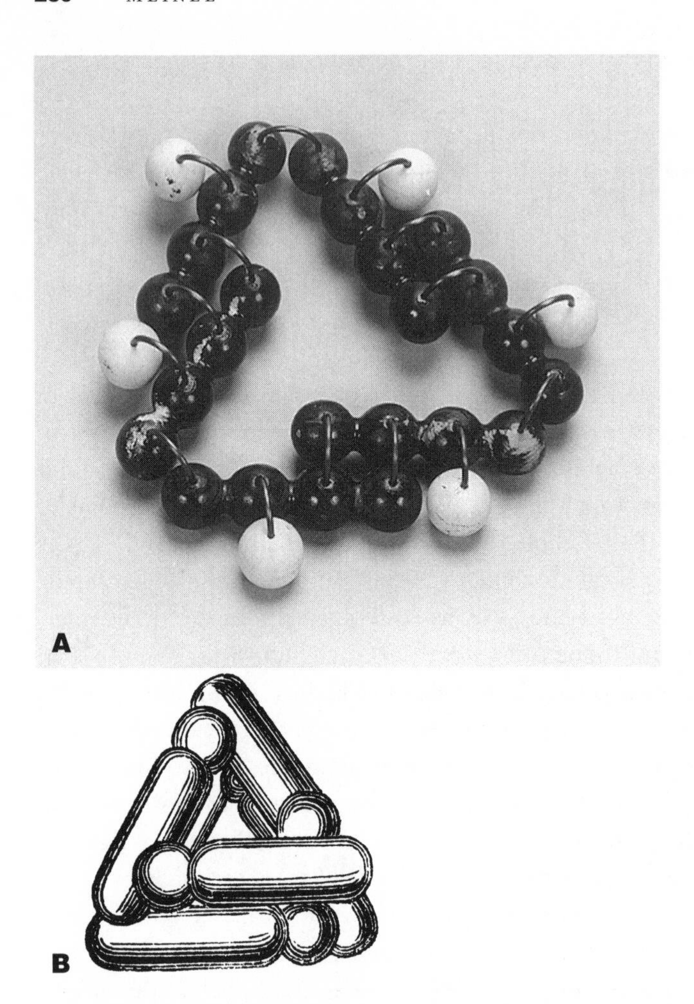

Figure 9.5 (A) August Kekulé's benzene model of 1865 and (B) Paul Havrez's graphic representation in Kekulé's textbook. The four-membered C-'sausages' are made from one piece of wood, the connecting tubes are of metal; total dimensions of the benzene ring circa 15 × 15 × 2 cm. Sources: Museum Wetenschap en Techniek, Ghent, inv. no. MW95/118; Kekulé, *Lehrbuch der Organischen Chemie*, vol. 2 (Erlangen, 1866), 515.

at the time: representations of molecular structure that would show spatial arrangements—if not necessarily in a realist sense. The English ball-and-stick models were three-dimensional in the restricted sense that they translated drawings on paper into a bigger, and hence more visible and more attractive, lecture-hall device. But with their planar valencies Hofmann's glyptic formulae only appeared to be spatial, whereas, if taken seriously, everything remained in the plane and the model offered no real advantage compared to the drawing (Kekulé 1867, 218).

ARMS, SOCKETS, EYES, AND THE CONSTRUCTION OF THREE-DIMENSIONALITY

In late 1866 Kekulé received a set of Dewar's brass-bar models from Playfair in Edinburgh, and arranged for the young author to spend the next summer in his Ghent laboratory. In the same year the university hired Theodor Schubart as a new mechanic to work for Kekulé. During the spring of 1867 Kekulé, Körner, and Schubart were busily experimenting with molecular models intended for teaching purposes. In April 1867 Körner wrote to Kekulé: "Your benzene models are all painted, and Schubart has finished the stand. . . . The little spheres are as colourful as Easter eggs, and I am waiting for your order to send them. They can easily be packed in a small cigar box".[8] These new models took Hofmann's glyptic formulae as their point of departure. But the latter were modified such that the valency arms of the black carbon balls were no longer in a planar orientation but pointed to the edges of a tetrahedron. When their lengths were appropriately chosen so that the free ends were at equal distances, double and even triple bonds could be realised without bending (and possibly breaking) the thin brass tubes. The touching arms were joined by a straight slit-tube fastener or, in the case of angular bonds, by small brass sockets that could be joined by a ring looped through little eyes drilled through their ends. The white hydrogen and green chlorine atoms received brass tubes of an appropriate size so that the valency arms of the carbon could be plugged into them. In order to put the entire molecule on the lecture-hall desk some of the carbon atoms had little sockets to fix them on a firm stand (Anschütz 1929, 1: 356).

It seems as if in the beginning the introduction of the tetrahedral carbon atom was merely a technical trick to improve the joining of their valency sticks and to give the model a spatial appearance. There were of course ideas about molecular symmetry in the background, and Kekulé may also have recalled that Alexander Butlerov, a colleague from his years in Heidelberg, had tentatively suggested a tetrahedral carbon five years earlier. Despite this, Kekulé's new model was not introduced in order to solve any theoretical problems. Rather, it offered a solution to a mechanical problem and to the quest for *Anschaulichkeit*. And it was to a partly non-chemical audience that Kekulé presented his model for the first time at the assembly of German naturalists and doctors in Frankfurt am Main in 1867 (*Tageblatt* 1867, 95, 114).

FROM TEACHING AID TO RESEARCH TOOL

Designed as didactic devices, Kekulé's molecular models soon turned into research tools that could be used to interpret and guide chemical reactions. The first application in a research context was Kekulé's own paper on trimethyl benzene, the formation of which from three acetone units could thus be demonstrated convincingly (Kekulé 1867, 218). This same paper presented Kekulé's new model for the first time in print (Fig. 9.6A), and one also learns that the formulae in the article were drawn after a physical model (Fig. 9.6B).

Once chemists had become used to this form of representation and learned to read spatial meaning, the new model could be applied to understanding unknown mechanisms and predicting possible reactions by giving them visual plausibility. As a consequence, we find these models being used by students and pupils of Kekulé who had been accustomed to looking at, and to thinking of, molecules in this new spatial manner.

As early as November 1867 Körner wrote enthusiastically to his master that, according to the model, there should be five isomers in a given compound instead of three as had been previously believed.[9] Two years later, he applied the model to a major deficiency of Kekulé's benzene formula. A hexagon with alternating double and single bonds would have yielded two different bi-substituted derivatives, depending on whether the substituents had a single or a double bond between them. Such isomers, however, do not exist and the formula needed to be reinterpreted. Körner solved the problem by altering the spatial configuration of the benzene nucleus. He abandoned the flat ring and linked each carbon atom with three others to yield a space-filling structure. The crucial innovation was the way he made use of the model. In order to test the steric possibility of the resulting structure, Körner demonstrated its feasibility by assembling his 3-D benzene from a ball-and-stick model (Koerner 1869; Schütt 1975; Paoloni 1992) (Fig. 9.7).

The step from interpretation to prediction by means of the model was achieved in the same journal in a paper by Emmanuele Paternò, Körner's friend and a fellow assistant in Stanislao Cannizzaro's chemical laboratory

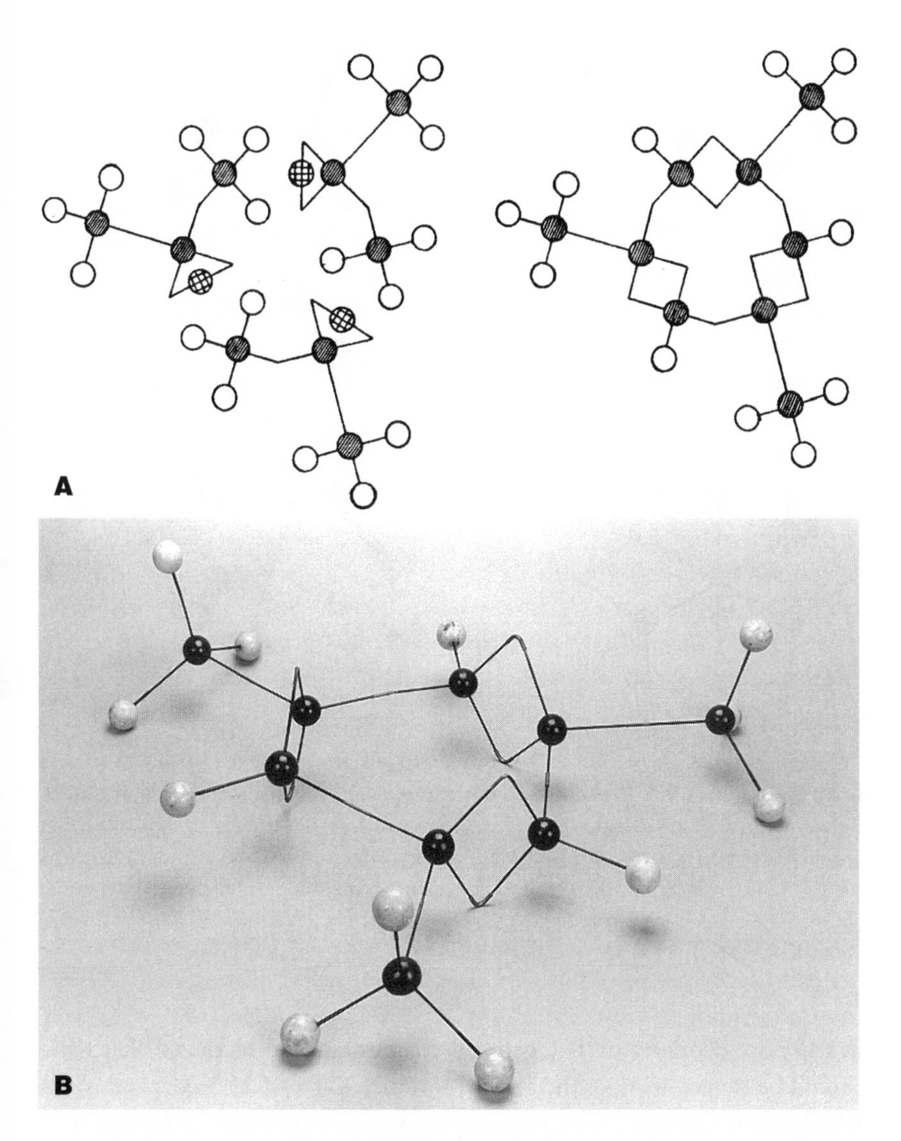

Figure 9.6 Mesitylene (1,3,5-trimethyl benzene), which is made of three molecules of acetone, as illustrated by August Kekulé using a tetrahedral carbon model. (A) Kekulé's figure, from "Über die Constitution des Mesitylens", *Zeitschrift für Chemie*, new ser., 3 (1867): 218. (B) An assembled mesitylene model, measuring circa 60 × 60 × 15 cm. Source: Museum Wetenschap en Techniek, Ghent, inv. no. MW95/116.

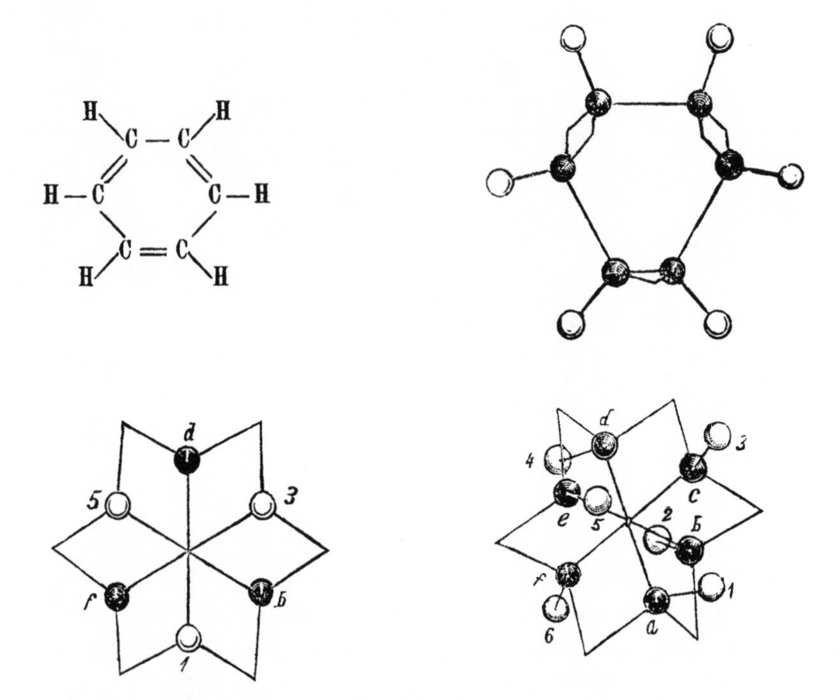

Figure 9.7 Wilhelm Körner's benzene model, presented in relation to that of Kekulé. Top: Kekulé's benzene formula realised by his tetrahedral carbon model. Bottom: Körner's proposal with the carbon atoms (dark spheres) roughly in what would today be a 'chair' conformation, three hydrogen atoms above and three below the C_6H_6 nucleus. Source: Koerner, "Fatti per servire alla determinazione del luogo chimico nelle sostanze aromatiche", *Giornale di Scienze Naturali ed Economiche di Palermo* 5 (1869): 237, 241.

in Palermo. Working on halogenated ethanes Paternò predicted that there should be two isomers of the general formula $XH_2C–CH_2X$, depending on whether the halogen atoms point in the same or different directions with respect to the tetrahedral carbon. If this were true the isomers could, at least in principle, be isolated by appropriate measures (Paternò 1869). In this argument, however, the use of the model was pushed much further than had been originally intended. The original ball-and-stick models were meant to represent only the atoms, their valencies, and their arrangement. They were essentially ball models, and the sticks just mere auxiliary means of denoting the number of valencies and joining balls that would otherwise have fallen apart. Paternò took the sticks seriously, so to speak, and imagined from their stiffness that the C–C bond would likewise be stiff. For his two isomeric 1,2-dibromoethanes would exist only if rotation about the carbon–carbon bond was restricted.

At a time, however, when the realist interpretation of a molecular model was still extremely controversial, no other chemist would have gone so far as to speculate about the rotation of bonds, which were, after all, not yet generally considered as physical entities. One should also bear in mind that the very publication of such speculations, done under Cannizzaro's patronage at the periphery of Europe, would have been much more difficult in the centres of nineteenth-century chemical orthodoxy such as Germany or France.

When Jacobus Henricus van't Hoff, who had been working with Kekulé in 1872–73, published his *Arrangement of Atoms in Space* in 1874 and thereby established stereochemistry, he could not only build upon an existing tradition of using 3-D models as didactic devices in chemical teaching, but may also have known about the first attempts to derive stereochemical consequences from them. We have no trace of an immediate influence, but it is evident that the problems van't Hoff was about to solve belong to the common context of understanding molecular complexity by introducing a new visual language to deal with the spatial properties of molecules. The solution he offered was different from the one Kekulé had proposed seven years earlier, but it is not unreasonable to assume that the model kits used in that group prepared him to think of chemical molecules in a visual way and in terms of three-dimensional structures.

New views do not emerge at once, nor do they spring from a single discovery. The chemists prepared 3-D models for teaching purposes, and in using them they learned to link the mind's eye with theoretical notions, with the manipulating hand, and with laboratory practice. Still, it took almost a decade before this gave birth to the new steric conception of matter. As there was surprisingly little discussion of epistemological questions and theoretical consequences of this approach, our story cannot be subsumed under the common nineteenth-century predilection for mechanical models in physical explanations; though these were a matter of vivid debate. The introduction of 3-D molecular models was not exclusively, nor even predominantly, part of a theoretical discourse, as is often assumed in the literature. Instead, the models were primarily used as tools for the creation of new types of *Anschauung* not only in the audiences taught, but also in the minds of those who developed these tools in struggling with the growing complexity of chemical constitution. And this approach was not peculiar to chemistry, but part of a more comprehensive change in perceiving the world and in dealing with it in a purposeful manner.

CONSTRUCTION THINKING AND THE CONQUEST OF SPACE

Reducing complexity to the simplicity of hidden structures and using structural thinking in a constructivist way was not restricted to science. The nineteenth century was a culture of construction (Ferguson 1992; Peters 1996; König 1997). Modes of thinking peculiar to engineers and architects established their primacy over traditional scientific or humanistic thought. Builders' thinking deals primarily with how to make; it mediates between concepts of form, methods of science, and practical ways of dealing with materials. Space is one of its main concerns, as the spectacular iron constructions of the time so impressively testify.

Beginning in the late 1860s a discussion began among British engineers and architects about the primacy of an 'aesthetics of construction' over an 'aesthetics of decoration'. London's new St Pancras Station, built in 1869 by George Gilbert Scott and W. H. Barlow for the Midland Railway Company with the widest span of any roof then in existence, was the prime example of a new type of structural and constructivist building by means of mouldings and standardised parts. At the same time the pages of the London weekly *The Building News and Engineering Journal* testify that, interspersed in between articles on gothic and neo-Palladian architecture, suddenly a new type of mostly anonymous article appeared, devoted to topics such as the primacy of construction, the use of moulds, or the visibility of natural laws in the structures of buildings as first and foremost exemplified by iron bridges and railway stations. In one of the rare programmatic contributions of this kind, Britain's prevailing architectural taste was criticised as lacking true creativity and beauty. This could only be based upon the principles of science and construction, and if their devotees

> were to begin by an experimental study of stones, and bricks, and timber, investigating their qualities, and making tentative efforts of combination, or exercising their minds with at least the simpler ideas of form and construction, investigating the properties of geometrical solids, cones and cylinders, . . . and other mechanical forms of construction, we should have a race of intellectual architects, whose minds, trained practically in the school of thought and

> invention, would outrival in their efforts all existing art by the very strength and vigour of their competitors ("Theory of the Arts" 1871, 209).

The builders' approach was by no means confined to architecture. A few years after the spectacular opening of Joseph Paxton's famous Crystal Palace, built for the Great Exhibition of 1851, and more than a decade before Hofmann invented his ball-and-stick model, toy boxes were sold in London and elsewhere, that enabled children to create a variety of polygonal forms by connecting peas or balls of wax by means of toothpicks (Brosterman 1997, 84) (Fig. 9.8).

Construction kits of this type originated with *Kindergarten* pedagogy. They were particularly designed to enable children to acquire a first notion of space (Brosterman 1997, 84; Ronge and Ronge 1855). According to Friedrich Fröbel's developmental psychology young children acquire their knowledge about the external world empirically by actively manipulating its particulars. Trained as an architect and later a student of physics, chemistry, and mineralogy, Fröbel was an assistant to Christian Samuel Weiss, the founder of modern crystallography, before he abandoned science and turned to pedagogy. In 1837 he opened the first *Kindergarten* for early childhood education. The *Kindergarten* movement spread rapidly through most of Protestant Germany, and after the failed revolution of 1848, liberal emigrée women brought it to Britain and the United States.

The foremost educational aids of the *Kindergarten* were geometric toys, meant to support the child's self-activity in exploring the world. Fröbel's typical *Kindergarten* 'gifts' (*Gaben*) were geometrical bodies, such as spheres, cylinders, and cubes, that would enable the child to apprehend and represent the external world and thus to train the mind's eye. By actively handling physical objects the invisible was thus to be grasped in a visible form, and an inner vision (*Anschauungsform)* of the world and the self would emerge (Fröbel 1974, 129).

Brick boxes made of little wooden cubes were particularly successful. From the 1840s Fröbel kits were commercially produced and became part and parcel of the *Kindergarten* movement.In the 1850s major international firms such as Myers & Co. in London and Milton, Bradley & Co. in Springfield, Massachusetts, started to produce them for a growing international market. Even more successful were Gustav Lilienthal and his brother Otto, the aeroplane pioneer, who, in 1875, invented a process for making artificially coloured stones for brick boxes that were mass produced and sold

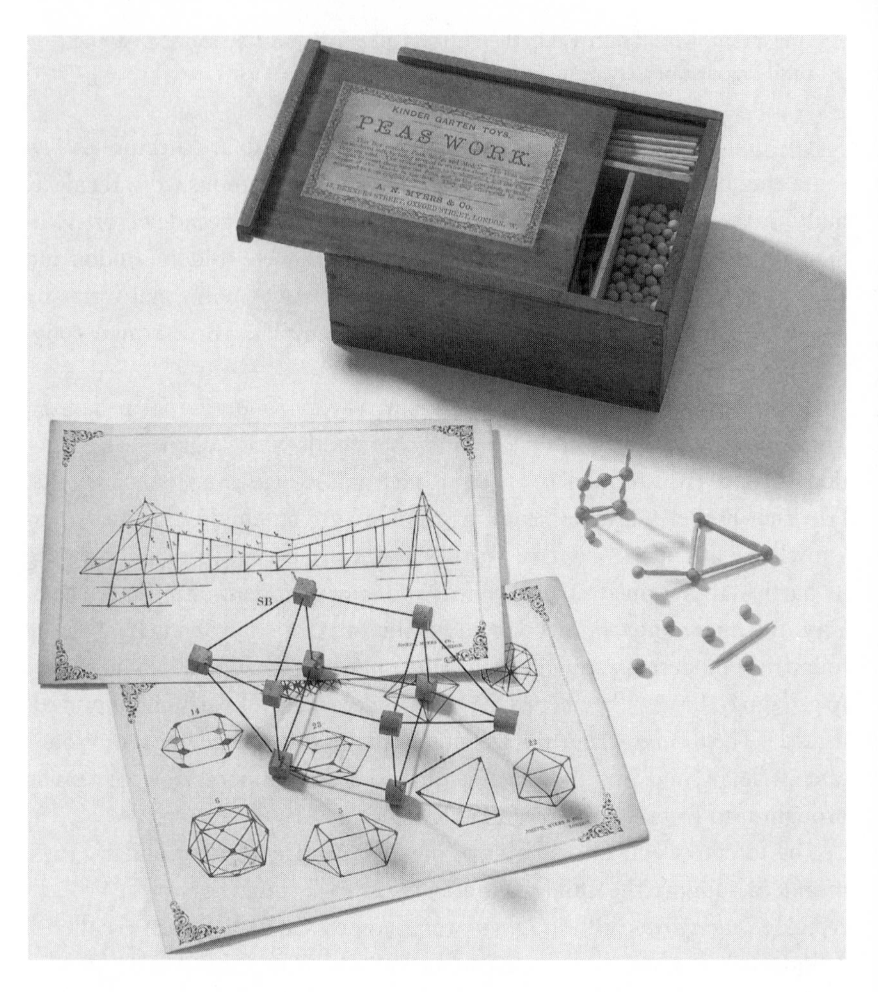

Figure 9.8 In the early 1850s the 'Peas Work' was added to the Fröbel gifts in order to build three-dimensional structures. Top: a box of peas and toothpicks made by A. N. Myers & Co. in London. Bottom: a more elegant version with little cubes of cork manufactured by Joseph, Myers & Co. in London, circa 1855. Source: Norman Brosterman, *Inventing Kindergarten* (New York, 1997), 84.

worldwide by F. A. Richter & Co. in Rudolstadt, Thuringia, under the trade mark *Anker-Steinbaukasten*—a precursor of today's *Lego* (Noschka and Knerr 1986; Leinweber 1999).

As far as appearance, use, and implicit philosophy are concerned, the correspondences between Hofmann's type moulds, his ball-and-stick models, and the Fröbel gifts are striking. Both were symbolic tools for exploring abstract structures and training the mind's eye by manipulating physical objects. Both had their own syntax built into their mechanical joints, and both carried an implicit message that went beyond epistemology: the bourgeois ideal of taking possession of space by constructing it.

Of course we do not need to assume that Hofmann and his fellow chemists took *Kindergarten* toys as their source of inspiration. Yet at the time it would have been almost impossible not to be familiar with their existence and meaning. The sheer explosion of assembly kits during the second half of the nineteenth century testifies to the spread of construction thinking through Europe. The Fröbel gifts and the molecular models were part of this same movement.

CONCLUSION

Models have various messages and can be put to various uses. Richard Buckminster Fuller, the architect who invented the geodesic dome and whose name is commemorated in the football-shaped $C_{60}H_{60}$ fullerene structure first identified in 1985, recalled having used Fröbel's 'Peas Work' haptically to discover structures and construction principles he had been unable to grasp otherwise because of his bad eyesight (Brosterman 1997, 84). Hofmann, the chemist who initially wanted to become an architect before he was drawn into chemistry by Justus Liebig, used painted croquet balls instead of vulgar peas to present Victorian gentlemen with the vision of he chemist as the builder of a new world out of man-made materials. Kekulé, he too a would-be architect in his early years, converted the planar ball-and-stick models into truly space-filling constructions, but the tetrahedral carbon atom, by which this was achieved, was initially a mere trick to improve both the appearance and the joining of the model. In all of these cases the models have a life of their own. They are neither mere representations of scientific theories or data, nor are they purely practical tools. This partial autonomy, which is partly embedded in their physical structure, is a tricky thing, for it may give birth to developments not intended by those who made these models. At the same time—and this seems to be peculiar to chemistry—they provide a material link between theoretical notions, chemical reactions, and the body and gestures of the chemist.

Throughout the period considered in this chapter, the meaning of these models remained liable to more than one interpretation and more than problematic from an epistemological point of view. Yet it seems plausible that this very ambiguity explains their success. For models mediate between audiences without dividing them as theoretical or ideological language would do. They mediate between the mind and the hand; they symbolically connect what the chemist thinks with the substances in his flasks and the operations he performs with them. But models also mediate between teacher and pupil, and between expert and general public. In doing so, they transmit a variety of messages, both explicit and hidden (Hoffmann and Laszlo 1991; Laszlo 1993; Schummer 1999–2000). And as their language was in accordance with the constructivist thinking prevailing at the time, these messages were easily understood even without being publicly discussed. In this way the molecular models invented by chemists in the 1860s created a symbolic and gestic space into which theoretical notions, bodily actions, cultural values, and even professional claims could be convincingly inscribed. The models could then be used to conquer new spaces of possibility by those second-generation students of chemistry who had become accustomed to thinking and working in three dimensions.

NOTES

1. The Science Museum, London, has ball-shaped atomic models said to have been made for Dalton, at his suggestion, around 1810 (inv. no. 1949-21; depicted in Thackray 1970, 266, fig. XII). There is another set of wooden balls of various sizes, believed to have been used by Dalton, with holes drilled to allow them to be joined by pins (Museum of Science and Industry, Manchester, acc. no. 1997.6.53).

2. No. 195 lists the items individually: carbon (black), hydrogen (pale blue), oxygen (scarlet), nitrogen (white), chlorine (pale green), sulphur (yellow), phosphorus (pink), light metals (gold bronze), heavy metals (copper bronze), organic radicals (blue and black), and neutral gases (brown).

3. For similar items, see Hill 1971, 62–63; and Ramsay 1974, fig. 4 and note 24.

4. A set given to Kekulé by Lyon Playfair in 1866 is said to have survived in Bonn until 1925 (Anschütz 1929, 1: 357), but none seems to be extant.

5. There is another surviving kit of circa 1870 from the laboratory of Thomas McLachlan, a London consultant. The black carbon balls are drilled with four holes giving four planar bonds; but one of the black balls has extra holes drilled into it to create a tetrahedral carbon atom (Science Museum, London, acc. no. 1964-495). The Science Museum also has a hybrid form of Hofmannian glyptic formulae with tetrahedral carbon atoms in a beautiful mahogany case made in Spain. The maker's plate inside the lid reads "Coleccion para demonstrar las Combinaciones Quimicas

según A. W. Hofmann, constructor Gonzalez Verdiguier, Madrid" (Science Museum, London, acc. no 1977-126; reproduced in Knight 1997, 385).

6. Wurtz to Auguste Scheurer-Kestner, 29 August 1876, Bibliothèque Nationale et Universitaire de Strasbourg, ms. 5983, Correspondance d'Auguste Scheurer-Kestner, fol. 466. Alan Rocke kindly pointed me to this source.

7. Kolbe to Frankland, 23 July 1866, as translated in Rocke 1993a, 314; see also Kolbe to Frankland, 9 July 1867, quoted in Russell 1996, 285.

8. Körner to Kekulé, 25 April 1867, Kekulé-Archiv, Technische Universität Darmstadt.

9. Körner to Kekulé, 4 November 1867, ibid.; see also Klooster 1953.

REFERENCES

Anschütz, Richard. 1929. *August Kekulé*, 2 vols. Berlin: Verlag Chemie.

Berthelot, Marcellin. 1860. *Chimie organique fondée sur la synthèse*. Paris: Mallet-Bachelier.

Brock, William H. 1967. *The Atomic Debates: Brodie and the Rejection of the Atomic Theory*. Leicester: Leicester University Press.

———. 1992. "Liebig's and Hofmann's impact on British scientific culture." In *Die Allianz von Wissenschaft und Industrie. August Wilhelm Hofmann (1818–1892)—Zeit, Werk, Wirkung*, ed. Christoph Meinel and Hartmut Scholz. Weinheim: VCH.

Brock, William H., O. Theodor Benfey, and Susann Stark. 1991. "Hofmann's benzene tree at the Kekulé festivities." *Journal of Chemical Education* 68: 887–88.

Brodie, Benjamin C. 1867. "On the mode of representation afforded by the chemical calculus, as contrasted with the atomic theory." *Chemical News* 15: 295–302.

Brooke, John H. 1976. "Laurent, Gerhardt, and the philosophy of chemistry." *Historical Studies in the Physical Sciences* 6: 405–29.

Brosterman, Norman. 1997. *Inventing Kindergarten*. New York: Harry Abrams.

[Crookes, William]. 1869. Letter to G. E. D. *Chemical News* 17: 84.

Crosland, Maurice P. 1962. *Historical Studies in the Language of Chemistry*. London: Heinemann.

Crum Brown, Alexander. 1864. "On the theory of isomeric compounds." *Transactions of the Royal Society of Edinburgh* 23: 707–19. (Originally submitted as an M.D. thesis to the University of Edinburgh; reprinted as *On the Theory of Chemical Combination*. Edinburgh: privately printed, 1879 [held in Edinburgh University Library].)

Dewar, James. 1866–67. "On the oxidation of phenyl alcohol, and a mechanical arrangement adapted to illustrate structure in the non-saturated hydrocarbons." *Proceedings of the Royal Society of Edinburgh* 6: 82–86.

Eliel, Ernest L. 1997. "Conformational analysis: The elevation of two-dimensional formulas into the third dimension." In *Pioneering Ideas for the Physical and Chemical Sciences: Josef Loschmidt's Contributions and Modern Developments in Structural Organic Chemistry, Atomistics, and Statistical Mechanics: Proceedings of the Josef Loschmidt Symposium, 25–27 June 1995*, ed. W. Fleischhacker and T. Schönfeld. New York: Plenum.

Ferguson, Eugene S. 1992. *Engineering and the Mind's Eye*. Cambridge, Mass.: MIT Press.

Frankland, Edward. 1866. *Lecture Notes for Chemical Students: Embracing Mineral and Organic Chemistry*. London: Van Voorst.

Fröbel, Friedrich. 1974. "Anleitung zum rechten Gebrauche der dritten Gabe [1851]." In *Ausgewählte Schriften*, vol. 3, ed. Helmut Heiland, 121–61. Düsseldorf: Küpper.

Gee, Brian, and William H. Brock. 1991. "The case of John Joseph Griffin: From artisan-chemist and author-instructor to business-leader." *Ambix* 38: 29–62.

Gillis, Jean. 1967. "De atoom- en molecuulmodellen van Kekulé." *Mededelingen van den Vlaamse Chemische Vereniging* 29: 175–80.

———. 1996. *Auguste Kekulé et son oeuvre réalisée à Gand de 1858 à 1867* (Académie Royale de Belgique, Classe des sciences, Mémoires, ser. 2, vol. 37, no. 1). Brussels: Académie Royale de Belgique.

"Glyptic formulae." 1867. *The Laboratory* 1: 78.

Görs, Britta. 1999. *Chemischer Atomismus. Anwendung, Veränderung, Alternativen im deutschsprachigen Raum in der zweiten Hälfte des 19. Jahrhunderts*. Berlin: ERS-Verlag.

Griffin, John Joseph. 1866. *Chemical Handicraft: A Classified and Descriptive Catalogue of Chemical Apparatus, Suitable for the Performance of Class Experiments*. London: Griffin.

Havrez, Paul. 1865. "Principes de la chimie unitaire." *Revue universelle des mines, de la métallurgie, des traveaux publics, des sciences et des arts* 18: 318–51, 433–48.

Heilbronner, Edgar, and Klaus Hafner. 1998. "Bemerkungen zu Loschmidts Benzolformel." *Chemie in unserer Zeit* 32: 34–42.

Heilbronner, Edgar, and Jean Jacques. 1998a. "Paul Havrez et sa formule du benzène." *Comptes rendus de l'Académie des Sciences de Paris*, ser. 2 c, 1: 587–96.

———. 1998b. "Paul Havrez und seine Benzolformel: Ein überraschendes Addendum zur Geschichte der Benzolformel." *Chemie in unserer Zeit* 32: 256–64.

Hill, C. R. 1971. *Chemical Apparatus*. Oxford: Museum of the History of Science.

Hoff, Jacobus Henricus van't. 1874. *Voorstel tot uitbreiding der tegenwoordig in de scheidkunde gebruikte struktuurformules in de ruimte*. Utrecht: Greven.

Hofmann, August Wilhelm. 1862. "On mauve and magenta." *Proceedings of the Royal Institution of Great Britain* 3: 468–83.

———. 1865. "On the combining power of atoms." *Proceedings of the Royal Institution of Great Britain* 4: 401–30. Reprinted in *Chemical News* 12 (1865): 166–90.
———. 1866. *Sur la force de combinaison des atomes*, ed. Abbé Moigno. Paris: Étienne Giraud. (Translation of Hofmann 1865.)
———. 1871. *Einleitung in die moderne Chemie*, 5th edn. Braunschweig: Vieweg. (First English edition: *Introduction to Modern Chemistry, Experimental and Theoretic: Embodying Twelve Lectures Delivered at the Royal College of Chemistry*. London: Walton & Maberley, 1865.)
———. 1890. "Die Ergebnisse der Naturforschung seit Begründung der Gesellschaft." In *Verhandlungen der Gesellschaft Deutscher Naturforscher und Ärzte: 63. Versammlung zu Bremen 1890*, 1–55. Leipzig.
Hoffmann, Roald, and Pierre Laszlo. 1991. "Chemical structures: The language of the chemists." *Angewandte Chemie International Edition in English* 30: 1–16.
James, Frank A. J. L., ed. 1987. *Chemistry and Theology in Mid-Victorian London: The Diary of Herbert McLeod, 1860–1870*. London: Mansell.
Kekulé, August. 1861–66. *Lehrbuch der Organischen Chemie oder der Chemie der Kohlenstoffverbindungen*, 2 vols. Erlangen: Enke.
———. 1865. "Sur la constitution des substances aromatiques." *Bulletin de la Société Chimique*, new ser., 3: 98–110.
———. 1866. "Über die Constitution der aromatischen Verbindungen." *Annalen der Chemie* 137: 129–96.
———. 1867. "Über die Constitution des Mesitylens." *Zeitschrift für Chemie*, new ser., 3: 214–18.
Klein, Ursula. 1999. "Techniques of modelling and paper-tools in classical chemistry." In *Models as Mediators: Perspectives on Natural and Social Science*, ed. Mary S. Morgan and Margaret Morrison, 146–67. Cambridge: Cambridge University Press.
Klooster, H. S. van. 1953. "Van't Hoff in retrospect." In *Proceedings of the International Symposium on the Reactivity of Solids*, 1095–100.
Knight, David M. 1993. "Pictures, diagrams, and symbols: Visual language in nineteenth-century chemistry." In *Non-Verbal Communication in Science prior to 1900*, ed. Renato G. Mazzolini, 321–44. Florence: Olschki.
———. 1997. "Drawing the unspeakable: Chemistry as visual art." *Chemistry & Industry*, 384–87.
König, Wolfgang. 1997. *Künstler und Strichezieher. Konstruktionskulturen und Technikkulturen im deutschen, britischen, amerikanischen und französischen Maschinenbau zwischen 1850 und 1930*. Frankfurt am Main: Suhrkamp.
Koerner, W. 1869. "Fatti per servire alla determinazione del luogo chimico nelle sostanze aromatiche." *Giornale di Scienze Naturali ed Economiche di Palermo* 5: 212–56.
Larder, David F. 1967. "Alexander Crum Brown and his doctoral thesis of 1861." *Ambix* 14: 112–32.

Laszlo, Pierre. 1993. *La parole des choses ou la langage de la chimie*. Paris: Hermann.

Leinweber, Ulf. 1999. *Baukästen. Technisches Spielzeug vom Biedermeier bis zur Jahrtausendwende*. Wiesbaden: Drei Lilien Edition.

Loschmidt, Joseph. 1861. *Constitutionsformeln der organischen Chemie in graphischer Darstellung*. Vienna: Carl Gerold's Sohn.

Lüthy, Christoph. 2000. "The invention of atomist iconography." *Max-Planck-Institut für Wissenschaftsgeschichte*, Preprint 141. Berlin.

Meinel, Christoph. 1992. "August Wilhelm Hofmann: 'Reigning Chemist-in-Chief'." *Angewandte Chemie International Edition in English* 31: 1265–82.

———. 1995. "'A World out of Chaos': Anschauung, Modellbildung und chemische Synthese bei August Wilhelm Hofmann." In *Inter Folia Fructus: Gedenkschrift für Rudolf Schmitz*, ed. Peter Dilg, 79–92. Frankfurt am Main: Govi.

Muspratt, J. S., and A. W. Hofmann. 1845. "Über das Toluidin, eine neue organische Basis." *Annalen der Chemie und Pharmacie* 54: 1–29.

Naquet, Alfred. 1865. *Principes de la chimie fondée sur les théories modernes*. Paris: Savy.

Noschka, Annette, and Günter Knerr. 1986. *Bauklötze staunen. Zweihundert Jahre Geschichte der Baukästen*. Munich: Hirmer.

Nye, Mary Jo. 1989. "Explanation and convention in nineteenth-century chemistry." In *New Trends in the History of Science*, ed. Robert P. W. Visser, H. J. M. Bos, L. C. Palm, and H. A. M. Snelders, 171–86. Amsterdam: Rodopi.

Paoloni, Leonello. 1992. "Stereochemical models of benzene, 1869–1875." *Bulletin of the History of Chemistry* 12: 10–24.

Paternò, Emmanuele. 1869. "Intorno all'azione del percloruro di fosforo sul clorale." *Giornale di Scienze Naturali ed Economiche di Palermo* 5: 117–22.

Peters, Tom F. 1996. *Building the Nineteenth Century*. Cambridge, Mass.: MIT Press.

Ramsay, O. Bertrand. 1974. "Molecules in three dimensions." *Chemistry* 47, no. 1: 6–9 and no. 2: 6–11.

———, ed. 1975. *Van't Hoff—Le Bel Centennial*. Washington: American Chemical Society.

———. 1981. *Stereochemistry*, *Nobel Prize Topics in Chemistry*. London: Heyden.

Rocke, Alan J. 1984. *Chemical Atomism in the Nineteenth Century: From Dalton to Cannizzaro*. Columbus: Ohio State University Press.

———. 1993a. *The Quiet Revolution: Hermann Kolbe and the Science of Organic Chemistry*. Berkeley and Los Angeles: University of California Press.

———. 1993b. "Rival visions, rival styles: Synthetic chemistry in France and Germany, 1845–1869." Paper presented at the Annual Meeting of the History of Science Society, Santa Fe, N. Mex., 12 November 1993.

Ronge, John, and Berta Ronge. 1855. *A Practical Guide to the English Kindergarten: Being an Exposition of Froebel's System of Infant Training.* London: Hodson.

Russell, Colin A. 1971. *The History of Valency.* Leicester: Leicester University Press.

———. 1987. "The changing role of synthesis in organic chemistry." *Ambix* 34: 169–80.

———. 1996. *Edward Frankland: Chemistry, Controversy, and Conspiracy in Victorian England.* Cambridge: Cambridge University Press.

Schiemenz, Günter Paulus. 1994. "Joseph Loschmidt und die Benzol-Formel. Über die Entstehung einer Legende." *Sudhoffs Archiv* 78: 41–58.

Schorlemmer, Carl. 1894. *The Rise and Development of Organic Chemistry*, rev. edn. London: Macmillan.

Schütt, Hans Werner. 1975. "Guglielmo Körner (1839–1925) und sein Beitrag zur Chemie isomerer Benzolderivate." *Physis* 17: 113–25.

Schummer, Joachim, ed. 1999–2000. "Models in Chemistry." *Hyle: International Journal for Philosophy of Chemistry*, special issues, 5: 77–160; 6: 1–173.

Spronsen, Jan W. van. 1974. "De ontwikkeling van het molecuulmodel in de organische sterochemie." *Chemisch Weekblad*, 26 April, 13–14.

Tageblatt der 41. Versammlung deutscher Naturforscher und Ärzte in Frankfurt am Main. 1867. Frankfurt am Main.

Thackray, Arnold. 1970. *Atoms and Powers: An Essay on Newtonian Matter-Theory and the Development of Chemistry.* Cambridge, Mass.: Harvard University Press.

"Theory of the Arts." 1871. *The Building News and Engineering Journal* 21 (22 September 1871): 209–11.

Torracca, Eugenio. 1991. "Teorie nuove e loro diffusione: A. W. Hofmann e la chimica degli anni '60 del secolo scorso." In *Scritti di storia della scienza in onore di Giovanni Battista Marini-Bettòlo* (Rendiconti della Accademia Nazionale delle Scienze detta dei XL, Memorie di scienze fisiche e naturali, ser. V, vol. 14/2, part 2), ed. Alessandro Ballio and Leonello Paoloni, 215–26. Rome: Accademia Nazionale delle Scienze.

Turner, Gerard L'E. 1983. *Nineteenth-Century Scientific Instruments.* London: Sotheby.

———, ed. 1991. *Gli strumenti*, vol. 1. Turin: Einaudi.

Tussen kunst en kennis: Een keuze uit de Gentse universitaire verzamelingen. 1992. Ghent: Archief Rijskuniversiteit.

Wotiz, John H., ed. 1993. *The Kekulé Riddle: A Challenge for Chemists and Psychologists.* Vienna, Ill.: Cache River Press.

Wurtz, Adolphe. 1864. *Leçons de philosophie chimique.* Paris: Hachette.

CHAPTER TEN

Mathematical Models

Herbert Mehrtens

In 1893 at the third annual meeting of the German Mathematicians' Association in Munich an "exhibition of mathematical and mathematical-physical models, apparatus, and instruments" was shown to the general academic public in the rooms of the Technical College. The show was seen by about a thousand visitors in the course of one month.[1] The terms 'models', 'apparatus', and 'instruments' are certainly not well defined in the mathematical sense. 'Models' is the first word in the list, followed by '*Apparate*', which is the most general word in German for something technical or somehow artificially organised to serve some function: machine, instrument, appliance, gadget, organisation, or equipment. At the end there are the 'instruments', tools with clear functions like a slide-rule or a logarithmic compass. The things to show something with come first, the things to do something with last, and everything else is in between. There were some photos of models on display, but the exhibition showed mathematics and mathematical physics, i.e., the most abstract science, in actual three-dimensional objects. Of these a large number were mathematical models, in their majority representing geometrical entities and characteristics, and they were made of cardboard, metal, wood, string, wire, or plaster. Figure 10.1 shows the geometry room of the 1893 exhibit. In the showcase at front left there are plaster models of geometrical surfaces. This is the type of model that is addressed in this chapter. My question is, what and how did these models represent? The question will not find a clear-cut answer.

The concept of 'mathematical model' has undergone a peculiar change. While by 1890 it would mean the physical 3-D representation of a

Figure 10.1 Geometry room with exhibit of models and instruments, Munich 1893. From Walter Dyck, *Katalog mathematischer und mathematisch-physikalischer Modelle, Apparate und Instrumente* (Munich, 1892–93), xiv.

mathematical entity, the recent *Encyclopaedia of Mathematics* (Vinogradov 1990, 128) reads: "*Mathematical Model*: A (rough) description of some class of events of the outside world, expressed using mathematical symbolism." Mathematical modelling in this recent sense is a technique of applying mathematics that has been developed mainly in the second half of the twentieth century.[2] The encyclopaedia also has the entries "model, regular", "model (in logic)", and "model theory", which all concern the representation of formal systems of logic and mathematics by other such systems. The concept of the physical model for visualising mathematical entities and characteristics appears to have survived solely in school mathematics, where it seems to be enjoying a revival.[3]

Physical models in mathematics slowly fell into oblivion after World War I. They were banished to cellars or attics, though in some places they remained visible, noted in passing as curiosities in their showcases. It was the return to visual representation through computer-generated images that probably brought about their rediscovery, most prominently in a two-volume German–English publication edited by Gerd Fischer in 1986, with photos of models (vol. 1) and mathematical commentaries by various specialists (vol. 2) (Fischer 1986). The books were given an illustrated review by one of the most prominent historians of geometry, Jeremy Gray, in the mathematicians' international tabloid, the *Mathematical Intelligencer*.[4] Gray's enthusiastic review praises "the power of geometry" and the "irresistible beauty" (Gray 1988, 69) of the photos. He takes the models to pose the question of what mathematics is about. His answer is that mathematics is about special cases and about general theory, "and it is in the overlap that mathematics is alive" (Gray 1988, 67). There is a problem with this overlap. The actual models are not the special cases themselves, but show figurative aspects of special cases. The delicacy of the relation between the models and the mathematics from which they are generated is part of the reason for the briefness of their career in mathematics. The neglect of 3-D models (and 2-D diagrams) during most of the twentieth century is closely related to the dominance of mathematical modernism with its preference for general theory, symbolic formalism, and the treatment of mathematical theories as worlds of their own without any immediate relation to the physical world around us (Mehrtens 1990).

With the computer, however, a new sort of world has been created, that of 'virtual reality'. Gray is convinced that the models "have a new life ahead of them", not only as objects of aesthetic pleasure but in mathematics (Gray 1988, 68–69). I am not sure whether he means the models or the geometrical surfaces they display. 'Virtual reality' is what Brian Rotman takes as the reality of mathematical objects and procedures. He suggests the "realization of the virtual, whereby mathematical objects, by being constructed in a computer, reveal themselves as materially presented and embodied" (Rotman 1995, 414). Rotman's vision—"little more than highly speculative extrapolation" he calls it—is connected with his appeal to bring "materiality, embodiment, and corporeality" back into mathematics (Rotman 1995, 414; see also Rotman 1993). While Gray is not clear about the possibility of a mathematical after-life for the corporeal models, Rotman appears to ascribe corporeality to the virtual reality of computer-generated images. The models, however, should obviously be interpreted in their corporeality in relation to the mind and the body of the mathematician. Yet mathematics is a disembodied science, so its embodiments in virtual reality or in plaster pose the question: is this mathematics or are these 'just' models?

In all its meanings, it should be added, 'model' is a relational term, indicating a form of representation of something, not a replication but an intentional selective construction of a new thing meant to stand for something else. I cannot attempt to develop a theory of representation and modelling here, but I wish to point out that a model shows us itself, something else, the intention of representation, and the specific form of representation. With the model a relation between itself and that which is modelled is created, and the model represents the act of representation. Physical mathematical models are fundamentally dissimilar to the usual representation of 3-D geometrical entities by concepts and formulae, possibly assisted by diagrams. This difference can be taken as one of ontology, and if ontology matters, the models are either insufficient attempts to visualise something that lies outside the sphere of visual imagination, or they make the statement that the entities represented are inside that sphere. Some mathematicians' use of models between the 1860s and the first decade of the twentieth century produced this ontological ambiguity. With the rise of mathematical modernism the ontological question found an answer that remained valid for most of the twentieth century: mathematics is about 'structures', developed in axiomatic theories, not about physical objects. Today this modernism is no longer modern, and the question arises anew.

The press that published the volumes on models mentioned above reissued four actual plaster models, among them 'Clebsch's diagonal surface' (Fig. 10.2).[5] Now, who would buy a mathematical model in 1986? Probably it would adorn a mathematician's desk or shelf, investing his science with an aesthetic and mysterious, maybe also an antiquarian touch. It is clearly the aesthetic value, the "irresistible beauty" of the models that makes them attractive (Gray 1988, 69). They look like abstract sculptures, and they did in fact play a role in twentieth-century art. Naum Gabo, one of the most important early constructivists in sculpture, used the techniques of mathematical models to represent spatiality; the surrealists made use of mathematical models in photography, visual quotations, and in exhibiting models from the mathematicians' collections. Rather than the power of geometry it is the power of the corporeality of conceptually created form, suggesting spatiality and temporality, that led from the models of academic mathematics to abstract sculpture. Artists were the heirs of the modelling geometers, but they did not produce mathematical models. Mathematical models, one might say, were illegitimate art. And they were, one might also say, illegitimate mathematics. Modernist purism excluded corporeality and visuality

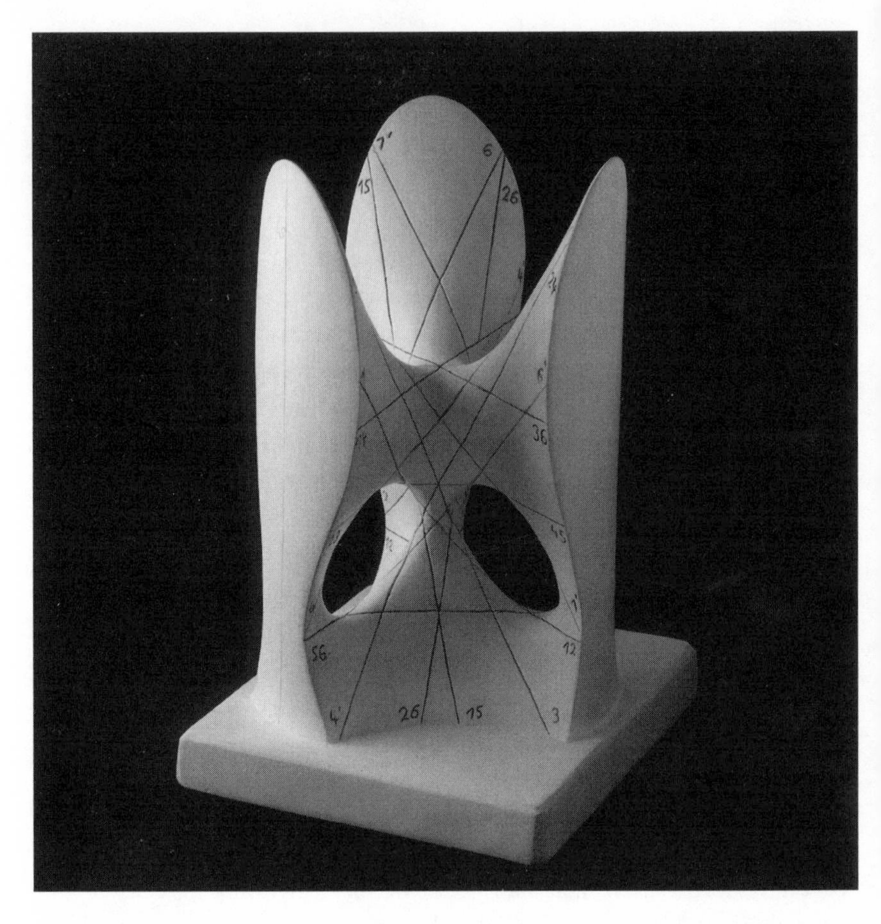

Figure 10.2 Clebsch's diagonal surface. Plaster, 17 × 15 × 24 cm. From Gerd Fischer, *Mathematische Modelle / Mathematical Models* (Braunschweig, 1986), vol. 1: fig. 10. Courtesy of Vieweg Verlag, Stuttgart.

from the discursive universe of mathematics. Modern conceptual mathematics had no use for any visualisable 'essence', while modernist purism in the arts introduced conceptuality with the help of the forms mathematical models provided as sculptural material.

ACTS OF REPRESENTATION

In order to introduce the concept of 'representation' and its use, and simultaneously to enter into the question, how do mathematical models represent, I shall read a photograph from the turn of the twentieth century (Fig. 10.3)

Figure 10.3 August Gutzmer. Photograph courtesy of Vieweg Verlag, Stuttgart.

as an act of representation, or rather as a complex system of such acts.[6] The photo shows a man with his left arm resting on a lectern, apparently standing on some sort of low table. In the background there is a dark vertical plane with various more or less intense white marks and shading. In the foreground is the lectern and on the other side, standing on the same table, there are two rather strange objects, one not completely visible, apparently made of wire, while the other is a solid white sculpture, arranged close to, but not on the vertical centreline of the picture; in its whiteness it corresponds in prominence to the face of the man, whose nose is almost exactly on the central vertical axis. The scene is lit mainly from the right, so that the right side of the man's face is in the dark, while his black moustache becomes prominent against the light left side of his face. The lighting also emphasises the three-dimensionality of the two objects and shows the man's hands, which are placed such that the white sculpture is between them. The hands are at rest, they do not touch the sculpture, but the diagonal between them cuts through its centre.

The knowledgeable 'reader' of the picture will decode it as the representation of a mathematician presenting himself on what is probably one of the postcards of university professors that were sold to students in late-nineteenth- and early-twentieth-century Germany. The background is a blackboard with a grid, carrying traces of previous use and an obviously mathematical diagram together with some mathematical signs to the right. Together with the lectern they put the man in the position of a mathematics professor. His beard and his clothing suggest that the picture is at least eighty years old. The two objects in the foreground, however, remain enigmatic. The object made of wire looks regular enough to have some scientific meaning. The white thing looks like an abstract sculpture. Today's reader, used to comic-strips and animated films, might see it as some sort of cartoon character carrying a ball or a disc. The reader of these lines, however, will already know that this is a mathematical model. But she may not know what it represents as a mathematical model. The knowledgeable contemporary viewer, like a student who might have bought such a postcard, might also have known nothing about the mathematics of this object but would probably have recognised it as some representation of higher mathematics. The object is enigmatic, and taken as a mathematical object it remains enigmatic except to the few who know the specific mathematics to which it relates. Thus the man in the picture is put in the position of the one who knows, and the position of his hands signals that he can handle the object and with it the knowledge it embodies. Mathematics is signified in the photograph in two distinct ways, as a specific type of knowledge encoded in diagrammatic and typographical signs on the blackboard (where there are no numbers: this is real mathematics) and as embodied in 3-D objects. The illumination in the photograph is arranged so that the little white thing is highlighted in its sculptural quality as strongly as the forehead of the man. Mathematics, the picture says, is head and hands, is signs and objects.

The photograph presents an image of the mathematician and of mathematics. It represents a set of pre-existing things, but it also produces the meaning of these things in relation to each other and in relation to the cultural code that the viewer is supposed to know. As the term is used in cultural studies, this representation of the mathematics professor in a postcard is a productive and not simply a reproductive act. The picture represents a series of representations of something that cannot be pictured but only performed, namely doing mathematics. And the photo is not the scene depicted. Very probably the scene was staged to produce a photo. In this representational act the model is merely an attribute signifying mathematics in general. The photo might even be a fraud, playing on the representational code, and the 'sculpture' a piece of art smuggled into the sphere of mathematics.

But there is no reason to suspect fraud: Fairly reliable information tells us that the man is the mathematician August Gutzmer, from 1900 full professor of mathematics at the University of Jena and subsequently in Halle. Gutzmer was highly interested in applied mathematics and also taught descriptive geometry, which was unusual in universities. He had planned a mathematical institute for Jena and realised one at Halle, completed in 1911–12, including a drawing room and a room for models (Lorey 1916, 247, 253, 326, 333–34). The 'sculpture' is the model of 'Kuen's surface', a special case of 'Enneper's surfaces', which in turn are a subclass of 'Joachimthal's surfaces' (Fischer 1986, 2: 40–41). In the catalogue of the Munich exhibition edited by Walther von Dyck it is described as a "surface of constant measure of curvature with plane lines of curvature according to Enneper" (Dyck 1892–93, 292–93). With this caption it is offered at the price of 18 marks in Schilling's mail-order catalogue of mathematical models (Schilling 1911, 17–18, 144–45) together with a treatise by Thomas Kuen, who had discovered this surface as a special case of Enneper's surfaces, exceptional in that it can be represented by elementary (trigonometric) functions (Kuen 1884). Dyck also gives the information that the model had been produced under the supervision of Alexander Brill at the Munich Technical College and was published by the firm of Ludwig Brill.

The identification of man and sculpture has led us into the historiographic representation of the two most interesting objects depicted in the photo as factually existing in the past, but we have lost the photo as mediator. We can get back to it by inferring that the model was neither Gutzmer's product, nor a subject of his research. The model is involved in his act of self-representation as mathematician as an attribute in an allegory. It had probably been bought for his personal or the institutional collection of models, and was chosen for its aesthetic quality.

Had Gutzmer given lectures on such surfaces using this model another act of representation would have taken place. The specificity of its shape would have come into view, being represented by formulae and diagrams (as in Fischer 1986, 2: 40–41), while the model would have served as the representation of this specific case, describable by comparatively simple formulae. So an intricate web of meanings would have been woven by imagined, immaterial things and material things like blackboards, chalk marks, and plaster objects, as well as by material processes like talking, writing, drawing, and gesturing. Representation here is obviously not a one-to-one relation, and nor is it unidirectional but rather mutual. The question about the model cannot be "what does it represent?", but rather must be about how it is involved in acts of representation and what meanings these produce. In a lecture course the model is probably a didactic aid, but it could also be interacting with the 'epistemic things' mathematicians handle.

'Epistemic thing' is a concept introduced by Hans-Jörg Rheinberger to point to researchers' irreducibly vague objects of desire around which experimental systems are arranged (Rheinberger 1992; 1997). Moritz Epple has used the concept for the history of mathematics (Epple 1999, 10–17). The immaterial things of mathematics, e.g., in Epple's case the mathematical idea of a knot, or in ours an algebraic surface, have to be represented, and modernist purism would accept typographical representations only. But even in modernist research a wider set of graphematic representations of epistemic things were used in attempts to achieve ultimate representations of accepted results, i.e., those at the sight of which epistemic desire dies. It seems that for a very brief period mathematical models were representations of epistemic things, i.e. representations used on the road to new knowledge. In the lecture course they might represent this road, if they were used to point out the questions or observations that led to the ultimate representation.

Kuen's surface made it to become the cover model of Fischer's books, in a photo on the first volume and as a drawing on the second (Fischer 1986). Here we find the two versions also representing difference and unity in the contents of the two volumes. The drawing (with shadings) indicates a 3-D object, but with far more distance than the photo. At the same time the drawing shows the model with its base, indicating its character as a model, while the photo cuts off the base and thus adds further symmetry. The two volumes present two spheres of representation, the aesthetic sphere of photography and the mathematical sphere of typographical symbols illustrated by diagrams. Photography stages the models and adds the mysterious quality of the frozen moment, in which point of view and lighting are fixed. The models are represented in this way as sculptures.[7]

Now mathematical models, Kuen's surface included, had already had a career as photographic models. Henry Swift, a not so well-known member of the American group 'f.64' (better known are Ansel Adams, Imogen Cunningham, and Edward Weston), presented three pictures of mathematical models in a group exhibition of 1932 (Heyman 1992, 67, 70, 71, 73). The 'straight photography' of the time represents things in their sculptural qualities, and I do not know by what chance Swift met with mathematical models. In any case, they fit well, for example, with Weston's well-known pictures of sea shells, which may well be taken to represent mathematicity (e.g., Newhall 1965).

Four years later, and in a somewhat similar vein, Kuen's surface appeared as one of twelve photographs of mathematical objects by the surrealist artist Man Ray that accompanied an essay on mathematics and abstract art in a special issue entitled "*L'objet*" of the Parisian journal *Cahiers d'art* (Zervos 1936; Werner 2002, chap. 4) (Fig. 10.4). In this context the primary level of representation was artistic production of a photographic image. The object itself was represented as a specific mathematical object seen and staged by the artist; the caption reads: "Enneper's surface of constant negative curvature derived from the pseudosphere" (*surface à courbure constant négative d'Enneper, dérivée de la pseudosphère*; Zervos 1936, 13). Hardly any reader of the journal would have been able to attach much meaning to these words. This special issue of the *Cahiers d'art* was published on the occasion of the *Exposition surréaliste d'objets*, which took place in May 1936 and showed, according to the title of the catalogue: "Mathematical, Natural, Found and Interpreted, Mobile, Irrational, Objects from America and from Oceania" (*Mathématiques, Naturels, Trouvés et Interprétés, Mobiles, Irrationels, Objets d'Amérique et d'Océanie)* (Werner 2002, 145). The exhibit did not present itself as art but as surrealism; it presented objects in their

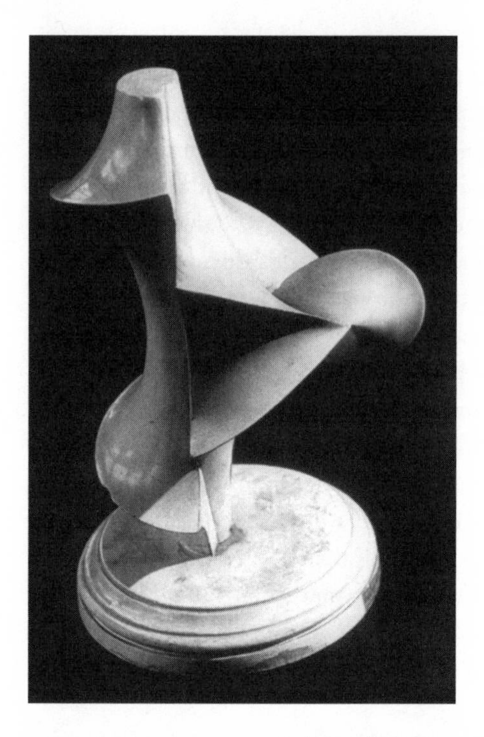

Figure 10.4 Kuen's surface. Plaster, 18 × 18 × 24 cm. Untitled photograph by Man Ray, circa 1935. © Man Ray Trust, Paris/VG Bild-Kunst, Bonn 2001.

sur-reality. André Breton's metaphor for it was the moving train: "In the windows, beings-objects (or objects-beings?) characterised by the fact that they are prey to a continual transformation and express the perpetuity of the struggle between the integrating and disintegrating forces fighting for true reality and life. The movement-master!"[8]

Here mathematical models or 'objects' become part of a web of representation, made to show the mystery of 'the object'. Gabriele Werner has given a thorough interpretation of the meaning of mathematics and mathematical objects in surrealism, a central thesis being that the objects represent "mathematicity" (Werner 2002). They are shown as coming from and embodying mathematics, nothing more. In the context of the exhibit they are related to other objects and take part in showing the fantastic side of reality. Man Ray's photos are about photography, the construction of bodies and images, and about mathematicity as a product of mental construction in relation to the female body as a product of mental construction as well (Werner 2002,

151–25; compare Krauss 1985). They are not about mathematics, and not about Kuen's surface, they are about objects and images of objects.

Through images, objects, and texts the surrealists discussed epistemology and the relations of art and science. That is not my topic, but in representing mathematics by mathematical models and images of models, surrealism points to something important about the models. Though they are objects like other objects, at the same time they are different. Observed or touched they are immediately involved in acts of representation, because they prompt questions about where they come from and what they mean. In the dark of a cellar they are mute and motionless beings; in the presence of an observing subject they become object-beings. In surrealist discourse they remain 'objects', for the word 'model' is not used. As 'models' they are related to something else and belong to a specific discourse, namely mathematics. It is their mere being in the world that is the basis for their potential representative and discursive functions; which functions these are is in the last analysis a matter of historical contingency. However, they show themselves as unnatural, as man-made, and therefore they persistently pose the question about their origin and their originally intended functions.

MATHEMATICS

What is mathematics about? This was Gray's question when confronted with the photos of the models, Kuen's surface being his figure and example number one: "a special case, doubtless, but the reward is a wonderful figure" (Gray 1988, 64). The plaster object has been produced to show its figure, its *Gestalt*. It represents a unique case of the surface in question. The model turns the virtual reality of the surface into physical reality. In the virtual realm of differential and algebraic geometry it is of indefinite size and extends into infinity. For a unique case to be realised physically all parameters have to be fixed such that the resulting object gets the symmetries and singularities that are intended to become visible. The physical object then incorporates mathematics, and used or observed by mathematicians it embodies mathematics. Elsewhere I have pointed to the fact that an artillery shell also incorporates mathematics (Mehrtens 1996, 112, 122). I called this "attached mathematics", which "ceases to be mathematics in the disciplinary sense of the word, i.e., it is the actual process of complex calculations and constructions for some extra-mathematical purpose" (Mehrtens 1996, 91).

The mathematical formulae and calculations that determine the shape of a body built for stable flight at supersonic speed disappear into this body, but can be 'read' by experts. Confronted with the models I find attached mathematics in use for mathematical purposes. The formulae disappear and what they represent reappears in another form of representation. The standard interpretation would be: but the models are not really mathematics. There is no corporeality in mathematics. The models are means of visualisation; they serve mainly pedagogical purposes and in some cases may have a heuristic function in research. But neither pedagogical devices nor heuristic tools are the real mathematics. Models just represent the mathematics.

This view in part explains the little interest models have found among historians of mathematics. It would have been my answer, too, until I was in the last stage of writing this chapter. But rethinking representation and rereading some of my sources I reconsidered Brian Rotman's argument about the realisation of the virtual. Thus let me tentatively extend the list of characterisations of mathematics—pure, applicable, applied, and attached—with 'embodied', to include the representation of the epistemic thing in a physical, 3-D realisation. Especially Felix Klein's introduction to the papers on "Intuitive Geometry" (*Anschauliche Geometrie*) in his collected works shows that models were more to him than a physical aid to mental imagery (Klein 1922a). Klein studied with the physicist and mathematician Julius Plücker at the University of Bonn and was his assistant from 1866 to 1868. Plücker constructed models of geometrical surfaces, and Klein had to assist him. In 1871 Klein himself "published" (*herausgeben*) a series of models in zinc, which, as he stresses, were not constructed "empirically" (as Plücker had done) but geometrically "by using the planes which touch the surfaces at the extension of complete conic sections". He then "chose a case with no special symmetries which would have facilitated the survey and simultaneously the construction" but soon abandoned this principle as inappropriate. Producing these models was "pioneering work" in a still "quite unknown area. The general comprehension of the figurative conditions [*gestaltliche Verhältnisse*] for this field ... has only gradually been accomplished" (Klein 1922a, 3). In his introduction Klein mentions a few other models, among them those of Rodenberg (see below) before moving on to his "organisational efforts". He notes that the movement of producing and working with models had declined since the turn of the century, and ends by stressing the importance of concrete geometrical figures and spatial imagination (*Vorstellung*) in his own work, and of spatial *Anschauung* for mathematics in general (Klein 1922a, 5–6).

In the late 1860s and early 1870s Klein was involved in the study of algebraic surfaces, and here models were for a while heuristic. In 1873 Klein published a paper on surfaces of the third order (i.e., defined by a polynomial of third degree) in general, starting from the diagonal surface Alfred Clebsch had studied (Fig. 10.2). The paper mentions models, it is to a large extent descriptive, and it aims at the "determination of the figures [*Gestalten*] of all surfaces of third order" (Klein 1922b, 12). Here we have possibly the decisive point, Klein's interest in what he calls *Gestalt*. It is the geometrical form that guides the mathematical interest and the characterisations given in typographical formalism. The models (and diagrams) were constructed to 'grasp' the form and to show the typical forms of a class of surfaces. If form or *Gestalt* is the epistemic thing around which the research circles, then the model is one of the representations of this thing, and not fundamentally different from the algebraic formula or the diagram. It is one of the many simultaneous and consecutive representations of the object. In this sense the model is part of doing mathematics, embodied mathematics. Research is the creative work of representing epistemic things. In this act of representation, I suggest, the model is real mathematics, embodied mathematics. But can a *Gestalt* be an epistemic object of mathematics?

In his famous treatise of 1872, called the *Erlanger Programm* (Klein 1974; Mehrtens 1990, 60–67), which attempts the unification of diverse geometrical theories and explicitly takes an "abstract mathematical point of view", Klein adds an endnote to a footnote, and, thus distanced from the text, he states:

> There is an essential [*eigentliche*] geometry, which does not only mean to be, as the investigations discussed in the text are, a visualised [*veranschaulichte*] form of abstract investigations. Here it is the task to grasp the spatial figures in their full figurative reality [*gestaltliche Wirklichkeit*], and (which is the mathematical side) to understand the relations valid for them as evident consequences of the principles of spatial intuition [*Anschauung*]. For this geometry a model—be it realised and observed or only vividly imagined—is not a means to an end but the thing itself (Klein 1974, 75).

For Klein the model indeed represents the (epistemic) thing itself, and both imagination and observation have access to its reality which is in the *Gestalt*. *Gestalt* is observed in the physical reality of the model and it has mental reality as a mathematical object. In this view the plaster model presenting its *Gestalt* represents mathematical reality. As an object of spatial intuition it is 'evidence' for the mathematical theorems about surfaces. Klein's *eigentliche Geometrie* is about figures taken as real in the way they relate mental imagery and physical shape. Figures like Clebsch's diagonal surface or Kuen's surface had not been seen before the models were produced. Klein nevertheless ascribes to them a reality arising from the principles of spatial *Anschauung*. I shall discuss the epistemological meaning of *Anschauung* below. For the moment the decisive points are, first, that the figures are taken as a preexisting reality and as essential to geometry, and second, that Klein makes no substantive distinction between the mental image and the visual image.

Yet in these very decades mathematics discovered its 'monsters', mathematical entities like space-filling curves or nowhere-differentiable functions that defied *Anschauung* completely. Klaus Volkert has analyzed what he calls 'the crisis of *Anschauung*' from a semiotic point of view, and, by the way, ignoring 3-D models completely. Volkert characterises Klein as taking *Anschauung* only in its heuristic function and as representing an empiricist conception, which takes *Anschauung* and observation to be the same (Volkert 1986, 242). In the early 1870s, however, Klein had still taken *Anschauung* as mathematical evidence. Klein's retreat is nicely visible in the lectures he gave at the mathematical congress on the occasion of the World's Columbian Exposition in Chicago in 1893. German mathematicians showed their models, and Klein gave a series of lectures, some of which were connected with specific models. One of these lectures was about surfaces and relates to his work of 1873; at the end Klein says: "these methods give us an actual mental image of the configurations under discussion, and this I consider as most essential in all true geometry" (Klein 1894, 32). In a later lecture he qualifies geometry as an "applied science" and then adds: "From the point of view of pure mathematical science I should lay particular stress on the *heuristic value* of the applied sciences as an aid to discovering new truths in mathematics" (Klein 1894, 46; his emphasis).

Anschauung had been disqualified as mathematical evidence. But Klein remained a staunch defender of *Anschauung* as essential to mathematics, and of geometry as the science of space and spatial mathematical objects, by declaring it to be 'applied' mathematics. In applied mathematics, which deals with the real world, real objects are allowed, so in this way the models would remain objects of 'real'—if only applied—mathematics.

It was not only the counter-intuitive constructions and results which devalued visual intuition. My impression is that in algebraic and differential geometry itself as it developed during the last quarter of the century (Gray 1989) the 'reality' of what could be presented by 3-D models did not play a role for long. Take Carl Rodenberg, mentioned above, who wrote his dissertation on surfaces of the third order and produced a well-known series of models in 1881 (Fischer 1986, photos 10–33). The corresponding article does not use any argument related to models, and only in an endnote does he mention his intention to produce a series of models "which will facilitate concrete comprehension" (Rodenberg 1879, 110). Klein's own posthumously published lectures on nineteenth-century mathematics (Klein, 1926) still stress the importance of *Anschauung* but mention models only in connection with descriptive geometry, an applied field in which geometrical means are developed for the representation of real physical objects. Finally, from the turn of the twentieth century geometry became 'pure mathematics' in Klein's sense, mathematical theory about things that come into existence by axiomatic stipulation and have no existence or reality outside the theory.

So the models were representations of the epistemic object only briefly and only for a few mathematicians. In Klein's sense of 'applied mathematics' they were epistemic things that found their ultimate form as a result and were then lost from interest. The shape of an artillery shell designed for supersonic speed is an epistemic thing for the applied mathematician only as long as it is not determined. Once it is, the formulae become merely technical apparatus and disappear as attached mathematics into the non-mathematical world of workbenches and means of destruction. In Klein's 'essential' geometry, the model together with a mathematical paper would be a result, a teachable mathematical object. But already in Rodenberg's presentation the result is in the paper, the models are only an additional didactic tool. And so they became pedagogical devices, collector's items and show pieces with certain functions for the mathematical profession.

The rise of interest in the geometrical *Gestalt* and the status of models in the development of algebraic and differential geometry clearly deserve closer research.[9] Klein is the most important figure in a group of young mathematicians who in the late 1860s got interested in models and were responsible for building the collections. It is interesting to note that of the older generation Plücker and Clebsch, both teachers of Klein, had also been working as physicists. Plücker had the idea of working with models from

Michael Faraday (Tobies 1981, 25) and Alexander Brill writes of an early model of a wave surface produced by the physicist Gustav Magnus in 1840 (Brill 1887, 74). At a time when physics and mathematics were drifting apart, thinking in terms of models in pure mathematics might possibly be understood as a residuum of physical thinking. Yet most of the older and of the younger mathematicians involved with models spent at least some time teaching in polytechnic schools, such as Hermann Amandus Schwarz in Zürich and Christian Wiener in Karlsruhe. Here descriptive geometry had been the dominant subject since the foundation of the École Polytechnique in Paris. Descriptive geometry was central to the education of engineers and involved training in drawing, including the use of models. For mathematicians, teaching this fairly elementary subject might have been a point of departure from which to rethink higher, abstract mathematics in terms of models and with a feeling for physical reality.

The collections of models that had been gathered by the end of the century were presented as pedagogical means for the academic teaching of mathematics. The function they served was still called *Anschauung*. The prominence of this word and its decline parallel those of the models. The concept itself is rather vague, relating to the philosophical concepts of cognition and intuition, to mental imagery, to the physical visualisation of abstract entities, and also denoting straightforward visual perception of form. *Anschauung* is a discursive event in the formation of the self-conception of professionalised, modern mathematics. I have written about this so-called foundational debate at length elsewhere (Mehrtens 1990) and shall briefly sketch the controversy.

The first lines of the entry in a contemporary encyclopaedia read:

> *Anschauung* in its proper sense means the perception by the sense of sight and simultaneously the mental conception of an object obtained in this way; in the wider sense generally the immediate cognition of an object in opposition to cognition by thought or that mediated by concepts; the former is also called intuitive, the latter discursive cognition (*Meyers Konversations-Lexikon* 1894, 646).

The opposition involved—intuitive versus discursive—is more partially expressed in the opening sentences of the first chapter of a book on the training of *Anschauung* in mathematics:

> For more than a hundred years the word *Anschauung* has become the code word for all those educational attempts which fight the empty trumpery of

> words and stand for a vivid grasp of reality and a clear and uncompromising objectivity. We just have to say the word and anybody will have a distinct inner image of an education replacing the dead letter with the colourful abundance of the phenomena (Timerding 1912, 1).

The judgement is rather sharp: *Anschauung* is vivid, clear, and objective, while words and letters are empty and dead. Later in the book the mathematical enemy is named "methods of logical inference" or "formalism" which is a "monster" (Timerding 1912, 42). The discursive event in mathematics is not the concept of *Anschauung* as such but rather its use as a battle cry against a clear enemy called 'formalism' or 'logic'. It was 'logical inference' that disqualified *Anschauung* as evidence in mathematics, and 'formalism' is the mathematics that restricts itself to work with formal systems of typographical symbols, using diagrams not at all or merely as illustrations.

'Formalism' in a stricter sense is the foundational position represented above all by the name of David Hilbert, Klein's colleague in Göttingen at the turn of the twentieth century; it holds that truth and existence in mathematics are founded only in the consistency (the freedom from contradiction) of mathematical theories. This sort of mathematics is usually called 'modern mathematics', and I have named the opposition to it 'counter-modernism', to point out that, as in many other cultural fields, here too it is a modern opponent of modernism. In the last analysis, I believe, their dispute is about authority and creativity. On the counter-modern side *Anschauung* and intuition (in a non-visual sense) stand for recourse to a 'higher' authority granting objectivity and truth while modernism claims creative freedom independent of any outward source: mathematics is "the free creation of the human mind" (Richard Dedekind, quoted in Mehrtens 1990, 35).

Mathematical models are creations of human hands and minds, but with Klein they stand for *Anschauung* and thus for a specific brand of counter-modernism. The models represent immaterial things in physical materiality and thus suggest an immediate, material relation between the human body and the bodies of mathematics. But the problem with algebraic surfaces is that they are not even the skin of a body. At the time of the decline of interest in models, Richard Baldus, in an inaugural lecture on *Anschauung* in mathematics, took a distanced position towards models. Mental imagery (*innere Anschauung*) in geometry is always directed by thought and systematically observes specific spatial characteristics, he argued, while "the mathematical model cannot take this into consideration and shows in its whole more and in its detail less than desired" (Baldus 1921, 9). *Anschauung* even as a pedagogical orientation was seen with more and more skepticism. In academic mathematics the foundational counterprogramme to formalism became L. E. J. Brouwer's 'intuitionism', which attempted to base all of mathematics in the primordial intuition of number or, more radically, of the one-after-the-other (Mehrtens 1990, chap. 3.4). There was no room here for geometrical models either.

MODELS IN THE PROFESSION

The dispute between moderns and counter-moderns in mathematics was by no means only an epistemological discourse about mathematical certainty. It was a matter of positioning the self of the mathematician in relation to the imagined community of mathematicians and their imaginary source of unity, mathematics. This characterisation relates to Benedict Anderson's definition of 'nation' as 'imagined community' (Anderson 1983). The 'nation' or 'mathematics' has no existence other than in practices, discourses, and imaginations. They can be symbolised in flags or models and performed in a pledge of allegiance to the flag or the exhibition and mathematical characterisation of models. The symbols and performances take place in historically specific cultural and socio-political environments and have meaning only within these environments. The (not quite) blind spot of the foundational dispute in mathematics was the social and cultural position of mathematics which was negotiated between the lines and the acts. Thus it is necessary to look at the other side of the coin, the social history of the models. The following brief sketch concentrates on the German story, attempting to point to elements of the social function of models.[10]

The production of models, it seems, began in France at the École Polytechnique, at the meeting point of technology and mathematics, in descriptive geometry (Brill 1887, 72–73; Klein 1926, 78), and led to an exhibit in the Conservatoire des Arts et Métiers, which Klein saw in 1870.[11] In the same context Klein mentions the technical colleges (*Technische Hochschulen*) of Karlsruhe and Darmstadt, where he got to know "the whole range of the work descriptive geometers had already done for the visual grasp [*das anschauliche Erfassen*] of higher spatial figures" (Klein 1922a, 3). Another occasion where Klein saw a model (made by Christian Wiener), was a meeting of mathematicians in 1868. Like so many scientific societies the German

Mathematicians' Association developed from regular meetings under the umbrella of the Society of German Naturalists and Physicians, founded in 1822. The first attempt at organising German mathematicians came from Clebsch and led to the meeting where Klein met Wiener and Alexander Brill, with whom he later worked closely. A further meeting took place in Göttingen in 1873, where again models were shown (Lorey 1916, 213–14), and the big show of which we have the catalogue (Dyck 1892–93) was planned for the second annual meeting of the association in 1892 and took place in 1893. Subsequent meetings regularly had exhibits of models (Lorey 1916, 336). Another previously prepared catalogue listed the models in the German mathematical exhibit at the World's Columbian Exposition of 1893 (Dyck 1892–93). Finally, I should mention the show of models put on at the third International Congress of Mathematicians in Heidelberg in 1904.

In addition to their role in the education of engineers the models had a function in the formation of the growing community of mathematicians. And when the profession was well established they were still put on show, representing mathematics, and even, in this era of high imperialism, cultural foreign policy. After World War I the shows were over. In this they appear to parallel the development of exhibitions and museums in general. On various levels, from the local collection of, say, natural history objects, to the World's Fairs, these served local community functions, were national or regional representations, and played their role in imperialist competition. As a medium of communication and representation exhibitions had their heyday from the mid-nineteenth century to the turn of the twentieth. The history of mathematical models and their display could and maybe should be told as a subhistory in this general field. It should be added that expositions lost their prominence for many reasons, one being the tendency to make them permanent, that is, to build museums. In the case of mathematics, Walter von Dyck became a central figure in the foundation in Munich of the German Museum for Masterpieces of Technology and Science (Deutsches Museum) which was provisionally opened in 1906 (Fischer 1994, xvii–xxi). Another aspect is the role of expositions for the infrastructure of the sites where they took place. Just as the Paris Universal Exhibition of 1900 was functional for the creation of the Paris *métro*, so models in mathematics were functional for the creation of institutes. In this brief sketch I cannot let the history of mathematical models represent and reflect the general history of expositions,

Figure 10.5 Graduation in mechanical engineering at the Hanover Technical College in 1877. Photograph courtesy of Universitätsarchiv Hannover.

but this context should be kept in mind. Nor shall I go into mathematics at the Deutsches Museum, because mathematics played too little of a role there, but institutes are highly significant for the formation of the profession.

Interest in models beyond their use in descriptive geometry arose at a time when technical education was being raised to the academic level. The École Polytechnique in Paris, and the polytechnics in Zürich and in Karlsruhe, where Clebsch and Christian Wiener taught around 1860, were the 'model' institutions. In Germany industrialisation had gained strong impetus only since the mid-nineteenth century, and technical education was instrumental in this process. During the 1870s the polytechnic schools were raised in status and renamed '*Technische Hochschulen*'. The engineering profession struggled for status, especially in the arena of higher education, and won recognition in 1899, when Kaiser Wilhelm II gave the Berlin Technical College the status symbols the universities enjoyed, most importantly the right to award doctorates. The other German states followed the Prussian example. Until the last third of the century drawing and models of machines and machine parts had played a big role in engineering education, training the technical imagination and teaching functionally correct sketching. And technical models were a means of self-representation too, as a photo of a graduation at Hanover shows (Fig. 10.5). But since the 1870s machine laboratories were established, and academic technology turned from being largely descriptive to using experimental methods. At the same time drawing became mainly oriented towards construction, and machine-models tended to go out of date more and more quickly (Mende 1994).

In technical colleges mathematical models were thus in a sympathetic environment until the 1890s, when an 'anti-mathematical movement' set in. Then the engineers attacked the unnecessary abstraction of mathematics and proposed that mathematics be taught by engineers. The struggle of mathematicians to keep their positions led to the invention of applied or 'practical' mathematics as a separate field, including descriptive geometry, but quickly numerical methods became much more important. In teaching mathematics for engineers certain sorts of models could play a role, but those were not the showpieces of higher geometry. The mathematicians could defend their position, and in their professional rhetoric *Anschauung* was a constantly used argument against logic and formalism. In general, however, it was not related to models. Ironically, the development of applied mathematics with separate positions helped to make room for pure mathematics to develop the purist modernist version that would dominate much of the twentieth century (Mehrtens 1990, chap. 5.2; Stäckel 1915; Hensel 1989).

Descriptive geometry was not usually taught in universities; models entered here with higher geometry. As Renate Tobies has pointed out, it was Felix Klein, the gifted organiser and politician of his discipline, who used the models to 'make room' for mathematics (Tobies 1992). Mathematicians had their 'seminars', but these were associations of teachers and students with membership but not necessarily a room, though if there was one it would in some cases also hold a small library (Lorey 1916, chap. 4). In Erlangen and Munich Klein asked for more. When Klein received his first chair in Erlangen he requested a reading room with a library and a collection of models. The reading room was realised, and here he established the first collection of models and instruments (including a calculating machine) in a German mathematical seminar (Tobies 1981, 38–39). When he accepted a call from the polytechnic in Munich in 1875, Klein asked for more, namely "rooms for a new mathematical institute to be established". He specifically pointed out the necessity of a workshop to produce models and another room for the collection, and demanded an assistant and a budget for this work (Tobies 1992, 759). In the same year a new second chair for higher mathematics was established and Alexander Brill called to it. Klein and Brill became the directors of the mathematical institute and started the extensive production of models. Brill's brother Ludwig, who had a printing works and a bookshop in Darmstadt, took on the publication of the models.[12] Next, in Leipzig, where Klein went in 1880, he established a 'mathematical institute' with two departments, the collection of models (with workshop) and the mathematical seminar (König 1981, 61–68). At the turn of the century mathematical institutes, as connected systems of rooms, were still rare, but every university had a more or less extensive collection of models (Lorey 1916, 311–34). Klein had paved the way. His own interest in models was deep, but in other cases the motive for the collection may just have been the quest for rooms and self-representation.

Klein's report on his early contacts with models relates to the initial attempts, mainly by Clebsch, then at Karlsruhe, to organise the mathematicians (Klein 1922a). At the first meetings no papers were read, but models were presented. We can only speculate about their communicative function. Even if not all of the mathematicians present would have been interested in the mathematics of the model, they would have been able to see, for example, that Prof. Wiener from the technical college did not do only elementary, descriptive geometry but serious mathematics. The models would in any case represent the mathematician, signal the reality of mathematics, and, as objects of aesthetic and intellectual pleasure, they were also gifts of a kind. As a gift they would help to establish communality, and by partaking of both physical reality and pure mathematics they bridged the problematic gap between the manual (technical college) and the intellectual (university). I suspect that models were indeed functional as a means of symbolic exchange between mathematicians, and that in the years before there was applied mathematics as an answer to the attacks from the engineers, this means of exchange was of special importance for mathematicians from the different institutional settings.

By the time the Mathematicians' Association was founded these relations were well established and where academic mathematics was, there was generally also a model collection. As in the photo of Gutzmer, models were representing mathematics and the mathematician. They could be used to introduce students and laypeople to mathematics in a way that made them see what it was about. Mathematics, which had always been a difficult subject for 'popular' science, could be shown to have an aesthetic value that appealed to non-specialists. In this way, that is with fairly elementary 'real' mathematics (the formulae), even the arch-modernist in mathematics, David Hilbert, in 1920–21 gave a popular lecture course on *Anschauliche Geometrie*, in which he used models extensively. The lectures were published with a considerable number of photos of models (Hilbert and Cohn-Vossen 1932).[13] But by that time Hilbert's modernist axiomatic geometry was the progressive research program in mathematics, and it did not care for *Anschauung* at all.

Felix Klein had his (more seriously mathematical) show with models at the Columbian Exposition in Chicago. The whole enterprise, in which the models again played their role as objects of symbolic exchange, was, however, clearly an act of cultural imperialism, as David Rowe and Karen Parshall have shown. They quote from Klein's report to the German government: "Without question, *at the present time and for the immediate future, America represents the richest and most promising object for scientific colonization*" (Parshall and Rowe 1993, 45, Klein's emphasis).

Models obviously had multiple functions in the formation of the mathematical profession. As embodied mathematics they (or images of them) could be traded over many borders and made sense not only to all sorts of mathematicians but also to physicists and engineers, and maybe philosophers. But this trade was perhaps too lively, and this raised questions about certainty and existence in mathematics. The further rise of modernism, and especially David Hilbert's formalist program of an axiomatic mathematics that ultimately had no other source of existence and authority than itself, can be interpreted as a next step in the formation of the profession. If mathematics rests in itself, no other science can legitimately discuss the mathematical principles of truth, existence, and value. A mathematics that takes *Anschauung* or intuition as essential will have to face questions from philosophers, psychologists, and maybe neurologists. Modernist purism restricted mathematics to the construction of strictly regulated worlds of meaning made from formal typographical sign systems. For other tasks there were the applied mathematicians (Mehrtens 1990, chaps 2.2, 5.4).

CONCLUSION

While there was no longer any place in modern mathematics for the models, they were taken over into modern art. I have mentioned Naum Gabo, who had studied natural science and engineering at the Munich polytechnic and had seen the mathematical models. In his sculptures he wanted to show that spatial form is not identical with volume, that the elements of artistic production—like the line, the plane, and the surface—have their own force, independent of the natural world (Gabo 1971). Another artist who transferred mathematical material into his work is Georges Vantongerloo, and later came Max Ernst, who appears to have discovered the models of the Poincaré Institute in Paris for art and showed them to Man Ray, who made his photographs. Ernst himself made extensive use of images from Schilling's 1911 catalogue in his collages (Vantongerloo 1986; Werner 2002, chap. 6). While Gabo and Vantongerloo took figurative and constructive principles from mathematical models as material for their artistic work, the surrealists took the models and their images themselves as material. Both groups argued about the relation of art, mathematics, the natural sciences, and their different relations to nature. To the artists the models were not art because they were the products of mathematicians meant to represent mathematics; and they were taken to be without relation to the irrational (Werner 2002, 118, on Zervos 1936). Used by artists, however, they could become part of art.

In mathematics the models had, for Felix Klein and for a short time, been epistemic things; later he interpreted them as applied mathematics. But by the end of the century mathematicians took the models as imperfect representations of geometrical entities that could be used as an aid in communication about mathematics. The models were not 'evidence' in any sense of the term. They were boundary objects, good for trading meanings and for showing figures but illegitimate as objects of pure mathematics and, except for Klein and maybe a few others, they were not applied mathematics either, because they were not of the real world but of mathematics.

In the early twentieth century the models crossed the border from mathematics to art. While the visual part of the mathematical discourse faded away, the discourse of art was enriched with new analytical forms of the representation of spatiality and temporality (Krauss 1981, chap. 2). The mathematicians had produced objects of aesthetic and intellectual interest, made to be shown and seen. They had used them for the self-representation of mathematics and mathematicians, and they had done this at a time when and in a social setting where no one would have identified such objects as art. But because of the intrinsic interest in geometrical figures and the possible forms of mathematical regularities in space, and because of the models' materiality, they are close to what we have come to understand as 'art'. These mysterious 'object-beings' came out of the geometrical imagination, the virtual reality of nineteenth-century geometry. Once present in the physical world they found uses of various sorts. They are not art, not mathematics, they are just mathematical models. Their shape can be explained mathematically, but the only way to explain their physical existence is to tell their history.

NOTES

1. A brief description is given by Gerd Fischer in his introduction to the reprint of the catalogue (Dyck 1892–93; Fischer 1994). A catalogue of 304 entries, many of which covered more than one object, had already been prepared for 1892, but the meeting was cancelled because of a cholera epidemic. For the Munich

exhibition a supplement with another 304 entries was published. The reprint contains both parts.

2. Some Leninist epistemology apart, I recommend the philosophical and sensitive historical discussion of applied mathematics and mathematical modelling in Paul and Ruzavin 1986.

3. The journal *Der Mathematische Unterricht* has had special issues on platonic solids (1991, vol. 37, no. 4) and physical 3-D models (1997, vol. 43, no. 5). One sign of a possible revival is the wide interest in Albrecht Beutelspacher's efforts to create 'mathematics you can touch' (*Mathematik zum Anfassen*); with undergraduate students he produced an exhibit of mathematical models (Beutelspacher 1998) and is currently planning a museum of mathematics (Walter 2000).

4. In its section 'The mathematical tourist' the *Intelligencer* also published a paper on the collection of mathematical models at the Mathematical Institute of the University of Göttingen (Mühlhausen 1993).

5. A leaflet in Fischer 1986, entitled "Mathematics in plaster", describes the model as 17 × 15 × 24 cm; the price was DM 198.

6. For the use of the concept of representation in cultural studies, see Hall 1997; for its use in history of science, Rheinberger 1997, chap. 7.

7. In a recent exhibition catalogue (Billeter 1997a) two photos of mathematical models (one of Kuen's surface photographed by Man Ray) appear, though without specific comment, as illustrations to a chapter titled "Sculpture which exists through photography" (Billeter 1997b, 45).

8. This is my very literal translation from the French as quoted in Werner 2002, 148–49.

9. It would, for example, be of interest to look at doctoral dissertations like Rodenberg's, to see to what extent models were taken as a legitimate 'result' of mathematical work.

10. For a survey of the situation of German mathematics at the end of the nineteenth century, see Mehrtens 1990, chap. 5.

11. Brill reports that mathematicians in France appear not to have developed much further interest in models (Brill 1887, 73). The Munich catalogue (Dyck 1892–93) lists only two French exhibitors; about 25% of the exhibitors were non-German, half of whom were from Britain, where Arthur Cayley and Olaus Henrici (a student of Clebsch) appear as the main figures in work with models (Brill 1887, 73). Dyck's introduction to the catalogue also mentions that mathematical models and instruments were included in the exhibition at the South Kensington Museum in London in 1876 (Dyck 1892–93, iii).

12. A recipe for the production of modelling clay from Munich is reprinted in Fischer 1986, viii. While in Erlangen, Klein had bought the models from a French publisher. Brill published models until 1899, when the publisher Martin Schilling took over; see the extensive catalogue, Schilling 1911. The last sign of Schilling's activities is a letter of 1932 to the mathematical institute in Göttingen, reporting

that "in recent years no new models have appeared .... [T]here are various new models in preparation. However, as a result of the bad and unpredictable market, we have held back their production" (Fischer 1986, ix–x).

13. One of the photos shows a very specific 'model', namely a bust of Apollo with parabolic curves on the face and the neck, which Klein had ordered to be made in order to see whether the "artistic beauty of a face had its cause in certain mathematical relations"—with no result (Hilbert and Cohn-Vossen 1932, 174 no.1, photo 175; also in Fischer 1986, 1: fig. 78).

REFERENCES

Anderson, Benedict. 1983. *Imagined Communities: Reflections on the Origin and Spread of Nationalism*. London: Verso.

Baldus, Richard. 1921. "Mathematik und räumliche Anschauung. Antrittsvorlesung gehalten an der Technischen Hochschule Karlsruhe am 15. Mai 1920." *Jahresbericht der deutschen Mathematiker-Vereinigung* 30: 1–15.

Beutelspacher, Albrecht. 1998: "Geometrische Modelle—Mathematik zum Anfassen." In *Mathematik im Wandel*, ed. Michael Toepell, 17–24. Hildesheim: Franzbecker.

Billeter, Erika, ed. 1997a. *Skulptur im Licht der Photographie. Von Bayard bis Mapplethorpe* [Exhibition catalogue]. Bern: Benteli.

———. 1997b. "Zur Ausstellung, 2. Teil. Der erweiterte Dialog." In *Skulptur in Licht der Photographie. Von Bayard bis Mapplethorpe*, ed. Erika Billeter, 37–57. Bern: Benteli.

Brill, Alexander. 1887. "Über die Modellsammlung des mathematischen Seminars der Universität Tübingen." *Mathematisch-naturwissenschaftliche Mitteilungen* (Stuttgart) 2: 69–80.

Dyck, Walter, ed. 1892–93. *Katalog mathematischer und mathematisch-physikalischer Modelle, Apparate und Instrumente*. Munich: Wolf. (Reprinted Hildesheim: Olms 1994.)

———. 1893. *Deutsche Unterrichtsausstellung in Chicago, 1893: Special-Katalog der mathematischen Ausstellung*. Berlin: n. p.

Epple, Moritz. 1999. *Die Entstehung der Knotentheorie. Kontexte und Konstruktionen einer modernen mathematischen Theorie*. Braunschweig: Vieweg.

Fischer, Gerd. 1986. *Mathematische Modelle / Mathematical Models*, 2 vols. Braunschweig: Vieweg.

———. 1994. "Vorwort." In *Katalog mathematischer und mathematisch-physikalischer Modelle, Apparate und Instrumente* [1892–93], ed. Walter Dyck, vii–xxii. Hildesheim: Olms.

Gabo, Naum. 1971. *Naum Gabo. Skulpturen, Gemälde, Zeichnungen* [Exhibition catalogue]. Berlin: Nationalgalerie.

Gray, Jeremy. 1988. [Review of Fischer 1986.] *The Mathematical Intelligencer* 10, no. 3: 64–69.

———. 1989. "Algebraic geometry in the late nineteenth century." In *The History of Modern Mathematics*, ed. David E. Rowe and John McCleary, vol. 1: 361–85. San Diego, Calif.: Academic Press.

Hall, Stuart, ed. 1997. *Representation: Cultural Representations and Signifying Practices*. London: Sage.

Hensel, Susann. 1989. "Die Auseinandersetzung um die mathematische Ausbildung der Ingenieure an den Technischen Hochschulen in Deutschland Ende des 19. Jahrhunderts." In *Mathematik und Technik im 19. Jahrhundert in Deutschland*, ed. Susann Hensel, Norbert Ihmig, and Michael Otte, 1–111 and appendix. Göttingen: Vandenhoek & Ruprecht.

Heymann, Therese Thau. 1992. *Seeing Straight: The f.64 Revolution in Photography*. Oakland, Calif.: The Oakland Museum.

Hilbert, David, and Stefan Cohn-Vossen. 1932. *Anschauliche Geometrie*. Berlin: Springer.

Klein, Felix. 1894. *The Evanston Colloquium Lectures on Mathematics*. New York: Macmillan.

———. 1922a. "Vorbemerkungen zu den Arbeiten über anschauliche Geometrie." In Felix Klein, *Gesammelte mathematische Abhandlungen*, vol. 2: 3–6. Berlin: Springer.

———. 1922b "Über Flächen dritter Ordnung." In Felix Klein, *Gesammelte mathematische Abhandlungen*, vol. 2: 11–56. Berlin: Springer.

———. 1921–23. *Gesammelte mathematische Abhandlungen*, 3 vols. Berlin: Springer.

———1926. *Vorlesungen über die Entwicklung der Mathematik im 19. Jahrhundert. Teil 1*. Berlin: Springer.

———. 1974. *Das Erlanger Programm. Vergleichende Betrachtungen über neuere geometrische Forschungen (1872)*, ed. Hans Wussing. Leipzig: Akademische Verlagsanstalt.

König, Fritz. 1981. "Die Gründung des Mathematischen Seminars der Universität Leipzig." In *100 Jahre Mathematisches Seminar der Karl-Marx-Universität Leipzig*, ed. Herbert Becker and Horst Schumann, 41–71. Leipzig: Deutscher Verlag der Wissenschaften.

Krauss, Rosalind E. 1981. *Passages in Modern Sculpture*. Cambridge, Mass.: MIT Press.

———. 1985. "Photography's exquisite corpse." In *In the Mind's Eye: Dada And Surrealism*, ed. Terry Ann R. Neff, 43–61. New York: Abbeville Press.

Kuen, Theodor. 1884. "Über Flächen von konstantem Krümmungsmaass." *Sitzungsberichte der mathematisch-physikalischen Classe der königlich bayerischen Akademie der Wissenschaften* 14: 193–206.

Lorey, Wilhelm. 1916. *Das Studium der Mathematik an den deutschen Universitäten seit Anfang des 19. Jahrhunderts*. Leipzig: Teubner.

Mehrtens, Herbert. 1990. *Moderne—Sprache—Mathematik. Eine Geschichte des Streits um die Grundlagen der Disziplin und des Subjekts formaler Systeme*. Frankfurt am Main: Suhrkamp.

———. 1996. "Mathematics and war: Germany 1900–1945." In *National Military Establishments and the Advancement of Science and Technology: Studies in 20th-Century History*, ed. Paul Forman and José Manuel Sánchez-Rón, 87–134. Dordrecht: Kluwer.

Mende, Michael. 1994. "Modelle und Zeichnungen. Anschaulichkeit als Prinzip der Ausbildung in der Maschinenlehre und Technologie während des 19. Jahrhunderts." *Dresdener Beiträge zur Geschichte der Technikwissenschaften* 23, no. 2: 126–134.

Meyers Konversations-Lexikon. Ein Nachschlagewerk des allgemeinen Wissens. 1894. 5th edn, vol. 1. Leipzig: Bibliographisches Institut.

Mühlhausen, Elisabeth. 1993: "Riemann surface—crocheted in four colors." *The Mathematical Intelligencer* 15, no. 3: 49–53.

Newhall, Nancy. 1965. *Edward Weston: The Flame of Recognition*. New York: Aperture.

Parshall, Karen V., and David Rowe. 1993. "Embedded in culture: Mathematics at the World's Columbian Exposition of 1893." *The Mathematical Intelligencer* 15, no. 2: 40–45.

Paul, Siegfried, and Georgij Ruzavin. 1986. *Mathematik und mathematische Modellierung. Philosophische und methodologische Probleme*. Berlin: Deutscher Verlag der Wissenschaften.

Rheinberger, Hans-Jörg. 1992. *Experiment—Differenz—Schrift. Zur Geschichte epistemischer Dinge*. Marburg: Basilisken Presse.

———. 1997. *Toward a History of Epistemic Things: Synthesising Proteins in the Test Tube*. Stanford: Stanford University Press.

Rodenberg, Carl. 1879. "Zur Classification der Flächen dritter Ordnung." *Mathematische Annalen* 14: 46–110.

Rotman, Brian. 1993. *Ad Infinitum.... The Ghost in Turing's Machine: Taking God Out of Mathematics and Putting the Body Back In*. Stanford: Stanford University Press.

———. "Thinking dia-grams: Mathematics, writing, and virtual reality." *South Atlantic Quarterly* 94: 389–415.

Schilling, Martin. 1911. *Catalog mathematischer Modelle für den höheren mathematischen Unterricht*, 7th edn. Leipzig: Martin Schilling.

Stäckel, Paul. 1915. *Die Mathematik an den Technischen Schulen*. Leipzig: Teubner.

Timerding, Heinrich E. 1912. *Die Erziehung der Anschauung*. Leipzig: Teubner.

Tobies, Renate. 1981. *Felix Klein*. Leipzig: Teubner.

———. 1992. "Felix Klein in Erlangen und München. Ein Beitrag zur Biographie." In *Amphora. Festschrift für Hans Wussing zu seinem 65. Geburtstag*, ed. Sergei S. Demidov et. al., 751–72. Basel: Birkhäuser.

Vantongerloo, Georges. 1986. *Georges Vantongerloo* [Exhibition catalogue]. Berlin: Akademie der Künste.

Vinogradov, I. M. 1990. *Encyclopaedia of Mathematics*, vol. 6. Dordrecht: Kluwer.

Volkert, Klaus Thomas. 1986. *Die Krise der Anschauung. Eine Studie zu formalen und heuristischen Verfahren in der Mathematik seit 1850*. Göttingen: Vandenhoeck & Ruprecht.

Walter, Stefanie. 2000. "Kurvendiskussionen sind schön. In Gießen entsteht ein Mathematik-Museum." *Frankfurter Allgemeine Zeitung*, no. 174 (July 29): 10.

Werner, Gabriele. 2002. *Mathematik im Surrealismus. Man Ray—Max Ernst—Dorothea Tanning*. Marburg: Jonas.

Zervos, Christian. 1936. "Mathématiques et art abstrait." *Cahiers d'Art* 1/2: 4–10.

CHAPTER ELEVEN

Science, Art, and Authenticity in Natural History Displays

Lynn K. Nyhart

NATURAL HISTORY MUSEUM EXHIBITS AS MODELS

Natural history group displays are among the best-known three-dimensional (3-D) didactic representations in the sciences. These displays, which came to be identified with natural history museums internationally in the twentieth century, come in many varieties. The simplest typically show several mounted animals of one species—usually including a male, female, and young—posed together in 'characteristic' positions and unencumbered by scenery. Others, usually called 'habitat groups', are more elaborate, placing the animals in meticulously reproduced naturalistic settings that include the features characteristic of their surroundings, such as particular plants or rocks, or a leaf-covered forest floor. The most elaborate of these forms, the habitat diorama, features a painted background that fades into the 3-D scene, so that the viewer's eye is drawn into the scene as a whole (Wonders 1993, 19–22).

These naturalistic representations are unlike most other models in science. Models of planetary motion, atomic structure, or DNA, for example, aim to render visible that which we cannot see (Meinel, Francoeur, de Chadarevian, this volume); a natural history display usually depicts something we already recognise as an animal, a plant, or a section of landscape. Nor do natural history group displays resemble physical embodiments of abstract ideas or theories, such as models of mathematics or economies (Mehrtens, Morgan and Boumans, this volume); as simple, 'realistic' slices of life, they appear, at least, to be profoundly non-theoretical. Nor are they just like anatomical or obstetrical models intended for teaching practitioners in a hands-on way how the body is structured or how its processes work (Mazzolini, this volume); traditionally, natural history museum exhibits are for looking at, not for manipulating. They are neither scaled up, like embryological models, nor scaled down, like archaeological models or ship models (Hopwood, Evans, Schaffer, this volume); they are life-sized, presenting organisms as they look in nature.

Natural history group displays may thus be said to occupy a very special niche among 3-D models in the sciences. They are full-scale, realistic representations of once-living creatures. Crucially, actual parts of original animals, plants, or landscapes form a prominent and necessary feature of the display, a characteristic that James Griesemer has felicitously captured in the term 'remnant models' (Griesemer 1990).[1] Indeed, the key quality of such models is not their analogical power, their manipulability, or their ability to render things on a human scale, but rather their authenticity. When a family stands before a natural history diorama or walks around a free-standing biological group, they are in the presence of 'the real thing'. No matter how complex the process of reconstruction, this experience of authenticity depends on the fact that the skins and feathers of the animals displayed once covered living creatures.[2]

These peculiar qualities of natural history group displays, which have been with us for so long as to seem almost inevitable, raise intriguing historical questions. Why and how did these kinds of displays, and the value of authenticity that they carried with them, come to play such an important role in natural history museums? What relationship did authenticity bear to notions of scientific validity? Given that considerable effort had to go into constructing such displays, what made a particular representation of nature 'realistic' or 'true to life'? Which products of art and artifice were to be allowed in the name of legitimate verisimilitude, and when was the line crossed into illegitimate artificiality? This chapter examines how museum workers in Germany answered these questions at the turn of the twentieth century, when such displays—known in Germany collectively as 'biological groups'—were a new and exciting but problematic feature of the museum world.

I begin by arguing that, given the novelty of the group display within public museums around 1900 and its close resemblance to more popular entertainments, its status as science was not automatic but had to be achieved. The following two sections examine the particular mixes of science, art, and authenticity expressed in biological groups at two very different museums, the sources for which lend themselves to different insights. The biological groups at the Altona City Museum near Hamburg generated a lively controversy in the professional museum literature; the programmatic aims of the curators at the Oceanography Museum (Museum für Meereskunde) in Berlin, by contrast, were not made explicit in the published record but are revealed in archival correspondence and unpublished reports. As the chapter's final section shows, practical considerations also played a significant role in shaping the construction of the displays.

MASS MUSEUM ANXIETY

In late 1905, the recently founded German journal for museum professionals, *Museumskunde* (*Museology*), published a review of the 1903–4 *Annual Report* of the Trustees of the Field Museum in Chicago. Rather than merely summarising for German readers the activities of one of the world's richest natural history museums, its author Benno Wandolleck, an assistant at the Dresden natural history collection, also criticised the group displays (or 'tableaux', as he called them) shown in several of the report's photographs.

> Although such tableaux captivate the lay public very much, one must always bear in mind that in fact they forsake the scientific foundation that a scholarly museum must under all circumstances maintain, coming dubiously close to those institutions that go by the name of 'panopticon'. Setting aside the fact that it requires an eminently artistic gift to prepare such tableaux so that they correspond to some degree to the truth, one must keep in mind that one is mostly forced to work with artificial materials; that one can use no real rocks, no real water; that the dried plants, despite the best painting, always give the impression of hay and the mannequins can never get rid of the stiffness and the lack of interest characteristic of mannequins. The slightest lapse makes the thing laughable, and even if all is correct, the groups will always have a tedious effect. They do nothing for the understanding of nature, nor will the public be educated in art (Wandolleck 1905, 236).

This critique neatly linked several key issues. In the name of democratic uplift and entertainment, the poor art of the tableau threatened to subvert scientific truth, substituting instead the sensationalism and cheap tricks of the panopticon, or commercial waxworks display. Because it was bad art and bad science, the tableau was doomed to fail as mass education as well.

Though couched as a criticism of a distinctly American form of exhibit, the review could also be read as a warning to the European museum community not to succumb to the dangerous allure of the tableau or biological group exhibit.

And indeed, that is how it was read by Otto Lehmann, director of the Altona City Museum, which when it opened in 1901 was viewed as a pioneer among German museums in the use of biological groups to educate the masses (Hinrichsen 2001). Responding directly to Wandolleck in *Museumskunde*, Lehmann expressed admiration for the American approach, which, he argued, was characterised by the healthy attitude that museums were for all the people, not just some of them. German museums should follow the American lead—after all, these public institutions were largely paid for through taxes collected from the entire citizenry. Moreover, in contrast to Wandolleck's claim of the laughable artificiality of such groups, Lehmann asserted that they *could* be truly lifelike. Admittedly, many were less realistic than they should be, but to reject the form itself because of its frequently imperfect execution would be foolish; when successful, a museum with such displays would be "for all levels of society the place that is repeatedly sought out as an ever-flowing source of interior joy and contemplative self-education" (Lehmann 1906, 66).

Wandolleck's review and Lehmann's reply expressed important tensions within the German museum community in the early twentieth century. Like other museums internationally, German natural history museums were in the midst of a dramatic reform movement that had begun in the late 1880s and would last well into the new century. Following examples set in the 1870s by the American Museum of Natural History in New York and prominently taken up in the early 1880s by the U.S. National Museum of Natural History and the British Museum, reformist curators in Germany sought to open up publicly funded museums to educate a much broader public than before.[3] They separated research collections from public display areas (placing researchers in the back rooms along with their study materials), increased the opening hours for casual visitors, improved the informational content of labels, and devoted considerable time, effort, and money to constructing public exhibits that would attract and educate visitors not yet versed in the basics of natural history.

Not all museum curators shared equal enthusiasm for catering to 'the people' (often equated with the working class; see *Museen* 1904).

Research-oriented curators at well-established natural history collections, such as those at Hamburg, Dresden, and at some universities such as Berlin and Breslau, while themselves engaged in important exhibit innovations, sought to strike a judicious balance between the attention and resources they devoted to their traditional research tasks of specimen collecting and analysis, on the one hand, and the newer, additional tasks associated with public exhibits and education, on the other. Striking the 'right' balance was particularly problematic for university-educated, research-oriented curators with strong academic ties because the status of their scientific subdiscipline, systematics, was simultaneously coming under challenge from laboratory-based scientists. To accept public education as their leading task seemed to many museum naturalists to accept a lower status as mere archivists and displayers of nature, rather than as researchers producing new knowledge.

Given this situation, it should not surprise us that the greatest enthusiasm for innovative public exhibits came not from the existing research museums but rather from new museums with particular reasons for wanting to attract the general public. The largest number of such museums were founded to serve local communities and valued broad participation above professional research. Many of these museums, especially the 'Heimat' (homeland) museums devoted to characterising the local region that began to spring up across Germany at the turn of the century, contained a mixture of exhibits on history, folklore, and nature designed to educate people on all aspects of their locality. Other museums conceived around novel projects, such as the museum of ethnology, natural history, and commerce in Bremen, which was structured primarily to display the economic benefits of colonialism, and the oceanographic museum in Berlin, which took as its purview the entirety of the ocean's properties and human interactions with it, also made especially bold and highly publicised efforts to draw in the public with innovative tableau-style exhibits. Both the Heimat museums and these other new museums used biological groups to show off the natural history of the region on display, be it local, distant, or even undersea.

Thus the new fashion for biological groups must be understood as it was at the time, as part of a profoundly challenging destabilisation of assumptions about the projected audiences, purposes, and structure of the museum. Within this context, the new turn toward the public posed the acute problem of how to open up to a new audience while maintaining the traditional dignity and status of the museum as a scholarly institution. Lehmann was an articulate spokesman for the democratic ideal of educating the masses in what Germans viewed as the 'American' style, and for an approach to education that did not divide knowledge along strict disciplinary lines, while Wandolleck was one of the more conservative defenders of the research role of the traditional natural history museum. Most museum professionals stood somewhere between the two, seeking a new balance that catered to a broad public while continuing to advance a research agenda within the museum. To the professional museum community, biological groups encapsulated both the possibilities and the dangers of opening up to the masses.

Museum officials acknowledged that the rows upon rows of taxonomically organised specimens that typified prominent natural history research collections held little appeal to lay visitors. Comparative anatomical exhibits, which set anatomical structures of different organisms next to one another for comparison, attracted perhaps wider interest, especially if they included human parts in their juxtaposed skulls, skeletons, and bottled brains. However, what clearly appealed most to visitors were more dramatic renderings of animals reconstructed in lifelike poses. Although birds had been exhibited in this way in some museums for some time, with family groups arrayed in and around a nest, before the 1890s few German museums displayed mammals in groups (Wonders 1993, 23–45; Lampert 1904, 23).

Group displays were widely visible in other settings, however. As Karen Wonders has pointed out, from the mid-nineteenth century, competitions in 'artistic taxidermy' (reconstructing creatures in lifelike, dynamic poses) became a feature of the international exhibitions and trade fairs that did so much to shape the burgeoning mass culture of Europe and America (Wonders 1993, 23–45). Those exhibitions also provided sites for related visual forms such as ethnographic displays of living men and women performing 'native' tasks and historical village reconstructions with Europeans dressed in old-fashioned clothing. Both these kinds of tableaux were echoed in non-living ethnographic and historical exhibits using wax models (Greenhalgh 1988; Bruckner 1999; Wörner 1999). Panoramas were a related illusionistic form that underwent a new boom beginning in the 1880s (*Sehsucht* 1993; Schwartz 1998, 149–76; Comment 1999, 66–76); their construction sometimes included not only the characteristically huge, circular paintings (often of historical events such as battles) but also natural materials such as grass and stone in the foreground to enhance the realism of the scene (Sternberger

1977, 9–10). Finally, the private commercial collections of waxworks and miscellaneous curiosities that in Germany went by the name of 'panopticon' particularly attracted the anxieties of museum men (König and Ortenau 1962; Bruckner 1999, 127).[4]

What all these display forms had in common was, first, they were essentially theatrical in their design; second, they all claimed to be 'realistic', at least in the sense of seeking anatomical accuracy; and third, they were extremely popular—they drew in the crowds. Thus when reformers sought to open museums to the masses, these various theatrical forms of display had obvious appeal.

But borrowing from popular commercial spectacle carried risks. Many museum professionals were deeply concerned to maintain the demarcation between their institutions, as sites of science and education, and what Lehmann called the "imbecilic panopticon that only satisfies sensual pleasures" (Lehmann 1906, 66). From the perspective of public museum men, such commercial establishments, although often claiming to demonstrate natural historical, anatomical, and ethnographic scenes and objects, existed mainly to titillate their viewers. Emil Eduard Hammer's three-storey panopticon in Munich, for example, offered such gruesome exhibits as "Buried alive" (*Lebendig begraben*), showing a horrified, long-haired, kohl-eyed maiden pushing off the lid of a coffin; a "Gorilla carrying off a farmer's daughter"; a polar bear attacking an underclad woman in the act of hanging herself inside a bear-pit (Fig. 11.1); and a piece titled "Nightmare", showing a small ape leering at an unconscious lace-garbed woman while perched on her stomach (König and Ortenau 1962).[5] No wonder advocates of scientific exhibits of animals sought to distance themselves from such commercial ventures.

Perhaps the most repeated concern among scholarly museum men was the possibility of losing, as Wandolleck put it, the "scientific foundation" that a museum must "under all circumstances maintain" (Wandolleck 1905, 236). It was in this connection that the panopticon was most often invoked and where the dubious truth value of group displays became most prominent. As Friedrich Dahl, assistant at the Berlin natural history museum, emphasised, "Truth is the first thing that the researcher owes the lay public, and the truth should therefore be the first principle that the director of a museum should make his duty." He continued,

Figure 11.1 "Lunatic beauty in the bear-pit." Late-nineteenth-century display by Emil Eduard Hammer in Hammer's Munich panopticon. From Hannes König and Erich Ortenau, *Panoptikum* (Munich, 1962), 111.

> It has been said in favor of panopticon-exhibits in museums that the public wants them, and that may be so. But should the researcher let himself be guided by that? I am persuaded that the public would stream in in even greater crowds if—at the cost of truth in the reproduction—all animals could be made to move through clockwork gears—if, for example, the lion would open his jaws and children were allowed to throw something in (Dahl 1908, 129).[6]

As Dahl's sarcastic comments suggest, museum curators faced a delicate problem. Although the spectacular group displays clearly attracted the public, this form needed in turn to be moulded into one suitable to convey both the seriousness of the scientific-educational enterprise and the truths of nature. How was this to be done? And what truths were the group displays to reveal? What aspects of science could the group form properly show? Although curators agreed that the most important quality of biological groups was that they represented a true reality, their commitment to realism did not solve the problem of what was actually to be displayed.

REFINING VISITORS' SENSIBILITIES AT THE ALTONA CITY MUSEUM

In 1903, Lehmann rehearsed his views on group exhibits at a conference on 'Museums as sites of popular education [*Volksbildungsstätten*]'. As a general Heimat museum covering the natural history, history, and folklore of the home region of Schleswig-Holstein, the broad aim of the Altona City Museum, according to Lehmann, was to teach 'the people' how to see—how to look thoughtfully at objects, whether of nature or culture, to take in their meaning, and to appreciate their beauty. For such a project, the visitors had to want to see, and to that end a homely scene—whether of a peasant's kitchen, a waxwork family dressed in traditional local costumes, or a beaver building its lodge—was far more gripping than a 'scientifically' organised array of cookpots, dresses, or stuffed rodents. That is, what is appropriate to display to such an audience is, as Lehmann put it, not "knowledge [*Wissen*] but living, fruitful views [*Anschauungen*] of nature". As he repeated later on, specifically with reference to the museum's natural history section, "It is not the intention simply to convey a certain quantity of facts, but to educate the museum's visitors into thinking humans; they should become accustomed to observing, in order to understand nature" (Lehmann 1904, 37, 41; for a similar view, see Schauinsland 1904, 30).

To exemplify his pedagogy, Lehmann described an exhibit depicting a group of elk being attacked by a pack of wolves (Fig. 11.2). "The way that the wolf hunts is expressed in these animals; the breathless hounding by the attacking animal", as well as the doe's defensive movements, the buck's turn of the antlers, and the stupidity of the background animal in not knowing how to react to the attack, are all present to the careful observer. According to Lehmann, the group scene presented a far better picture of the natural history of the elk than a single stuffed elk alone. The visual information offered by the

Figure 11.2 Elk and wolf exhibit, Altona City Museum. From Otto Lehmann, *Festschrift zur Eröffnung des Altonaer Museums* (Altona, 1901), 14.

group display was enhanced by the accompanying label, which, in keeping with the historicist and nostalgic tenor of the museum as a whole, explained that elk had been common in the German primaeval forest in Caesar's time, but the last wild one had been caught in 1746; the last wild wolf in central Germany was shot in 1838 (Lehmann 1904, 42–43).

In explicating the advantages of such biological groups, Lehmann distinguished his educational form from commercial or spectacular ones by rearticulating what might be seen as a sensationalistic appeal to drama within the pedagogical vocabulary of instruction through *Anschauung*, or teaching through the direct perception of real objects, which had often been advocated in natural history teaching. Drawing on the philosophy of perception as it had been developed in educational psychology, late-nineteenth-century pedagogical theorists argued that the most basic level of mental intake above that of raw sense perception, but preceding that of 'understanding', was

Anschauung, an intuited sense of the world mediated through vision. Educators had long taken the cultivation of *Anschauung* to be essential to elementary education and a necessary first step to developing higher levels of comprehension (Deussing 1884; Harder 1891; see also Secord, Mehrtens, this volume). Like other museum educators, Lehmann adapted the rhetoric of *Anschauung* to teach 'the people' by stimulating their imagination and emotions through visual impressions. As he put it, biological group exhibits would convey to the naïve visitor a quantity of information about causal relationships that would otherwise be mystifying: "[Y]ou poor thing, you can't know why nature allowed these forms, these colours, to be, why they fulfill their tasks in only such and such a way—but see, here the animal lives under these conditions—thus must it seek its food—these enemies threaten it—thus must it protect itself from them—and now all at once the form is understandable, and, what's more important, the presentation offers a lively idea [*einen lebendigen Begriff*] that will bear a hundred fruits" (Lehmann 1906, 62). Thus education would be served by the very same means that might, under other conditions, simply satisfy a visitor's thirst for spectacle.

To Lehmann, then, it was important that the display of activity be both realistic and visually and emotionally charged. Dramatic scenes rendered abstract scientific concepts like 'the struggle for existence' or 'the conditions of existence' comprehensible by the masses at the intuitive level of *Anschauung*. The use of drama could also help develop the aesthetic sensibilities. At the same time, Lehmann also developed a new, sparer aesthetic for his biological groups. Like many other museum educators in Germany and beyond, he shuddered at what he called the 'horribilia' of elaborately faked woodlands or flower-filled fields.[7] He advocated a minimalist, dull-coloured background with just a few suggestions of the physical setting, which kept the attention focused on the animals and their behavioural interactions—the essentials that he was trying to convey (Lehmann 1906, 64). To allow visitors to study the exhibit closely, he preferred open, free-standing group displays that visitors could walk around, rather than placing the animals in the more theatrical diorama-style glass-fronted niche; in addition, this choice entailed dispensing with any background painting. These decisions also distanced the display visually from the stage-mounted style of contemporary commercial displays, which frequently placed the figures before a painted backdrop or a theatre-like curtain (see illustrations in König and Ortenau 1962, and Schwartz 1998).

In 1906 and 1907, Lehmann's approach to making the biological group more scientifically respectable came under sustained attack by his earlier opponent Wandolleck, who charged that the biological groups at Altona did not tell the truth about nature, but rather sacrificed it for drama. Considering the elk and wolf display, he pointed out that since both animals were now extinct, no living person had ever seen such a scene, certainly not the artist who constructed it. So it was a product of the imagination and had completely left the realm of solid (inductive, observational) science. Worse yet, in Wandolleck's judgement, this scene "shows us the characteristic features, that is, the mode of life, of neither the elk nor the wolf"; at best the viewer enjoys a moment of horror. And, he noted, the extinction of the elk in Germany (a fact emphasised in the museum's guidebook and on the signage) was not actually due to the depredations of the wolf, as one might reasonably infer from the display (Wandolleck 1906, 645).

Wandolleck also objected to Lehmann's pedagogical claims concerning the display's appeal and its ability to teach the uninstructed visitor. The freestanding structure of displays like that of the elk and wolves, for one, was highly problematic. Open to the depredations of dust and to discolouration by exposure to the light, the scenes were unlikely to appeal for very long. This form also meant that the visitor's gaze was not absorbed into the scene but all too easily distracted by other viewers standing opposite, further inhibiting the concentration required for any serious self-education. In addition, the very poses of the animals themselves, stiffly captured in a moment of high action, called attention to their unreality (Wandolleck 1906, 647). All in all, to Wandolleck, the public would be neither elevated aesthetically nor educated in scientific truth by the means Lehmann used.

The dispute between Lehmann and Wandolleck highlights the intricate ways in which judgements concerning scientific truth and aesthetic merit intertwined in the evaluation of museum displays. From a scientific perspective, museum scientists had to decide what standard of evidence was required to determine the validity of a scene. Did a scientist have to vouch for its actual existence, or just its plausibility? Did the artist or the preparator have to have seen it? Were photographs enough? For Wandolleck, if the scene were to be true, surely *somebody* had to have seen the scene depicted—otherwise it was a product of the artist's imagination, and subject to the dangerous slide away from truth and toward the panopticon. Lehmann, by contrast, was untroubled by a need for Wandolleck's level of exactitude, possibly in part because he disputed Wandolleck's insistence on the sharp boundary between scientific truth and artistic imagination. In contrast to Wandolleck, in Lehmann's view the aim of the biological group display was not merely to convey a correct representation of nature, or *Naturrichtigkeit*, but rather what Lehmann called 'the truth of nature', *Naturwahrheit*. "The slavishly working preparator strives for correctness, the artist for truth, and I am persuaded that biological groups can be constructed by an artist's hand to be true to nature [*naturwahr*] without forsaking the scientific foundation" (Lehmann 1907, 87). The difference between Lehmann and Wandolleck over the scientific value of 'artistic' renderings was intensified by their disagreement over the tastefulness of showing animals in motion: did such dramatic moments pull the visitor in or call attention to their own artificiality? Would catering to the crowd's desire for drama help elevate their sensibilities or would it simply lower the scientific value of the exhibit? Here, too, judgements concerning what constituted 'truth' are difficult to extricate from judgements concerning aesthetic appeal.

Collectively, the various disagreements between Wandolleck and Lehmann point to two different views of authenticity. Although concerned to convey true information about nature, Lehmann focused primarily on the experience of the visitor: an 'authentic' experience was one that was felt to be true in the breast of the viewer. Wandolleck's 'authenticity', by contrast, derived entirely from the authority of the exhibit's scientist-producer, who vouched for the veracity of the display based on his own empirical experience with, and scientific knowledge of, the objects in it. The ambiguity contained in the idea of authenticity is laid bare in their debate, revealing a tension over where the authorship of authenticity is to lie—in the scientist or in the visitor. In the context of the upheaval in museums at the time, this tension may be understood to have had a political dimension as well: whose interests, ultimately, was the museum to serve—the scientist's or the visitor's? Whose truths should be striven for? Much of the work done by the makers of museum exhibits revolved around finding ways to mediate that tension or to satisfy both needs.

Public discussions of the relationships among science, art, and truth, such as those found in the Lehmann–Wandolleck debate, were relatively rare in the German museum world, but the same concerns were regularly addressed implicitly by museum men designing their exhibits. One of the most detailed sets of surviving documents relating to these questions was produced by the staff at the Berlin Oceanography Museum, Germany's first in this area. Examining the fortunes of the biological exhibits of this museum offers more direct insight into the interplay between science, aesthetics, and practical issues than we usually obtain, as well as a set of concerns somewhat different from that revealed in the Altona museum.

REPRESENTING MARINE ECOLOGY AT THE OCEANOGRAPHY MUSEUM

Founded in 1900 and opened in 1906 as a highly atypical part of the University of Berlin, the Museum and Institute of Oceanography was dedicated to promoting a total science of the sea that would encompass physical oceanography, naval science and navigation, the technology and culture of fishing, and the study of life in the seas.[8] Within this broad plan, the study and presentation of life in the ocean and along its shores received a significant portion of the space and budget. The biological section of the museum was organised geographically and ecologically, using biological groups to depict the effects of climate and physical conditions on the life groupings there. Although the pedagogical assumptions about *Anschauung* may have been similar to Lehmann's, the more explicitly scientific agenda that stood in the foreground was tied to a different set of assumptions about audience, aesthetics, and authenticity. Rather than reconstructing scenes of a vanishing domestic nature threatened by civilisation, as Lehmann did, exhibits at the Oceanography Museum connected the visitor to a distant larger world she or he was unlikely to experience directly. Rather than focusing the viewer's attention on dramatic scenes of animal behaviour, these displays portrayed their animals as sharing a common physical environment and interacting with it, often without being engaged with each other in any visually obvious way. And rather than seeking mainly to evoke a visceral or intuitive connection to the drama of nature as a means of raising the viewer's aesthetic sensibilities, these exhibits worked to legitimate a new science through the reconstruction of the sites of its research. Indeed, the museum guide made a point of noting that scenes such as that of a coral reef on the Red Sea (Fig. 11.3) and of the South Pole (Fig. 11.4) represented distant places where marine scientists pursuing their new research agenda had actually gone to study, places that could not be represented by living creatures in zoos or aquaria because of their delicate or extreme living requirements (*Führer* 1907, 97–98).

Figure 11.3 Diorama of coral reef on the Sinai coast at El Tor, Oceanography Museum, Berlin. From *Führer durch das Museum für Meereskunde in Berlin* (Berlin, 1913), 27.

The displays at the Oceanography Museum differed from those in Altona in both form and content. In contrast to Lehmann's preference for free-standing groups unencumbered by background scenery, the curators at the Oceanography Museum considered background painting and artificial props legitimate, even necessary to reconstructing an authentic, actual scene. For some groups they embraced the form of the large niche-diorama, with its many requirements for complete reconstruction of the scene (Fig. 11.4); for other groups they employed an intermediate form with three glass-walled sides projecting into the room and a painted back wall (Fig. 11.3). These forms created different experiences for visitors. Instead of having the visitors walk around free-standing animals, which placed them almost in the same space and offered a feeling of physical immediacy, these versions of the biological group separated the viewer from the scene by large frames of glass, while protecting the specimens from dust and curious hands. The niche form also reduced the angles from which one could see—a situation especially important to the success of the illusion, which required blending the

Figure 11.4 South Pole diorama, Oceanography Museum, Berlin. From *Führer durch das Museum für Meereskunde in Berlin* (Berlin, 1907), 108.

two-dimensional background scene with the 3-D elements in the foreground (Wonders 1993, chap. 5).

The South Pole diorama illustrates how the museum's various agendas and forms mingled. This exhibit, placed in a corner of one of the biology rooms, grouped Antarctic fauna in front of a painting that depicted the research ship 'Gauss' of the German South Pole expedition. Naming both the ship and the expedition in its text, the 1907 museum guide noted that the specimens had been donated to the museum by the expedition. This was the nation's most recent large-scale project of scientific exploration, and the attention given to it in the text and background painting reinforced both the nationalistic naval themes of the museum as a whole and its scientific emphasis on oceanographic research.

The text went on to call attention to the ways in which the Weddell seals and two species of penguins in the diorama showed their adaptation to life in the ocean: their smooth, rounded forms; their closable ear and nose openings; the placement of the flippers on the seals and legs on the penguins so as to accommodate swimming. The background painting, the text noted, even showed how the penguin could use its beak to help pull itself out of the water (*Führer* 1907, 107–9). Thus this display modelled both basic scientific concepts about adaptation and a site of scientific research, while also underlining German imperial ambitions. The illusionistic scene, though less complete than classic dioramas because of the visible ceiling panels, was clearly meant to draw the viewer in to a 'real' place and moment.

From the start, the scientific aims of the curators intermingled with a fairly consistent set of commitments concerning art and authenticity. In particular, the museum's scientists were intent upon representing real scenes that could also be characterised as 'typical'. As the assistant in the Biological Division, Leopold Glaesner, wrote in a 1912 report, "One can, of course, place as many and as varied animals as one wants from a certain location into a museum case together, but as soon as one shows them in so-called natural surroundings, one must keep rigorously in view that the supposed 'slices of nature' really are such; indeed it is not sufficient that the exhibited combination *could* perhaps be found once, but it must reflect a normal, typical picture."[9] In another report the same year, Glaesner and the head of the Biological Division, Thilo Krumbach, underlined the point. "The biotic communities [*Lebensgemeinschaften*] are thinkable only as slices of nature [*Naturausschnitte*]. Any stylising must be avoided."[10]

Glaesner and Krumbach understood their 'slices of nature' literally. For example, in early 1912 Krumbach conceived of an exhibit that would show the effects of wind and salt on the rocky coast of Istria, near the northern end of the Adriatic. Krumbach, who directed the Biological Station at Rovigno (coming to the Berlin museum only during the winter months), spent much of his time collecting specimens in the Adriatic for shipping to European universities.[11] While collecting for these purposes along the Istrian coast in 1912 and 1913, Krumbach kept his eyes open for the perfect section to cut out. Unfortunately, the places he originally had in mind turned out to be unsuitable—"too large for the space in the museum, not completely typical, too heavy in their effect or misleading, etc., etc." Only in the summer of 1913 did he find the perfect spot, one that was protected enough from the wind to allow the foliage to grow right down near the water, yet that jutted out far enough to show the effects of the wind on the bushes. "This spot has the further ideal characteristic that—cut out of its surrounding—it offers so much stimulation to the imagination that the latter will fill in the features of the coast correctly and without difficulty."[12] Krumbach gained permission to cut out the cliff and did so.

Hacking out a large chunk of rock from an island in the northeastern Adriatic and sending it to Berlin was no minor matter. Here again, the value of authenticity—embodied in a real site, but one meeting a variety of practical and aesthetic requirements—seems to have been paramount in justifying the labour and expense involved. And yet not all of the exhibit could be original. Krumbach had already decided that this 'slice of nature', to be effective, would take the form of an illusionistic scene in a diorama-like niche. To achieve a true-to-life effect, foliage had to be reconstructed and a painted background was required. Krumbach took photographs to guide the painter, who would be hired in Berlin, and collected the original foliage from the site.

Krumbach's careful activities to document and reconstruct an original site suggest the centrality of the scientist's role in authenticating an exhibit and making sure it would be 'true to nature'. It was the scientist who had seen the original and who would supervise the construction and care of the exhibit. As Krumbach wrote a decade later, "According to the principle on which our collections are built, namely, that they are exactly reconstructed slices of nature, it is entirely out of the question that they should be assigned to the care of someone who does not know these things from their very own observation."[13] This vouching for the truth was particularly important in depicting scenes or animals from distant locations, for which visitors would be unlikely to have any independent means of judging their authenticity. It also placed the scientist in the key role as the authority who validated the scene as 'true', thereby aligning the new discipline with exhibits primarily reflecting the scientist's perspective rather than the artist's (or the visitor's).

But this form of scientific authorisation had to be matched by the craft with which the realistic scene was portrayed. If a feature was visibly false, it undermined both the visitor's illusion of reality and the claim to scientific authenticity as well. Poor taxidermic methods were universally deplored, and indeed the standards of accuracy scientist-curators demanded of preparators were rising rapidly. Even a badly stuffed moose, however, still clearly derived from an original real thing, and therefore embodied to a certain degree its own authenticity. More problematic were props or elements that were partly or entirely artificial. Thus Glaesner, after a trip to other German museums to study their biological groups in 1912, criticised in Bremen "the snow in the polar animal group, which even from a great distance is recognisable as cotton wool".[14] Lehmann's solution in his free-standing exhibits in Altona was to minimise the use of such props, leaving the scenery largely to the visitor's imagination. Krumbach and Glaesner chose instead to seek out innovative illusionistic methods that would bring the reconstructed scene as close as possible to reality. That is, they would surround the authentic elements with artificial ones to enhance the realism of the scene as a whole.

For example, during his 1912 museum tour Glaesner was especially interested in other museums' representations of water—a feature crucial to his museum's exhibits and particularly difficult to simulate. In Altona, he noted, "a group of gulls is displayed hovering over rough seas, of which the waves are modelled in plaster and painted over. Although the form of the waves turns out well this way, their opacity has a disturbing effect." About Bremen's efforts he sighed, "How cautious one must be with the dry display of free-swimming animals is shown by the sea turtles and sharks, which hang in a glass case over a coral bottom. They give the impression that they are flying through the air, and create an absolutely laughable effect." In Hamburg, by contrast, he complimented the successes they had achieved by substituting gelatin for water, "a method that seems to me especially suitable for cases where coloured or turbid water is to be imitated."[15]

In the event, most of the biological dioramas at the Oceanography Museum showed shoreline scenes in which the water was represented only in the backdrop paintings, suggesting that one crucial role for the paintings was not simply to convey 'background', but more specifically some elements, like water, that were difficult to model in 3-D form. By contrast, in the large coral-reef display, which had been produced earlier by Ludwig Plate, a predecessor of Krumbach and Glaesner, it appears that the waterline was marked by frosted glass, over which the viewer could still get a good look at the corals (Fig. 11.3). In this case the background painting served to situate the scene as a real location, showing on the left the distinctive coastal features of "the Egyptian town of El Tor, in the middle a quarantine station for pilgrims to Mecca, and on the right a Bedouin village" (*Führer* 1907, 98). Thus, like the South Pole diorama, the painting helped to bring almost photographic specificity to the scene. Here we see how the products of artifice—in this case, background paintings—served to reinforce two aspects of authenticity. The scientists' claim that the scene *is* the 'real thing', strengthened by visual references to specific locations, was brought together with the feeling that was intended to be experienced by the visitor, that the scene looked and felt emotionally like 'the real thing'.[16]

At the Oceanography Museum there appears to have been a greater willingness than in Altona to use props, paintings, and other artificial devices, but these were saved from potential accusations of lack of realism by the insistence that they be used only to represent actual sites and by striving to make the artificial elements of extremely high quality. Part of ensuring the truth value of the exhibit was the scientist's own experience of the site; conversely, the representation of that research site also served as a means of garnering support for the project of marine life science itself. With regard to meeting both the scientists' and visitors' needs for authenticity, one would have to say that in this museum, the curators' efforts to satisfy the visitors' need for authenticity derived from a deeper desire to win visitors over to the project of ocean science.

PRACTICAL REALITIES

If the form and content of biological group exhibits were determined in part by a combination of scientific decisions about what realities to portray and aesthetic decisions about the best ways to portray them, these decisions themselves still had to be in the realm of the possible. The most notable contrast between the great American natural history museums, which pioneered the use of scientific biological groups, and the German museums was their funding. Much to the envy of German museum men, who worked for public institutions, the coffers of American museums at the turn of the century were enriched by a new scale of private philanthropy unknown to German culture (Wandolleck 1905). This had significant consequences: these American museums could send out their own collecting expeditions, and they had highly trained in-house preparators and artists. As a result, the control over exhibits at these large museums was very tight. German museums, with their much more limited budgets, could rarely afford to foot the bill for in-house collecting expeditions and were thus far more often dependent on donations of specimens (or funds to purchase specimens) from museum benefactors. Similarly, although a few larger museums had in-house taxidermists for preparing skins for scientific purposes, with the construction of biological groups it was far more usual to send materials to outside contractors.

The story of Krumbach's Istrian cliff exhibit illustrates in rare documentary detail the kinds of constraints that funding limitations could place on exhibits. In July 1913, when Krumbach found his cliff, he wrote to the museum's director Albrecht Penck asking not only for funds to hire a stonemason but also requesting permission to send his plant materials to the artificial-flower factory of Christine Jauch in Breslau (now Wrocław) for reproduction. Penck granted the request, but advised Krumbach "that the means of the museum are beginning to get somewhat tight, and that we must not go too far in the imitation of nature. The things that we display in large cases don't all need to be replicated exactly in all their details, since no one can ever see them up close. If one takes this point into consideration, one can work substantially more cheaply." Penck also wanted to be sure that the exhibit would be ready in time for the planned opening of the museum's new wing in the late fall.[17]

Penck's worries about cost and time were justified. On 22 September, Frau Jauch wrote that she had received from Rovigno a large myrtle bush (2.5 m high and 60 cm wide), in good condition, with directions to reproduce it. The method of reproduction, she wrote, was to keep the original main trunk and branches, replacing the leaves with artificial foliage exactly where the originals had been. "Even if I put c[irc]a 10 girls onto this work, I can't get it completed in 5–6 weeks." Before she started anyone on a job this big, she wanted confirmation that they still wanted it done. Time was clearly an issue, but it was also a question of money, as is indicated by the note Krumbach scribbled on the letter after it was forwarded to him: "such work is carried out in masses in the zoological museum in Breslau, and the great Oceanography Museum in Berlin is supposed not to have the means for it?"[18]

Krumbach sought the go-ahead, but apparently received bad news, for the next letter expressed to his Berlin co-worker Glaesner disbelief and dismay. "Your letter brought me the bitterest disappointment," he wrote. "Just imagine how I have been working purely in the service of the museum for weeks on end, how our people, our motorboat have been working day in and day out *only* on these materials, and then you will grasp my feelings when I hear that all this was for nothing. . . . What is to be done now is unclear to me. The myrtle bush belongs together with the piece of cliff over which the sea-wind formed it. The cliff-piece lies downstairs in the courtyard, ready to be packed. This block, too, was carved out for nothing, if you throw out the bush."[19]

It appears that the rock was indeed sent, though the myrtle bush was not reconstructed (Fig. 11.5). Instead, smaller (and presumably cheaper) plants appear in the exhibit that was finally mounted, and the effect of the wind on the coastal plant-forms—like the water—was shown primarily in the background painting. 'Reality' had to accommodate a budget.

The story of Krumbach's myrtle bush reminds us that in seeking to understand the forms of realistic biological group exhibits presented in German museums in the early twentieth century, we need to keep in mind not just the varying ideals of scientific truth and emotional authenticity expressed by museum curators but also the practical realities of conveying those truths. The early twentieth century was an enormously innovative time in museum exhibit production, as curators strove to find new and appropriate ways to represent reality, but they were limited in many ways: by the materials available, by the skills of craftsmen and artists, and perhaps above all, by their budgets.

Figure 11.5 Istrian shoreline diorama, Oceanography Museum, Berlin. Photograph (Q VII 1) reproduced with permission from archives, Museum für Verkehr und Technik, Berlin.

MODELS, NATURAL HISTORY DISPLAYS, AND AUTHENTICITY

It is remarkable that Germans at the turn of the century never wrote about their full-sized reconstructions of natural scenes as 'models'. Although the new and reformed museums housed many models, this word was reserved for objects that were entirely products of artifice. Curators at the Oceanography Museum, for example, conducted correspondence discussing scale models of boats, harbours, and the ocean floor; glass models of plankton and sea anemones; plaster models of deep-sea fishes and giant squid; and even the process of 'modelling' living creatures in some other medium. But models were clearly distinguished from natural objects, as is suggested by an early statement that "The collection D. I [showing the different ocean zones] is composed of naturalia; the collection D. II [showing the forms of the ocean floor] consists in models, reliefs, and profiles."[20] Almost invariably, talk of 'models' concerned artificial objects that were displayed as illustrative material outside of biological group displays. They helped visitors see or understand some didactic point better, but they were not typically understood as 'real', or even, in most cases, as realistic stand-ins for the original. Only rarely did the curators talk about modelling in connection with the biological group displays, as in the interesting borderline case of 'modelling' the myrtle bush's artificial leaves, or when Krumbach mentioned the need "to identify and to model" [*zu bestimmen und zu modellieren*] the medusae and fish in another diorama.[21] (Here, too, the term probably referred to creating a wholly artificial object, since these creatures were notoriously hard to preserve.) Dioramas and other biological groups that used natural materials were not called models; they were instead considered authentic representations of the real thing—even if they used some artificial materials.

That biological groups were not considered 'models' in the language of turn-of-the-century museum curators—that they were not understood as artificial constructs—suggests two important features of the values of natural history. First, it indicates a powerful need among naturalists to preserve authenticity as central to their practice of science and representation of nature and to divide it sharply from artifice. Naturalists worked with 'the real thing'. Indeed, the appeal of authenticity appears to have been great enough that the existence of natural items such as rock and fur- or feather-covered skin in a diorama—even when mingled with man-made objects had the ability to allow the reconstructed whole to represent scientific truth rather than humbug or fantasy. Yet, precisely because many man-made elements were used to convey 'natural' reality, a great deal of work had to be done to sustain the truth value of the diorama. In this chapter I have suggested that the intensity of this effort was connected with anxieties about the status of natural history museums as scientific institutions. Many museum scientists equated the natural with high-toned 'truth' and the artificial with popular 'humbug' (or perhaps, in a more positive light, with 'artistic genius'), thus setting up an array of linked dichotomies: nature versus artifice, truth versus humbug, and perhaps even science versus art. These dichotomies seem to have been associated in turn with deeper attitudes about class and the role of public museums in developing taste: if private, commercial establishments catered to the crass, unenlightened tastes of the masses for the sake of profit, the job of the public museum was to educate their sensibilities toward a more middle-class valuation of truth—and an appreciation of science as a source of that truth. Biological group displays, with all their ambiguities of production, continually threatened to breach these dichotomies.

This problem was not unique to museum scientists: in just the same period, a parallel controversy erupted among nature writers in the United States concerning the validity (or danger) of certain kinds of 'true stories' about animals (Lutts 1990), and the nascent nature film industry faced related quandaries about staging events in nature and about leaving boring or overtly sexual footage on the cutting-room floor (Mitman 2000). It appears, then, that museum scientists' deep concern with authenticity mirrored larger bourgeois anxieties about the manipulability of nature's truths in an age of mass culture and artificial reproduction. But the claims to authenticity of 3-D representations were more direct than those of nature writers or film-makers, because the skins and feathers had undeniably once clothed living animals.

Conversely, the heavy weight of authenticity borne by natural history group exhibits sets them apart from other kinds of 3-D representations in the sciences. While surely one of the claims made for many scientific models is their ability to simulate reality (in at least some aspect), simulation is not the same as authenticity. Authenticity, as I hope to have shown, involved at once the scientist's authority in saying something was true and accurate, the preparator's or artist's skill in reproducing and evoking that truth, and the viewer's intuitive experience of the scene as true. Naturalists at the turn of the twentieth century—and through most of the century since then—held fast to the belief that the experience of authenticity required contact with at least a remnant of an 'original thing' in the display. Ultimately, that conviction—and the ability of museum curators to impress it on visitors—is what has given the natural history group display its immense and long-lasting power.

NOTES

I wish to thank Nick Hopwood for the many fruitful intellectual exchanges we have had on this chapter and for pointing me to numerous useful references. Research was carried out with support from the American Council of Learned Societies, the Howard Foundation, and the National Science Foundation.

1. Griesemer's analysis focuses mainly on the ways physical specimens, field notes, and other materials represented theoretical understanding in the ecological work of Joseph Grinnell; the question of authenticity, which is my focus here, is not a central concern of his essay.

2. Some anatomical models of humans also used original remnants such as hair (Jordanova 1989, especially 44–47) and even skin (Mazzolini, this volume). These, too, linked models to their originals, though the illusion of a living being tends not to be enhanced by the use of skin, which does not preserve well. Thus the links—at least for skin—may be more symbolic than illusionistic.

3. Winsor forthcoming; Conn 1998; Rainger 1991; Haraway 1989; Rupke 1994; Gunther 1975. On the importance of international comparison to German museum developments, see Köstering 2003; and Penny 2002.

4. Scholarship on the history of German panopticons is limited; even descriptions are hard to come by (but see Geist 1983; Letkemann 1973; and Röhrich 1977). The ways in which panoramas and waxwork museums worked together with other realistic spectacles such as the morgue, the newspaper and, finally, the early cinema to create a mass visual culture in Paris is the subject of Vanessa Schwartz's outstanding book (Schwartz 1998). My thinking owes much to this monograph.

5. The "Nightmare" was a three-dimensional adaptation of a famous painting of the same title by the late-eighteenth-century painter Henry Fuseli, which was much adapted and caricatured through the nineteenth century (Powell 1972, 59–60).

6. See also Fritze 1907–8, 37. For a different nineteenth-century attitude toward the appropriate relations between science and commercial display, see Secord, this volume.

7. For parallel concerns in America, see Wonders 1993, 118.

8. On the holism of ocean science, see Schlee 1973; and especially Schefbeck 1991, 24–26.

9. "Bericht über eine Dienstreise nach Altona, Hamburg, und Bremen, ausgeführt von Dr. L[eopold] Glaesner zum Studium der dortigen Museen", Berlin, 10 June 1912 (hereafter Glaesner Bericht), Archiv der Humboldt-Universität Berlin:

Akten des Instituts für Meereskunde an der Universität Berlin betr: die Biol. Abteilung. Allgemeines [B.3.1] (29 July 1903 – 1942) (unpaginated; hereafter IfMBiol. 1).

10. [Leopold] Glaesner and [Thilo] Krumbach to [Albrecht Penck], 3 February 1912: Plan für die Biologische Abteilung des Museums f. Meereskunde, IfMBiol. 1.

11. On the Rovigno station and Krumbach's role in it, see Geheimes Staatsarchiv Preußischer Kulturbesitz (Berlin): Rep. 76 V[c] Sekt. 1 Tit. 11 Teil II No. 29. Ministerium für Geistes-, Unterrichts- und Medizinalangelegenheiten I. Unterrichts-Abteilung. Acta betr: die Zoologische Station in Rovigno, Allgemein. Bd. 1 (January 1911 – June 1916). See also Zavodnik 1996.

12. Thilo Krumbach to [Penck], 18 July 1913. Archiv der Humboldt-Universität Berlin: Akten des Instituts für Meereskunde an der Universität Berlin. B.3.II: betr. die Biologische Abteilung. Erweiterung (9 April 1901 – 26 October 1937) (unpaginated; hereafter IfMBiol. 2).

13. "Betrifft: Museum für Meereskunde und Meeresbiologie", IfMBiol. 1 (undated); content indicates that document is from Krumbach to the new director Merz, near the end of 1923 or early 1924. Emphasis on the critical role of the scientist as mediator between the original scene in nature and its reconstruction in the museum was not unique to Krumbach or the Oceanography Museum; see Schauinsland 1904, 31.

14. Glaesner Bericht, IfMBiol. 1.

15. Ibid.

16. For the way this issue played out in early nature films, see Mitman 2000, especially chapter 2.

17. Krumbach (Zoologische Station Rovigno) to [Penck], 18 July 1913; Penck to Krumbach, 30 July 1913 [Nr. 1312], both in IfMBiol. 2.

18. Christine Jauch to Director of the Institut für Meereskunde [Penck], 22 September 1913, ibid.

19. Krumbach to Glaesner, 27 September 1913, ibid.

20. "Denkschrift zur Errichtung eines Marinemuseums in Berlin", fol. 9–12v, Geheimes Staatsarchiv Preußischer Kulturbesitz: Rep. 76 V[a] Sekt. 2 Tit. X Nr. 158. Acta betr. das Institut für Meereskunde an der hiesigen Universität. Bd. I: die Begründung eines oceanographischen Instituts nebst Meeresmuseum in Berlin (February 1899 – November 1900).

21. Glaesner and Krumbach to [Penck], 3 February 1912: Plan für die Biologische Abteilung des Museums f. Meereskunde; Krumbach (Zoologische Station Rovigno) to [Penck], 18 July 1913. Both in IfMBiol. 1.

REFERENCES

Bruckner, Sierra Ann. 1999. "The tingle-tangle of modernity: Popular anthropology and the cultural politics of identity in imperial Germany." Ph.D. dissertation, University of Iowa.

Comment, Bernard. 1999. *The Panorama*. London: Reaktion.
Conn, Steven. 1998. *Museums and American Intellectual Life, 1876–1926*. Chicago: University of Chicago Press.
Dahl, Friedrich. 1908. *Kurze Anleitung zum wissenschaftlichen Sammeln und zum Konservieren von Tieren*, 2nd edn. Jena: Fischer.
Deussing, Gottlieb Gustav. 1884. *Der Anschauungsunterricht in der deutschen Schule von Amos Comenius bis zur Gegenwart: Historisch-critische Darstellung*. Frankenberg i. S.: Rossberg.
Fritze, Adolf. 1907–8. "Biologische Gruppen." In *Jahrbuch des Provinzial-Museums zu Hannover*, 34–38. Hanover: Das Museum.
Führer durch das Museum für Meereskunde in Berlin. 1907. Berlin: Mittler.
Geist, Johann Friedrich. 1983. *Arcades: The History of a Building Type*. Cambridge: MIT Press.
Greenhalgh, Paul. 1988. *Ephemeral Vistas: The Expositions Universelles, Great Exhibitions and World's Fairs, 1851–1939*. Manchester: Manchester University Press; New York: St Martin's Press.
Griesemer, James. 1990. "Modelling in the museum; on the role of remnant models in the work of Joseph Grinnell." *Biology and Philosophy* 5: 3–36.
Gunther, Albert E. 1975. *A Century of Zoology at the British Museum Through the Lives of Two Keepers: 1815–1914*. London: Dawsons.
Haraway, Donna. 1989. *Primate Visions: Gender, Race, and Nature in the World of Modern Science*. New York: Routledge.
Harder, Friedrich. 1891. *Handbuch für den Anschauungsunterricht mit besonderer Berücksichtigung des Elementarunterrichts in der Realien*, 10th edn revised by J. F. Huttmann. Hannover: O. Goedel.
Hinrichsen, Torkild, ed. 2001. *In Ottos Kopf. Das Altonaer Museum 1901 bis 2001 und das Ausstellungskonzept seines ersten Direktors Otto Lehmann*. Hamburg: Dölling und Galitz.
Jordanova, Ludmilla. 1989. *Sexual Visions: Images of Gender in Science and Medicine Between the Eighteenth and Twentieth Centuries*. Madison: University of Wisconsin Press.
König, Hannes, and Erich Ortenau. 1962. *Panoptikum. Vom Zauberbild zum Gaukelspiel der Wachsfiguren*. Munich: Isartal.
Köstering, Susanne. 2003. *Natur zum Anschauen. Das Naturkundemuseum des deutschen Kaiserreichs, 1871–1914*. Cologne: Böhlau.
Lampert, Kurt. 1904. "Die naturhistorischen Museen." In *Die Museen als Volksbildungsstätten. Ergebnisse der 12. Konferenz der Centralstelle für Arbeiter-Wohlfahrtseinrichtungen*, 20–26. Berlin: Carl Heymann.
Lehmann, Otto. 1904. "Das Altonaer Museum." In *Die Museen als Volksbildungsstätten. Ergebnisse der 12. Konferenz der Centralstelle für Arbeiter-Wohlfahrtseinrichtungen*, 36–46. Berlin: Carl Heymann.
———. 1906. "Biologische Museen." *Museumskunde* 2, no. 2: 61–66.
———. 1907. "Die Aufgabe der Museen. Erwiderung." *Zoologischer Anzeiger* 31: 87–93.

Letkemann, Peter. 1973. "Das Berliner Panoptikum. Namen, Häuser und Schicksale." *Mitteilungen des Vereins für die Geschichte Berlins* 69: 317–26.

Lutts, Ralph. 1990. *The Nature Fakers: Wildlife, Science, and Sentiment*. Golden, Colo.: Fulcrum.

Mitman, Gregg. 2000. *Reel Nature: America's Romance with Wildlife on Film*. Cambridge, Mass.: Harvard University Press.

Die Museen als Volksbildungsstätten. Ergebnisse der 12. Konferenz der Centralstelle für Arbeiter-Wohlfahrtseinrichtungen. 1904. Berlin: Carl Heymann.

Penny, H. Glenn. 2002. *Objects of Culture: Ethnology and Ethnographic Museums in Imperial Germany*. Chapel Hill: University of North Carolina Press.

Powell, Nicolas. 1972. *Fuseli: The Nightmare*. New York: Viking Press.

Rainger, Ronald. 1991. *An Agenda for Antiquity: Henry Fairfield Osborn and Vertebrate Paleontology at the American Museum of Natural History, 1890–1935*. Tuscaloosa: University of Alabama Press.

Röhrich, H. 1977. "Die Wachsbossierer, Hersteller anatomischer Lehrmodelle in München." In *La ceroplastica nella scienza e nell'arte. Atti del I Congresso Internazionale, Firenze, 3–7 giugno 1975*, vol. 1: 433–41. Florence: Olschki.

Rupke, Nicolaas A. 1994. *Richard Owen: Victorian Naturalist*. New Haven: Yale University Press.

Schauinsland, Hugo. 1904. "Das Städtische Museum für Natur-, Völker- und Handelskunde in Bremen." In *Die Museen als Volksbildungsstätten. Ergebnisse der 12. Konferenz der Centralstelle für Arbeiter-Wohlfahrtseinrichtungen*, 27–35. Berlin: Carl Heymann.

Schefbeck, Günther. 1991. *Die österreichisch-ungarischen Tiefsee-Expeditionen 1890–1898*. Graz: Weishaupt.

Schlee, Susan. 1973. *The Edge of an Unfamiliar World: A History of Oceanography*. New York: Dutton.

Schwartz, Vanessa. 1998. *Spectacular Realities: Early Mass Culture in Fin-de-Siècle Paris*. Berkeley and Los Angeles: University of California Press.

Sehsucht: Das Panorama als Massenunterhaltung des 19. Jahrhunderts. 1993, ed. Kunst- und Ausstellungshalle der BRD GmbH. Frankfurt am Main: Stroemfeld/Roter Stern.

Sternberger, Dolf. 1977. *Panorama of the Nineteenth Century*. New York: Urizen Books.

Wandolleck, Benno. 1905. Review of Field Columbian Museum, *Annual Report* of the Director to the Board of Trustees for the Year 1903–4. *Museumskunde* 1: 235–36.

———. 1906. "Die Aufgabe der Museen." *Zoologischer Anzeiger* 30: 638–53.

Winsor, Mary P. Forthcoming. "Natural history museums." In *The Cambridge History of Science*, vol. 6: *The Modern Biological and Earth Sciences*, ed. Peter J. Bowler and John Pickstone. Cambridge: Cambridge University Press.

Wörner, Martin. 1999. *Vergnügung und Belehrung: Volkskultur auf den Weltausstellungen 1851–1900*. Münster: Waxmann.

Wonders, Karen. 1993. *Habitat Dioramas: Illusions of Wilderness in Museums of Natural History*. Uppsala: Uppsala University.

Zavodnik, Dušan. 1996. "Ein Jahrhundert des Aquariumsgeschäftes in einer wissenschaftlichen Anstalt—der ex-Zoologischen Station des Berliner Aquariums in Rovinj (Adriatisches Meer)." *History and Philosophy of the Life Sciences* 18: 107–22.

PART THREE

New Media and Old

CHAPTER TWELVE

Models and the Making of Molecular Biology

Soraya de Chadarevian

In 1976 the new research assistant for the chemical collection at the Science Museum in London, Ann Newmark, came across dismembered parts of a large molecular model made up of Meccano-type bits in the museum's store. The model carried no tag, but investigations revealed the "mystery object" to be John Kendrew's 'forest of rods' model of myoglobin, the protein that carries oxygen in muscle.[1] Based on X-ray analysis, this was the first atomic model of a globular protein ever built. Herman Watson, Kendrew's former colleague at the Laboratory of Molecular Biology in Cambridge, came from Bristol to reassemble the 6- by 8-foot structure (Fig. 12.1).

If the museum records give no clues to how the bulky model came into the museum store,[2] other molecular models found their way into the same premises more officially. Only a few months before the large atomic model was rediscovered, Kendrew had presented an earlier and much handier low-resolution model of myoglobin, also known as the 'sausage model', to the museum as a gift (Fig. 12.2). It was accompanied by the original three-dimensional electron-density map, a representation of the crystallographic analysis which preceded model-building (see below).[3] All three objects became part of the *Living Molecules* exhibition, staged in the museum in 1987. Dedicated to celebrating forty years of research at the Laboratory of Molecular Biology in Cambridge, the display bore witness to the overriding importance of models in the public presentation of the science. Besides the myoglobin models, visitors—moving around skilfully illuminated perspex cases—could admire the intricacies of other, even larger, protein models. Also on display were various nucleic acid models, including a replica of the

Figure 12.1 'Forest of rods' model of myoglobin (1959). Source: Medical Research Council, Laboratory of Molecular Biology, Cambridge.

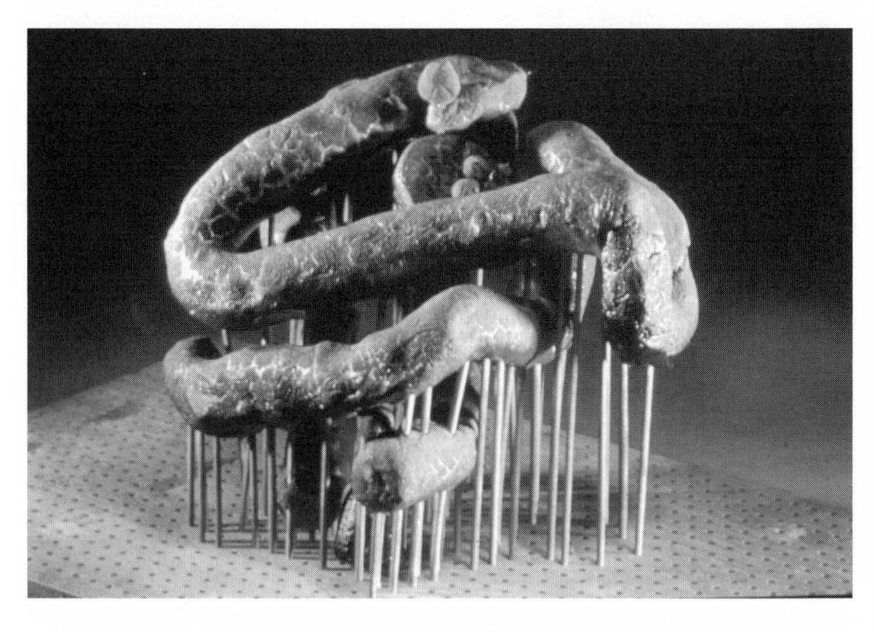

Figure 12.2 'Sausage model' of myoglobin (1957). Height of the model circa 8 inches. Source: Science Museum/ Science & Society Picture Library, slide no. SCM/PHY/C100369A.

Watson and Crick double helix, as well as models of virus structure and a mechanical model of the sliding filament mechanism in muscle fibres. Pushing a button made Max Perutz, Kendrew's colleague at Cambridge and co-Nobel laureate, appear on a small screen, surrounded by some of the models on show, to explain their function.[4]

Most of the objects now safely guarded in their cases were not originally built as museum exhibits. Their place was in the laboratory where they were handled, discussed, measured, tested against data, corrected, refined, and—if superseded—dismantled and the pieces used for other projects. This was even the fate of the most well-known of the models, Watson and Crick's double helix, which was built in the same Cambridge laboratory. The troubled 'life story' of this model (here I mean the physical model, not its ubiquitous iconographical representations) nicely illustrates the place of models in a research laboratory and other roles that models can assume.[5]

The first double-helical model of DNA, built of brass models and specially cut metal bases, was the result of James Watson and Francis Crick's protracted attempts to interpret the X-ray diffraction data of the molecule. Further turns of the helix were added later to impress visitors. One of the very few existing records of the 'original' model, which was never published as such, is a photograph taken by Antony Barrington Brown, a freelance photographer, for a report in *Time* magazine that never appeared. The model itself—once the object of intense discussion, but soon superseded by more refined models built at King's College, London—slowly fell to pieces and was eventually disassembled. Some of the original bases, the only parts which, because they were purpose-built, could later be identified as belonging to the original model, ended up in Bristol, when they were packed together with other model-building parts for the move of Herman Watson (no relation to James) from Cambridge. By the 1970s, however, Watson and Crick's work was much more widely recognised and had acquired symbolic meaning as the origin of the new science of molecular biology. Watson's best-selling discovery account was published in 1968, and in 1974 *Nature* dedicated an issue to the model's twenty-first birthday.[6] Thus, when in 1976 Newmark travelled from London to Bristol to discuss with Herman Watson the reconstruction of the myoglobin model and on that occasion discovered the original metal plates, she arranged for the pieces to be acquired by the Science Museum. Farooq Hussain, a research student at King's College, London, volunteered to build a facsimile of the model, incorporating the few original bases. This model, later found to be taller than the original, has since served as prototype for further replicas, one of which (made to the correct size) was included in the *Living Molecules* exhibition. Hussain's facsimile, held in the Health Matters gallery of the Science Museum in London and correctly labelled "The nearest there is to the original model ... ", has meanwhile undergone conservation, sealing its status as a museum object.

While much about models can be learnt from this exemplary tale, in the following I propose to focus not on the double helix, but on the protein models, and especially on the models of myoglobin and (to a lesser extent) haemoglobin, that were also based on X-ray diffraction studies. Today the double-helical model of DNA has become the most powerful symbol of molecular biology, but in the 1950s and 1960s the protein models, among the first of their kind, played as central a role in the making of the new science.[7] They were not only crucial tools for the study of molecular structure–function relationships. Being suited for display, they also made their entrance into many arenas outside the laboratory and the lecture hall, participating in the postwar advertisement of science and creating a public image of the new molecular biology. The prominence of models in representations of molecular biologists, from informal photographs to commissioned portraits, confirms the extent to which they too identified themselves—and were identified—with these objects.

Despite their ubiquity and high visibility, models have been neglected by a historiography biased towards scientific theories and texts. Even the more recent interest in scientific representations has focused on flat inscriptions and images (Lynch and Woolgar 1990). Protein crystallographers, however, always insisted that their results were hard to convey in words or pictures. Building, looking at, and manipulating models were crucial to appreciating the results. The same models could be used to good effect in all arenas in which manipulation and display was called for. Focusing on models then, their production, circulation, and uses, allows an analysis of scientific practice and its material culture that moves from the laboratory to the marketplace, the exhibition hall, the television studio, and the museum.

What I propose is not a history of molecular biology illustrated with the models, but a history of the models themselves and their role in the making of molecular biology. The chapter has three parts. I shall look first at the place of models in crystallographic practice and at modelling techniques in investigations of protein structure. In the second part I shall investigate how the results embodied in 3-D models were published and circulated. I shall discuss attempts to overcome the limitations posed by the flat pages of scientific journals and explore parallel channels of communication and 'publication' which developed around models. In the last part I shall deal with the uses and display of models in public arenas and their role in the advertisement of the new science of molecular biology, especially (though not only) on television.

MODELS AS RESEARCH TOOLS

Model-building has always been an integral part of crystal structure determination, a practice pioneered by the British physicists William and Lawrence Bragg in the 1910s. In the trial-and-error method commonly used by X-ray crystallographers, probable structures were built and then tested against the diffraction data (Watson and Crick also proceeded in this way). The protein models I shall be discussing were built by 'direct' structure determination, that is by using crystallographic data only. Here the models were the last step of a series of transformations involved in the interpretation of the highly complex diffraction pictures. The models could then be used for further refinements of the structure. Even in this case, then, model-building was part and parcel of the research process. A quick run through the steps involved in the X-ray analysis of protein structure will make this point clearer.

Analysis started from X-ray diffraction pictures, gained by exposing single protein crystals to a beam of X-rays. The result was recorded on photographic plates. From the intensities, indicated by the darkness of the diffraction spots on the picture, and the (reconstructed) phases of the same spots, the distribution of electron density in the molecule could be calculated. This step involved an immense amount of tedious measurement and calculation, for which protein crystallographers made pioneering use of electronic computers.[8]

In electron-density (or contour) maps the electron distribution is represented by contours, much like altitude on topographical maps. This technique was first introduced by Lawrence Bragg in the 1920s (Bragg 1929). To capture the three-dimensional distribution of electron densities, sections were laid through the molecule and contour maps plotted on translucent sheets which could be superimposed and illuminated. The series of sections

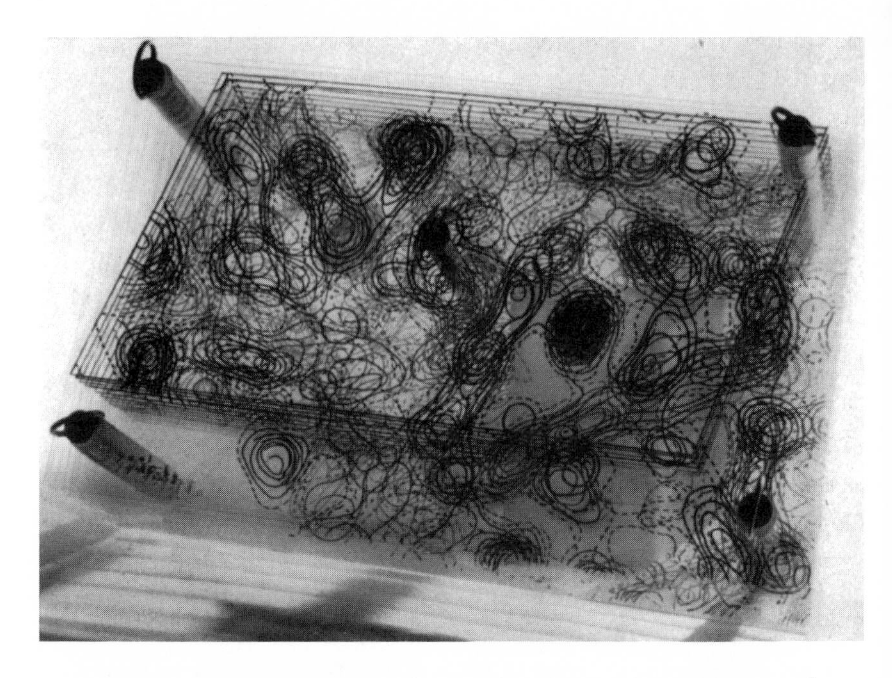

Figure 12.3 Three-dimensional electron-density map of myoglobin at 6-Å resolution. Source: Medical Research Council, Laboratory of Molecular Biology, Cambridge.

through the molecule could be seen "rather like a set of microtome sections through a tissue, only on a thousand times smaller scale" (Perutz 1964, 660) (Fig. 12.3).[9] But in 3-D contour maps the configuration of the protein molecules could not easily be made out: the structure needed to be 'pulled out'. Model-building was intended to do exactly that.

While model-building was an established practice in crystallography, protein crystallographers had to be inventive in their attempts to translate maps of electronic densities into molecular models. Modelling techniques from a whole range of disciplines besides chemistry as well as from outside the sciences provided clues, while building materials derived from the most diverse sources.[10]

Kendrew built his first 6-Å resolution model of myoglobin out of plasticine following the course of the polypeptide chain through regions of high electron density in the 3-D map (Fig. 12.2).[11] Looking at this first ever model of a globular protein derived by direct structure determination, Kendrew expressed surprise at the unexpected twists the protein chain was performing (Kendrew et al. 1958). Later visitors regularly described it as "visceral".[12]

Perutz produced the first model of haemoglobin at a similar resolution by reproducing in plasticine the shapes of the high-density areas in the various sections of the 3-D contour map and assembling the discs in the right order. When this first model collapsed, Perutz resorted to more unusual material. The contour map was reproduced on sheets of thermosetting plastic from which the shapes could be cut out, stacked together in the right order, and baked ("like a cake") to set permanently.[13] This resulted in the well-known black-and-white disc model of haemoglobin, which resisted the wear and tear of time (Perutz et al. 1960) (Fig. 12.4). Haemoglobin was built of four subunits, which resembled the myoglobin molecule. To be able to demonstrate the similarity, Perutz built a model of the myoglobin molecule, using the same technique.

The interpretation of the 2-Å map of myoglobin with its 1,260 atoms (excluding hydrogens) represented a fresh challenge. Kendrew invented a new modelling technique which took its inspiration from the toy construction kit Meccano. Steel rods, of the kind used in Meccano for the wheel axes of toy cars, but six foot tall, were positioned at the same intervals chosen for the mathematical analysis of the structure, so as to form a vertical grid consisting of about 2,500 rods. Coloured clips, used in Meccano to fix the wheels, indicated the electron density along the rods (the brighter the colour, the higher the density). Skeleton-type atomic models giving the exact position for each atom could then be built between the rods, following the density indicated by the clips. Once the main chain was inserted, the density of the side chains could be seen emerging at the appropriate intervals. Careful observation of the lumps in the map combined with model-building allowed Kendrew to identify many more side chains than anticipated. Again, modelling was a crucial step in determining the structure. The atomic coordinates deduced from the model with a plumb-line were further used to refine the structure by mathematical methods (Kendrew et al. 1960; Kendrew 1961).

The scale of the model (5 cm: 1 Å) was chosen so as to allow a human hand to reach in and fix the clips and model bits. The latter were designed by Kendrew, perfecting an idea first developed by Charles Bunn, research chemist and self-taught crystallographer at ICI. The 'Kendrew models' differed from conventionally used skeletal models in the connector system, which consisted of a barrel with two screws fitted into grooves in the model

Figure 12.4 Max Perutz's first model of haemoglobin (usually protected by a perspex case) in front of two gold-framed portraits of the benefactors of the Laboratory of Molecular Biology in Cambridge. Facing the entrance door, the model became the 'obligatory passage point' for every person entering the building. The entrance area has recently been redesigned, but Perutz's model still guards the main door. Source: Medical Research Council, Laboratory of Molecular Biology, Cambridge.

rods. This tight fit was essential to build large structures like proteins. First constructed by the laboratory's own workshop, they were later commercially produced by Cambridge Repetition Engineers, a small precision-engineering firm started by John Rayner, a hobby model-builder and car mechanic, with a business partner. The models were widely used, and were only superseded

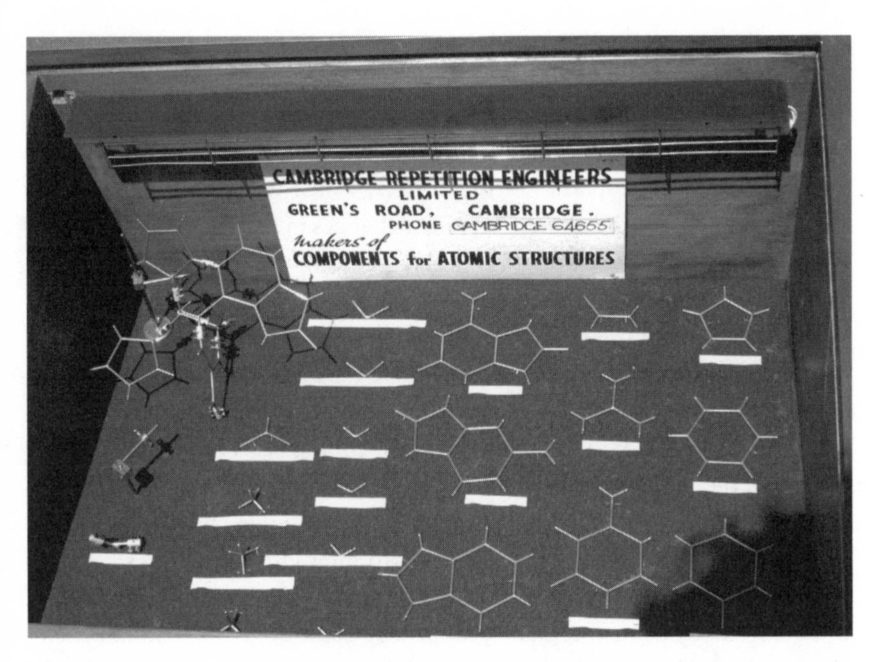

Figure 12.5 Cabinet with 'Kendrew type' model parts, produced by Cambridge Repetition Engineers. This cabinet was displayed in the firm's workshop, but could also be carried around to clients or exhibited at trade shows. Photograph by the author.

by computer modelling in the 1980s (Fig. 12.5). For over two decades Cambridge Repetition Engineers employed three people full-time to cover the demand. The success of the business is an indication of the impact Cambridge protein crystallographers had on the field. Many, if not most, orders came from researchers who, at some point in their careers, had been connected to the Cambridge laboratory.[14] The replication and marketing of the model parts also compensated for the difficulties of circulating the models in print, a point to which I will return.

The lady (inevitably a lady) who, on the occasion of a high-profile royal visit to the Laboratory of Molecular Biology in Cambridge, when shown different protein models, including ball-and-spoke models, exclaimed, "I had no idea that we had all these little coloured balls inside us", has entered the anecdotal history of the laboratory (Perutz 1987 and elsewhere). To crystallographers, models are conventionalised ways of representing the spatial relations of atoms. The large variety of available model parts further underlines the conventional character of model-building. This, however, did not preclude crystallographers from expecting their models to reveal the 'real' structure of the molecules and that studying the models would give clues about how the molecules work.[15] Models shaped the way crystallographers talked and 'thought' about molecules. Their explanations made use of the mechanical properties of the models, which were looked at from all sides, manipulated, and altered to predict changes in structure and functions. A few examples will illustrate this point.

Kendrew never pushed the functional analysis of the myoglobin molecule very far, but Perutz used the model to predict the structure of haemoglobin, composed of four subunits nearly identical to the myoglobin molecule (Perutz 1965). Haemoglobin carries oxygen from the lungs to the tissues and helps to transport carbon dioxide back to the lungs. On the basis of crystallographic studies of the molecule in oxygenated and deoxygenated forms, Perutz proposed a mechanism for the dual function of haemoglobin which he described as a "clicking back and forth between two alternative structures" (Perutz 1978). Similarly, once an atomic model of chymotrypsin, the second structure to be 'solved' by protein crystallographers at Cambridge, was available, biochemist Brian Hartley, who had provided the sequence data for the molecule, was allowed to "play" with it. Building the sequence changes of elastase, another protein of the same family, into the model, he predicted the 3-D structure of the molecule and confirmed that proteins with similar sequences had similar structures.[16] Looking at the chymotrypsin model, Hartley and David Blow, who built it, also found that one amino acid in the active centre had an odd position and was possibly wrong. Re-doing that bit of sequence Hartley quickly confirmed that there was indeed an error, and this correction allowed them to come up with a reaction mechanism. It was based on the stereochemical relationships the model had revealed, and was confirmed by chemical and kinetic data (Blow, Birktoft, and Hartley 1969). The work was celebrated as further important confirmation of the power of the crystallographic approach for the investigation of protein structure and function. Later Blow was ready to concede that to understand how enzymes "really" work, thermodynamic considerations needed to be taken into account. These were not represented in the models the crystallographers built and did not enter their theoretical considerations, at least not in the first instance.[17]

Model-building, this brief survey shows, was an integral part of protein crystallographers' investigative practice and, an essential step in their attempts to interpret the diffraction pictures of their molecules. In its final form the model embodied the 'result' of their investigations and itself formed the starting point of further steps of refinement, comparisons with other structures, and studies of mechanisms and structure–function relations. As we shall see in more detail below, models were powerful rhetorical tools in communication between scientists as well as with other publics. However, the fact that crystallographers best grasped molecular structures and their implications by handling physical models posed problems with respect to the circulation of results in the established channels of scientific publication.

MODELS AND THE SCIENTIFIC PAPER

The results embodied in the protein models were not easily conveyed, either in words or pictures, on the flat pages of scientific journals. Hardly any paper on the subject was published that did not refer to this problem. "The whole question of publication is difficult", Lawrence Bragg emphatically summarised the issues involved, for "how can one convey all the information in a huge molecule.... The 'paper' sent to colleagues should be a model. There seems to be no simpler way of conveying the information".[18] The outburst followed Bragg's repeated attempts to convince Kendrew finally to publish the results of the last step of his X-ray diffraction studies of myoglobin, i.e., the structure at 1.4-Å resolution. "It is extraordinary that [the results] have never yet been written up", he confided to the Secretary of the Medical Research Council, whom he kept informed on the protein work funded by the agency. "[Kendrew] got his Nobel Prize in effect for unpublished work. The world ought to know about the structure, and many people who worked with him ought to get credit for what they did."[17] As Bragg himself suggested, laziness was not the (only) reason for procrastination. The positions of atoms in the 'forest of rods' model were continually adjusted as the data were refined, but the full structure was not easily transferable into a text.[19]

Various solutions were found to convey at least some of the information the models revealed to the trained eye and hand. Photographing the models was the first, but not very satisfying, solution.[20] Kendrew's first low-resolution model of myoglobin had already appeared in print in a series of photographs, when Kendrew set out to find the help of an artist. "[Y]ou said that you know a draughtsman, experienced in anatomical drawing, who might be able to make a drawing of my myoglobin for me", Kendrew wrote to the Secretary of the Medical Research Council. "Since everyone is immediately reminded by my model of viscera, it seems a very promising way of going about it."[21] The London-based artist Frank Price undertook to draw 'front' and 'back' views of the model, stripping it of all constructive elements and rendering only the course of the polypeptide chain (Bodo et al. 1959, 100, figs 22a and b). Fellow scientists agreed that the drawings "[show] up the run of the chain better than photographs of the model".[22] It soon became clear that, as in the original model, the haem group had the wrong inclination. Despite this error, precisely the drawing which showed the position of the haem group, and in which the problem was therefore more visible, was frequently reproduced (e.g., Kendrew et al. 1960, 426, fig. 5; Kendrew 1964, 683, fig. 3; Kendrew 1966, pl. 19), while the second one disappeared from circulation. The inaccuracy of the drawing was pointed out in the 1960 *Nature* paper, but not again. Clumsy verbal descriptions accompanied the drawing. These were Kendrew's best: "[E]mbracing [the haem group], so to speak, are 2 segments of polypeptide chain arranged like the letter V." And: "The polypeptide chain is shown as an irregular rod winding its way around the molecule" (Kendrew 1966, captions to figs 18 and 19).[23]

Most artists worked only occasionally on molecules and it is as hard to find documentation about them and their work as about their counterparts in earlier centuries.[24] One artist, however, has made his name and his career by drawing 3-D protein structures: the architect and illustrator Irving Geis. His career as 'molecular artist' started with the commission to draw Kendrew's molecular model of myoglobin for *Scientific American* in 1960. The drawing took six months to complete, cost the journal more than any illustration before, and met with the contempt of the editor who stamped it "a mass of crumpled chicken wire" (Gaber and Goodsell 1997) (Fig. 12.6). This devastating critique notwithstanding, Geis continued to depict molecular structures, including several more for *Scientific American,* for the next 40 years of his life. The production of each picture was a laborious process. Geis would travel to the model (even if this meant crossing the Atlantic), spend several weeks sketching and photographing it, and then proceed to "conceptualize . . . and finally paint it".[25] Numerous conversations and correspondence between the scientist and the artist accompanied the process. In his renderings of 3-D protein structures, drawn to exact coordinates, Geis practiced what he called "creative lying", by which he understood the introduction of small distortions and exaggerations to resolve overlaps and create an understandable image, in spite of the complexity (Gaber and Goodsell

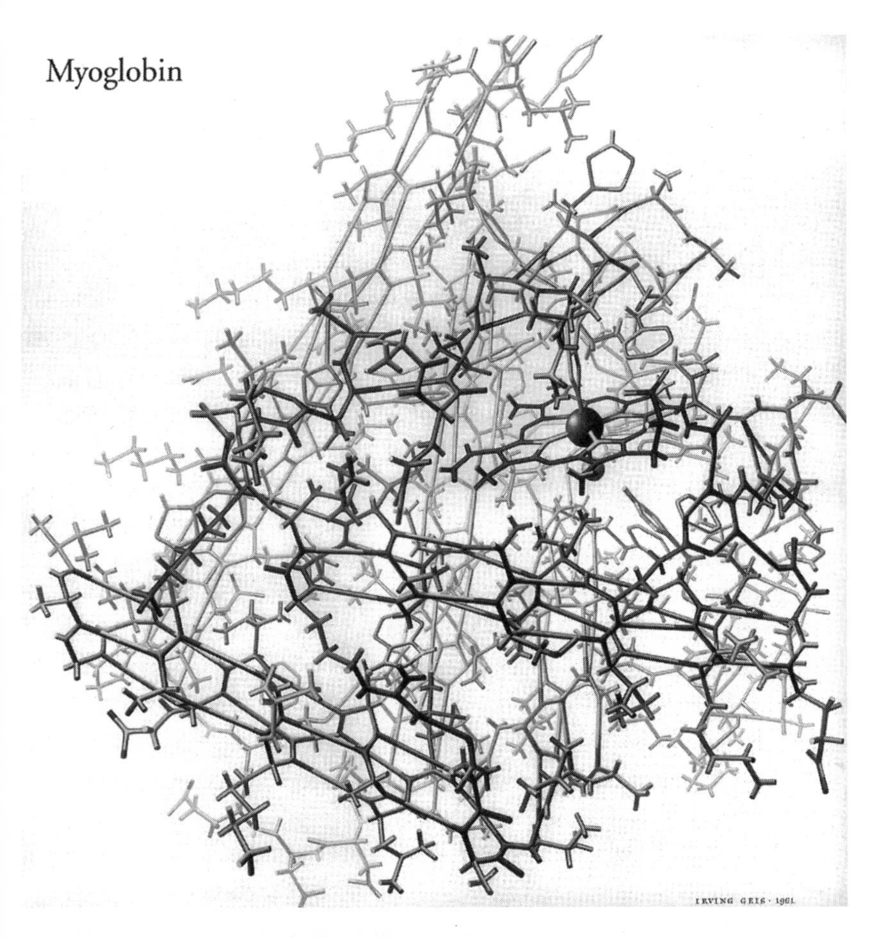

Figure 12.6 Drawing of the atomic model of myoglobin by Irving Geis. Source: The Geis Archives [presentation booklet]. Original in *Scientific American* 205 (1961) 98–99. Rights owned by Howard Hughes Medical Institute. Reproduced by permission of Sandy Geis.

1997, 58). According to Geis it was this capacity which distinguished the human artist and gave his or her products an advantage over photographs or, later, the mindless drawings of a computer (ibid., 59).

Geis's drawings deeply influenced the conventions for depicting and viewing intricate protein structures. The extent to which scientists came to depend on the artist's skills is documented by the fact that his name appeared as co-author on what has become a classic in the field: Dickerson and Geis's

The Structure and Action of Proteins, a heavily illustrated introduction to proteins (Dickerson and Geis 1969). Geis's illustrations in this volume were reproduced in many biochemical textbooks, introducing a whole generation of biochemists to his style of representation. According to the biochemist Richard Dickerson, who collaborated with the artist on two further book projects, Geis "taught us all how to look, how to understand and how to show others what we saw" (Dickerson 1997b, 1249).[26]

In his later drawings Geis shifted from an atom-to-atom representation to an increasingly abstract and 'artistic', though always stereochemically correct, rendering of molecular structures or to what a critic called the production of an "understandable metaphor for molecules" (Gaber and Goodsell 1997). This shift was an expression of his attempt to illustrate the function (as well as the structure) of the molecules, but was probably also his response to the development of molecular computer graphics, which borrowed his ribbon-and-arrow style of representation. These later drawings also appeared in textbooks, copyrighted by Geis. Placing himself in a long and venerable line of tradition, he saw his role as that of a "molecular Vesalius" who, like his Renaissance predecessor for the new human anatomy, used art to teach the modern public the equally new "molecular anatomy" (Dickerson 1997a, 2484).

Protein crystallographers resorted to stereoscopic drawings as an alternative way of conveying spatial information in print (e.g., Watson 1969). These were produced by manually retracing photographs of models taken at different angles or using data of atomic positions generated by early mathematical model-building programmes, a task requiring much patience and skill. In 1981, a whole *Atlas of Molecular Structures* with stereodiagrams of the myo- and haemoglobin molecules was published. To facilitate stereoscopic viewing the diagrams were printed in red and green, and a foldable pair of red–green spectacles was included in the back of the book (Fermi and Perutz 1981).[27]

Recommendations, and finally 'rules' approved by the International Union of Biochemists, were set up to regulate the description and graphical representation of polypeptide conformation (Edsall 1966; IUPAC-IUB Commission 1970). These allowed researchers to describe the position and orientation of every atom in a precise and standardised manner, greatly aiding communication and the comparison of different structures.

Yet all these efforts could not replace the importance of directly viewing, manipulating, and actively building models to grasp the structure and function of the molecules. Parallel to publication, models were mobilised in other ways. Every new structure attracted many visitors who came to appreciate "the body 'in the flesh'",[28] and to the extent to which the construction allowed it, models were carried along and showed around. With the complete set of coordinates (which were not always published but were generally available on request), the right model parts, and some experience, models could also be replicated. In the case of the myoglobin model this became easier once Cambridge Repetition Engineers started offering myoglobin component sets, complete with frames, clamps, angle gauges, and calipers, calibrated at the same scale as the model parts, allowing direct measurements in Ångstrøms.[29] The kits were mainly used to build models for demonstration or in model-building tutorials. Following growing demand in the mid-1960s, Kendrew also arranged for the commercial production of ball-and-spoke models of myoglobin, built to his specification by Alexander Barker of the Department of Engineering at Cambridge, who ran a small private business. Like other demonstration objects, these were sturdier and more colourful than the research models, but less accurate.[30]

TELEVISUAL LANGUAGE

The same models that it was so difficult to compress between the pages of a scientific article proved highly effective in all contexts and arenas in which display and manipulation were not only possible, but called for. Molecular biologists were keenly aware of this possibility and made effective use of their models not only among scientific colleagues and in lecture halls, but also in public arenas. Through these displays they created a public image and an institutional space for their science.

Television proved a particular congenial medium through which to present the new science to a wider public and promote the subject. The historian Edward Yoxen first noted that "the postwar period in which television rapidly emerged as a major medium of mass communication, was also the period in which molecular biology emerged as a new speciality in the life sciences" (Yoxen 1982, 138). With their models, Yoxen claimed, molecular biologists developed a "televisual language" which was most effective in attracting interest and mobilising public support for their enterprise (Yoxen 1978, 237). I shall pursue this suggestion, focusing on the first science series of British television.

The BBC series *Eye on Research*, launched in 1957, proclaimed as its aim to show what "the backroom boys of science", employed in industry, government, and the universities, did in terms of capital investment in knowledge for the future.[31] The programme, the most ambitious to that date, was supported at a high political level and, like most science information initiatives in the postwar era, was aimed at the appreciation rather than a critical assessment of the scientific enterprise (the audience was invited on "a voyage of discovery") (Lewenstein 1992). Among the first to compliment the series was the Director of Public Relations of the Atomic Energy Authority, commenting that if the public could be "thoroughly impregnated with the idea that research is a fascinating but a long and painstaking business", life would be "a lot easier for all of us in the science-information business".[32] Congratulations also came from the chemist Alexander Todd, Cambridge Nobel laureate and Chairman of the Advisory Council on Scientific Policy, who had been consulted on the series from the earliest stage. The programmes, which reached an audience of several million, appeared to Todd "a very valuable means of educating the general public in the methods and development of science and in giving them a proper appreciation of their implications".[33] It brought considerable prestige to BBC Television in academic and scientific circles and succeeded in 'selling' television as a means of "putting across scientific information in a reasonably entertaining way".[34] Even if British scientists did not depend directly on the public for funding, interest in science legitimated funding from tax money and was welcomed by scientists and science administrators alike.

In considering possible topics for a series, visual quality, next to intellectual content and treatment, was of overriding importance. Several academic subjects were *a priori* excluded as being too much of "the pencil and paper kind" to provide material for television. From the beginning, a programme on the models of large molecules, even if certainly a complex subject for a programme targeted at a "middle-brow audience", figured among the subjects under consideration. Series producer and interviewed scientists agreed on this choice. In the round of discussions preceding the first series, for instance, Todd excluded chemistry, his own subject, because he saw it as "impossible to represent extensively in visual terms", but agreed that crystallography "might well be made visual with the aid of models".[35]

Examination of the production files for the first programme on the work of the Cambridge protein crystallographers, broadcast in May 1960, confirms the central role attributed to the models and their 'performance' in making a good programme. Commenting on a draft of the script for the programme, the producer lamented the "prosaic nature" of the series so far, and encouraged scriptwriter and cameraman to make maximum use of the models, which could produce "exciting visuals" and bring up the reaction index. "I hope", he intimated to his staff, "you are planning to do that and not just show a lot of table top stuff. It is not sufficient to explain the subject in an interesting and exciting way." After the production he most deplored the lack of "decent close-ups of the models".[36]

While better effects might have been achieved with more imaginative camera work and better use of lenses, as Anbrey Singer suggested, watching the programme today it is striking how many scenes revolved exclusively around the models.[37] Long sequences show only the models with hands turning them around, pointing to, taking apart, and refitting parts of them—scenes and views only television, despite its 'flat' screen, could present. Time and effort was also spent on the choice of the most 'telegenic' models and on preparing them for the show. For instance, to create a better effect on the black and white screen, a 2-Å myoglobin model was equipped with fluorescing atoms in two nights' overtime by the staff of the Royal Institution before being escorted to Cambridge to be filmed (Fig. 12.7).[38]

The audience (8% of the television public, compared to 25% for the 9.25-p.m. News and 32% for *Play of the Week*, both broadcast at the same time) reacted rather positively to the programme, even if—models notwithstanding—most of the explanations went over the heads of many viewers.[39] Most complimentary, however, were the crystallographers themselves. They thanked the BBC for the "magnificent effort" by the production team in putting a difficult subject across to a lay audience and for the chance they had been given to participate in the programme.[40]

For neither Perutz nor Kendrew was this the first experience with the media and lay audiences. They had presented their work in various science magazines, including *Endeavour* and *New Scientist*, all of which had entered the market during or after the war. They also had experience with radio. Yet

Figure 12.7 Preparing the model room for the BBC programme *Shapes of Life* (1960). Source: Medical Research Council, Laboratory of Molecular Biology, Cambridge.

television proved much more congenial to their subject. Indeed, on at least one occasion the suggestion that Perutz should talk about his research on the structure of the haemoglobin molecule on a radio programme was turned down on the grounds that this was too dry and complicated a subject for that medium.[41] That was at the beginning of Perutz's research. But by the 1960s even Crick, who had consistently refused to appear on TV but did participate in radio science broadcasts, reckoned that molecular biology had become too difficult to be presented on radio, that is, without the aid of models.[42]

The 1960 *Eye on Research* production on the protein models was particularly timely. It coincided with the conclusion of lengthy negotiations with the Medical Research Council and the University of Cambridge, leading to the construction of a new Laboratory of Molecular Biology for the group including Perutz, Kendrew, and Crick, who had been working in the Physics Department, and some associates, especially from Biochemistry (de Chadarevian 2002). In preparation for the move the group, as the first institution world-wide, had incorporated the term 'molecular biology' in its name. The programme was a welcome chance to promote the new subject which was presented as a fundamental quest into the structure and function of 'living molecules' (a DNA model, even if not directly involved in the storyline, was present in the background).

The programme on the models of myo- and haemoglobin was only one in a long series of TV programmes on molecular biology. The double helix had its first appearance on TV in an earlier *Eye on Research* programme.[43] In 1964 a 10-week programme, *The Thread of Life*, authored by Kendrew, was dedicated exclusively to molecular biology, with the action again revolving around models of proteins and nucleic acids. These early and later productions, up to the BBC film *Life Story* (1987), a dramatised history of Watson and Crick's DNA model, played a crucial role in promoting molecular biology and creating a public image of the science tightly bound to its models.

Television programmes were important, but not the only occasions on which models were deployed to promote the subject. When, in 1958, Perutz went to London to present the case for a new laboratory of molecular biology to his patron, the Medical Research Council, he brought with him much the same set of models used in the BBC programme, making sure that a "large table ... on which the exhibits could be arranged was available".[44]

Figure 12.8 John Kendrew (in academic robe) showing Queen Elizabeth II (sitting) a small-scale wire model of myoglobin on the occasion of the official opening of the Laboratory of Molecular Biology in 1962. The object on the table is the 3-D electron-density map of myoglobin built from perspex sheets. Source: Medical Research Council, Laboratory of Molecular Biology, Cambridge.

Models were shown to high-level visitors to the laboratory and travelled to exhibitions (Fig. 12.8). In the late 1950s, which were crucial years for the formation of the discipline in Britain, Barker built a series of gigantic models, including a huge plaster replica of Kendrew's first myoglobin model.[45] The models were shipped to Brussels to be displayed in the International Science Pavilion at the World's Fair. In the hot phase of the Cold War and the atomic arms race, this exhibition, which carried the Atomium with its shining spheres as its symbol, was to foster a peaceful image of science and of nuclear physics in particular (Schroeder-Gudehus and Cloutier 1994; Lambilliotte 1959, 1961).[46] In the six months it was open to the public, it attracted over 40 million visitors. Fifteen countries, including both the United States and the Soviet Union, contributed to the realisation of the Science Pavilion. The exhibits were organised under the four headings of

'Atom', 'Crystal', 'Molecule', and 'Living cell', linking the physical and the biological sciences in a quest for the fundamental structures of inanimate and living matter. In his introduction to the Science Pavilion in the official report, Bragg stressed the disinterested nature and the fascination of scientific research (Bragg 1959).

The representation of the sciences at the Brussels World's Fair encapsulated the appeal of molecular biology and its models in the postwar era. With nuclear physics, molecular biology shared the aura of a 'fundamental' science. At the same time it carried the attraction of a 'physics of life' rather than a 'physics of death' (Rasmussen 1997; de Chadarevian 2002). Its intricate models promised applications in the medical field, rather than in destructive weapon systems. Indeed, despite the dominant rhetoric, the notion that once the structure of 'living molecules' was known it would also be possible to reconstruct and 'repair' them was never far away. In the *Living Molecules* exhibition in the Science Museum, mentioned at the beginning of this chapter, this connection was exemplified by a button. When pushed, a red light turned on in the haemoglobin model, indicating the position of the single amino acid that distinguishes sickle cell haemoglobin from 'normal' haemoglobin. If the cause of a complex disease picture can be pinned down to a small chemical difference, this exhibit seemed to suggest, a cure can be found. In more recent commercial advertisements concerning molecular probes, technologies, and drugs, the technical possibilities and medical benefits of manipulating molecules form the central message.

The models of molecular biologists did not only embody results, but were objects that could be manipulated, marketed, and displayed. Often the same models which served as research tools were also the ones which circulated outside the laboratory, appearing in lecture and exhibition halls, on television programmes, and before funding agencies. Other times the models were scaled up or down and replicated using different model parts, adapted to their new uses as didactic tools, travelling models, or exhibition objects. They were commercially produced and circulated widely. A few of the original research models have found their way into museums, where they are stored and preserved. By that time they have long ceased being research tools for the scientists, but they still work as objects of display, where they become embedded into new stories about molecular biologists and their discoveries.[47] When Kendrew's former colleague, in charge of rebuilding the 'forest of rods' model found in the museum store in London, against his brief, levelled the heights of some of the rods to give the model a more 'perfect' and 'finished' look, he prepared it for that new role as he understood it.

NOTES

I thank Robert Anderson, Antony Barrington Brown, Robert Bud, John Barker, Francis Crick, Richard Dickerson, Michael Fuller, Sandy Geis, Richard Henderson, Georgina Hooper, John Kendrew, Annette Lenton (neé Snazle), Hilary Muirhead, Ann Newmark, Kirsty Nott, Nicola Perrin, Max Perutz, and Paul Rayner for crucial information on the history of the models discussed in this paper as well as for making printed and photographic material available to me. Many thanks also to Richard Henderson, Martin Kusch, and Jim Secord, the referees of this volume, and participants at seminars in London and Cambridge for their many useful comments on earlier versions of this chapter, and to Nick Hopwood for our continuing conversations about models.

1. R. Anderson to A. Newmark, 23 March 1976; file T/1977/219, Science Museum, London.

2. Other evidence suggests that in 1968 Kendrew himself offered the model to the museum when his colleagues pushed for space and he faced the alternative of either dismantling it or finding it a new home.

3. J. Kendrew to F. Greenway (Keeper of Chemical Collection), 14 October 1975; file T/6762, Science Museum, London.

4. Originally planned to remain on display for a few months, the exhibition was only recently dismantled. Most of the models have been integrated into other displays in the museum.

5. On representations of the double helix and its place as a cultural icon, see de Chadarevian and Kamminga 2002; and Nelkin and Lindee 1995. On the fate of Watson and Crick's original model, see de Chadarevian 2003.

6. Watson 1968; *Nature* 248, no. 5451, 26 April (1974).

7. This is particularly true of Britain, where molecular biology grew out of an alliance of protein crystallographers, molecular geneticists, and protein chemists (de Chadarevian 1996; 2002). The new field was defined as the study of the structure and function of proteins and nucleic acids, i.e., the contested borderline between molecular biology and biochemistry was drawn according to the "politics of small and large molecules" (Abir-Am 1992).

8. The number of reflections to be included in the synthesis, and thus the volume of calculations, depended on the resolution of the analysis. To pass from one resolution to another twice as good, eight times as many spots had to be measured. In the analysis of myoglobin, Kendrew moved from a resolution of 6-Å, involving the measurement of the intensities and the calculation of the phases of 400 reflections per crystal, to an analysis of 2-Å and then 1.4-Å resolution,

including the data of 10,000 and 25,000 reflections per crystal, respectively. Data from several heavy-metal derivatives needed to be included in every stage of the analysis, further increasing the work. Data collection and calculations for each stage took several years to complete. The work would have been impossible without the help of many hands, the increasing automatisation of data collection and the use of the then most powerful computers (Kendrew 1964).

9. This is only one of many references by protein crystallographers to anatomical practices (see below). On modelling techniques based on microtome sectioning in nineteenth-century embryology, see Hopwood 1999. Generally, protein crystallographers did not refer to the contour maps as models, but reserved this term for physical models built on the basis of the electron distribution data.

10. In this, too, protein crystallographers followed tradition, as a note by William Bragg, written in 1925, confirms: "After trials of many ways of making models of atomic structure and of many substances I find that two have real merits: Balls representing the atoms may be made of hard dentists' wax, which softens in boiling water and can then be pressed into proper shape in metal moulds made for the purpose, just as we used to remake our golf-balls in the old days .... Gramophone needles made good connectors: the balls ... being drilled to receive them ... " (Bragg 1952, 231). Even when modelling kits became commercially available, molecular biologists still used a large variety of materials ranging from ping-pong balls to corks and baby shoes. Children's (that is boys') construction kits were always favourites. On the early history of physical modelling in chemistry, see Meinel, this volume.

11. Kendrew presented this model made of black plasticine to the Science Museum, describing it as "the first model of a protein" (J. Kendrew to F. Greenaway, 14 October 1975, Science Museum, file T/6762). How literally this should be understood remains open to interpretation. The model differs from the one which first appeared in print in a series of photographs and has recently returned to the Laboratory of Molecular Biology from Kendrew's private collection. This (much better preserved) model is covered with glossy white paint except for the oxygen-binding haem group, which is represented by a red disc. The structure is not sitting on pegs but supported by a few metal rods inserted between its bends. At the laboratory, priority is given to this model, but the available records give no definite answer. The 'white' model also exists in a few large-scale reproductions used for demonstrations and exhibitions, and certainly had a wider circulation at the time.

12. Dr. Norton, "Visit to the Molecular Biology Research Unit, Cambridge, 2 December 1957", unpublished internal report, file FD12/291 (E243/29), Public Record Office, London, and, for reactions more generally, Kendrew Papers, MS Eng. c.2408 (C.278), Bodleian Library, Oxford. It is interesting to speculate to what extent the different aesthetic appeal of the first protein model and the double-helical model of DNA affected their later careers.

13. Interview with Max Penitz, Cambridge, 1 July 1998. The disc structure of the model has been retained in the later decomposable demonstration models of haemoglobin that are part of many teaching collections.

14. Business correspondence, 1959 to the present, Cambridge Repetition Engineers, Cambridge. Around 80% of the shipments went to America, where many researchers settled after spending their postdoctoral years at Cambridge. In America orders could also be placed with the Ealing Corporation, which, however, marketed the models at a much higher price. The parts were produced semi-automatically. Brass rods were cut with special machines (normally used to build parts for sewing machines) and soldered by hand using gauges. While the 'Kendrew models' found wide distribution, Kendrew's 'rod and clip' method was soon superseded by the invention of an ingenious optical device, known as 'Fred's folly' (after the crystallographer Frederic Richards) which projected an image of the electron-density map onto the model while it was being built, greatly accelerating the task (Richards 1968). On the design and use of other molecular model kits as well as on molecular modelling in the practice and historiography of the chemical sciences more generally, see Francoeur 1997. On first attempts in the 1960s to use interactive graphics programs for protein modelling on the screen, see Francoeur and Segal, this volume.

15. When Perutz, in his well-rehearsed discovery story of the structure and function of haemoglobin, produced his structural models of the molecule, he told his audience: "That is what it really is"; see M. Perutz, *Science Is No Quiet Life*, Peterhouse Kelvin Club Video 1996.

16. Interview with B. Hartley, Elsworth, 28 September 1992.

17. Interview with D. Blow, London, 29 September 1992.

18. W. L. Bragg to H. Himsworth, 18 January 1968, file FD1/9033 (S18/1), Public Record Office, London.

19. A full description of the structure, including the complete set of coordinates, was finally published by Kendrew's colleague, Herman Watson, in 1969 (Watson 1969). The problem encountered by protein crystallographers of conveying their results in print was shared by other researchers whose results were embodied in 3-D models; see especially Hopwood, this volume.

20. Photographs nonetheless remained central documents for artists to work from (see below). They were also published in parallel with schematic drawings. For discussion of a similarly paired use of photographs and diagrams in electron microscopy see Lynch 1990. Photographic techniques were crucial for the recording of the diffraction patterns, and diffraction photographs were a fixture of crystallographic papers until the advent of automatic diffractometers. From the mid-1960s the use of 3-D photography was explored to reproduce the spatial qualities of models, but these remained isolated attempts; see R. A. Harte (American Society of Biological Chemists) to J. Kendrew, 21 November 1966, with attached a 3-D photograph of a space-filling model of myoglobin; Kendrew Papers MS Eng. c.2410 (C.305, as well as further correspondence on the topic in folder C.306), Bodleian Library, Oxford.

21. J. Kendrew to H. Himsworth, 17 December 1957, Kendrew Papers, MS Eng. c.2593 (M.1), Bodleian Library, Oxford.

22. G. H. Haggis to J. Kendrew, 6 June 1960; Kendrew Papers, MS Eng. c.2408 (C.281), Bodleian Library, Oxford.

23. When Perutz's model of haemoglobin became available, he used the services of A. Kirkpatrick Maxwell, an artist who worked on a freelance basis for the School of Anatomy at Cambridge, to produce a drawing of the molecule which is still widely used in textbooks. I thank Barbie Wells, librarian at the School of Anatomy at Cambridge, for her help in reconstructing Maxwell's name from his initials. The constant association of protein crystallography and the structures it produced with (molecular) anatomy and the use of a corresponding set of representational techniques, including models, drawings, and atlases of molecular structures, deserves further elaboration.

24. Biographical notes on British medical artists of the last half century, including Price and Maxwell, have been collected in Archer 1998.

25. S. Geis to author, 20 July 1998. I thank Sandy Geis for generously providing information on her father's activity and for making available material from the Geis Archives.

26. On Geis's molecular drawings, see also Dickerson 1997a and c.

27. At the Cambridge Laboratory of Molecular Biology many of the early stereoscopic drawings were done by Annette Snazle, a young assistant. On late-nineteenth- and early-twentieth-century atlases as expressions of a new mechanical objectivity and a new moral stance of scientists, see Daston and Galison 1992. On the production of late-twentieth-century atlases, see Galison 1998.

28. M. Perutz to H. Himsworth, 6 April 1953, inviting him to see the model of DNA; file FDI/426, Public Record Office, London.

29. Cambridge Repetition Engineers Price List (undated) and Ealing Catalogue 1975–76. Besides myoglobin component sets, Cambridge Repetition Engineers also offered model kits of ribonuclease, lysozyme, chymotrypsin, DNA, and bacterial cell wall.

30. Note that research and demonstration objects were not always neatly separated. Often the research objects themselves (or replicas thereof) served for demonstration or public display, as for instance in the *Living Molecules* exhibition and in the television films discussed below. The rhetoric of presentation used for different audiences as well as the reception of the (same) models by different audiences would warrant more detailed study.

31. The initiative for the series came from the BBC itself. The producer of the series was Aubrey Singer. The acquisition of a specialist scientific advisor to the production team was discussed, but finally dismissed on the ground that this would be a "rather useless individual, especially if he had no knowledge of television" (A. Singer to O.B.O. Tel [date missing], file T14/1502/1, BBC Written Archives Centre, Reading). Most of the scripts for the series were written by Gordon Rattray Taylor, Cambridge science graduate and wartime reporter for the BBC. He later became editor of the science programme *Horizon*, broadcast on BBC2, and wrote various books on science-related topics. There is not yet any extensive historical

study of science on television in Britain. In Brigg's official history of the BBC, science is hardly mentioned; the same is true for Seymour-Ure's introductory volume on *The British Press and Broadcasting since 1945* (Briggs 1979; Seymour-Ure 1991).

32. E. Underwood (Director of Public Relations, UK Atomic Energy Authority) to A. Singer (BBC), 13 February 1958; file T14/1502/1, BBC Written Archives Centre, Reading. Among those invited for the "end of series party" (i.e., the end of the first series; many more were to follow) were various scientists in leading administrative positions as well as high-level administrators from the Ministry of Supply, the Department of Scientific and Industrial Research, the Atomic Energy Research Establishment, and the Industry Research Association; A. Singer [internal memo], 27 January 1958; file T14/1502/1, BBC Written Archives Centre, Reading.

33. A. Todd to Sir Ian Jacob (BBC), 4 March 1958; file T14/1495/10, BBC Written Archives Centre, Reading.

34. Miscellaneous correspondence in file T14/1502/1, BBC Written Archives Centre, Reading relating to the first *Eye on Research* series in general.

35. G. Rattray Taylor, TV Science Series, 20 November 1957 [Note on discussion with Sir Alexander Todd, Professor Meyer Fortes, and Dr. Robin Marris], file T14/1502/1, BBC Written Archives Centre, Reading. See also "Note on a meeting with Lord Adrian [Vice Chancellor of the University] at Cambridge, 29 October", same file.

36. A. Singer to P. Daly [date missing] and 4 May 1960, file T14/1499/5, BBC Written Archives Centre, Reading. The general notes for production of the programme confirmed: "As much as possible should . . . be made of the finished models. This is what the lay viewer wants to see: The 'answer' to the problem"; G. Rattray Taylor, "Perutz Program (*Eye on Research* 1960)", 27 December 1959, same file.

37. *Eye on Research: Shapes of Life*. Script by Gordon Rattray; interview with Max Perutz and John Kendrew by Raymond Baxter, broadcast on 3 May 1960.

38. Miscellaneous notes and correspondence in file T14/1499/5. Even more discussion surrounded the choice, preparation, and staging of the DNA model in its first appearance on television in a 1958 *Eye on Research* production (*The Thread of Life: DNA*, 4 February 1958). In the end, a battery of models was used, one of which was placed on a revolving table, where for maximum effect it kept turning in the background throughout the production; notes and correspondence in file T14/1495/10, BBC Written Archives Centre, Reading.

39. See Audience Research Report for *Eye on Research*, 5—Shapes of Life, file T14/1499/5, BBC Written Archives Centre, Reading.

40. M. Perutz to P. Daly, 6 May 1960, file T14/1499/5, BBC Written Archives Centre, Reading.

41. W. L. Bragg to V. Alford (BBC), 1 May 1944 and response; Talks, W. L. Bragg, file 1, 1938–62, BBC Written Archives Centre, Reading. Bragg, who had introduced Perutz to the BBC, regularly talked on radio on scientific issues. In this

he followed his father's example. See for instance the wartime series of broadcast talks on the sub-visible universe coordinated by the elder Bragg, in which his son also participated (*Science Lifts the Veil* 1948).

42. F. Crick to M. G. Rhodes (BBC), 19 May 1966; Francis Crick, contracts–talks, file II, 1963–67, BBC Written Archives Centre, Reading. The trend of presenting science on television rather than on radio may not be irreversible. As an article in *The Guardian* recently argued, "most modern scientific equipment is both inscrutable in appearance and incomprehensible to the average viewer". The answer to this situation, the reporter suggested, is to revert to the radio, where people are left to imagine the details and feel less cheated. Intriguingly, the article featured a radio with one of its buttons operating the display of a chemical model (Green 1999).

43. See note 38.

44. M. Perutz to C. Norton, 22 February 1958, file FD12/292 (E243/109 I), Public Record Office, London.

45. This impressive model still decorates the landing of the laboratory staircase in Cambridge. The crystal structure models, especially the ball-and-spoke models, produced for the Brussels Fair were built according to standard scales and colour. The scale chosen was 2.5 cm: 1 Å.

46. The official report includes a description and photograph of the giant DNA model (Lambilliotte 1959, 71 and photo 25). This is but one example in which the double helix was used to redeem the darker face of nuclear physics. Under the title "A different harvest", Snow's *The Physicists* includes an account of the discovery of the double helix, which follows a chapter on nuclear fusion with gruesome pictures of the destruction caused by, and victims of, the atomic bombs dropped on Japan, and one on the prospects of nuclear war (Snow 1981).

47. On "objects of knowledge" in museums and for a compelling reading of the interwoven stories of production and viewing of a museum exhibit, see Jordanova 1989; see also Haraway 1989.

REFERENCES

Abir-Am, Pnina G. 1992. "The politics of macromolecules: Molecular biologists, biochemists, and rhetoric." *Osiris* 7: 210–37.

Archer, Patricia. 1998. "A History of the Medical Artists' Association of Great Britain, 1949–1997." Ph.D. dissertation, University College, London.

Blow, David M., Jens Birktoft, and Brian S. Hartley. 1969. "Role of a buried acid group in the mechanism of action of chymotrypsin." *Nature* 221: 337–40.

Bodo, G., H. M. Dintzis, J. C. Kendrew, and H. W. Wyckoff. 1959. "The crystal structure of myoglobin: V. A low-resolution three-dimensional Fourier synthesis of sperm-whale myoglobin crystals." *Proceedings of the Royal Society A* 253: 70–102.

Bragg, William. 1952. *Concerning the Nature of Things: Six Lectures Delivered at the Royal Institution.* New York: Dover.

Bragg, W. Lawrence. 1929. "The determination of parameters in crystal structures by means of Fourier series." *Proceedings of the Royal Society of London A* 123: 537–59.

———. 1959. "Introduction au Palais International de la Science." In *Le Mémorial officiel de l'Exposition universelle et internationale de Bruxelles 1958,* vol. 6: *Les Sciences*, ed. M. Lambilliotte, 17–21. Brussels: Établissements généraux d'Imprimerie.

Briggs, Asa. 1979. *The History of Broadcasting in the United Kingdom*, vol. 4: *Sound and Vision.* Oxford: Oxford University Press.

Daston, Lorraine, and Peter Galison. 1992. "The image of objectivity." *Representations* 40: 81–128.

de Chadarevian, Soraya. 1996. "Sequences, conformation, information: Biochemists and molecular biologists in the 1950s." *Journal of the History of Biology* 29: 361–86.

———. 2002. *Designs for Life: Molecular Biology after World War II.* Cambridge: Cambridge University Press.

———. 2003. "Relics, replicas, and commemorations." *Endeavour* 27: 75–79.

de Chadarevian, Soraya, and Harmke Kamminga. 2002. *Representations of the Double Helix*, rev. edn. Cambridge: Whipple Museum of the History of Science.

Dickerson, Richard E. 1997a. "Irving Geis, molecular artist, 1908–1997." *Protein Science* 6: 2483–84.

———. 1997b. "Irving Geis, 1908–1997." *Structure* 5: 1247–49.

———. 1997c. "Molecular artistry." *Current Biology* 7: R740–41.

Dickerson, Richard E., and Irving Geis. 1969. *The Structure and Action of Proteins.* New York: Harper and Row.

Edsall, John T., Paul J. Flory, John C. Kendrew, A. M. Liquori, George Némethy, and G. N. Ramachandran. 1966. "A proposal of standard conventions and nomenclature for the description of polypeptide conformations." *Journal of Molecular Biology* 15: 399–407.

Fermi, Giulio, and Max Perutz. 1981. *Atlas of Molecular Structures in Biology*, vol 2: *Haemoglobin and Myoglobin*, ed. D. C. Phillips and F. M. Richards. Oxford: Clarendon.

Francoeur, Eric. 1997. "The forgotten tool: The design and use of molecular models." *Social Studies of Science* 27: 7–40.

Gaber, Bruce P., and David S. Goodsell. 1997. "The art of molecular graphics. Irving Geis: Dean of molecular illustration." *Journal of Molecular Graphics and Modelling* 15: 57–59.

Galison, Peter. 1998. "Judgment gainst objectivity." In *Picturing Science, Producing Art*, ed. Caroline A. Jones and Peter Galison, 327–59. New York: Routledge.

Green, Dave. 1999. "Listen with prejudice." *The Guardian.* 1 December, G2: 17.

Haraway, Donna. 1989. "Teddy bear patriarchy: Taxidermy in the Garden of Eden, New York City, 1908–1936." In *Primate Visions: Gender, Race, and Nature in the World of Modern Science*, 26–58. New York: Routledge.

Hopwood, Nick. 1999. "'Giving body' to embryos: Modelling, mechanism, and the microtome in late nineteenth-century anatomy." *Isis* 90: 462–96.

IUPAC-IUB Commission on Biochemical Nomenclature. 1970. "Abbreviations and symbols for the description of the conformation of polypeptide chains." *Journal of Molecular Biology* 52: 1–17.

Jordanova, Ludmilla. 1989. "Objects of knowledge: A historical perspective on museums." In *The New Museology*, ed. Peter Vergo, 22–40. London: Reaktion.

Kendrew, John C. 1961. "The three-dimensional structure of a protein molecule." *Scientific American* 205 (December): 96–110.

———. 1964. "Myoglobin and the structure of proteins. Nobel lecture, December 11, 1962." In *Nobel Lectures Including Presentation Speeches and Laureates' Biographies: Chemistry, 1942–1962*, 676–98. Amsterdam: Elsevier.

———. 1966. *The Thread of Life: An Introduction to Molecular Biology Based on the Series of B.B.C. Television Lectures of the Same Title*. London: G. Bell and Sons.

Kendrew, J. C., G. Bodo, H. M. Dintzis, R. G. Parrish, and H. Wykoff. 1958. "A three-dimensional model of the myoglobin molecule obtained by X-ray analysis." *Nature* 181: 662–66.

Kendrew, J. C., R. E. Dickerson, B. E. Strandberg, R. G. Hart, D. R. Davies, D. C. Phillips, and V. C. Shore. 1960. "Structure of myoglobin: A three-dimensional Fourier synthesis at 2 Å resolution." *Nature* 185: 422–27.

Lambilliotte, M., ed. 1959. *Le Mémorial officiel de l'Exposition universelle et internationale de Bruxelles 1958*, vol. 6: *Les Sciences*. Brussels: Établissements généraux d'Imprimerie.

———, ed. 1961. *Le Mémorial officiel de l'Exposition universelle et internationale de Bruxelles 1958*, vol. 8: *Synthèse*. Brussels: Établissements généraux d'Imprimerie.

Lewenstein, Bruce V. 1992. "The meaning of 'public understanding of science' in the United States after World War II." *Public Understanding of Science* 1: 45–68.

Lynch, Michael. 1990. "The externalized retina: Selection and mathematization in the visual documentation of objects in the life sciences." In *Representation in Scientific Practice*, ed. Michael Lynch and Steve Woolgar, 153–86. Cambridge, Mass.: MIT Press.

Lynch, Michael, and Steve Woolgar, eds. 1990. *Representation in Scientific Practice*. Cambridge, Mass.: MIT Press.

Nelkin, Dorothy, and M. Susan Lindee, 1995. *The DNA Mystique: The Gene as Cultural Icon*. London: Freeman.

Perutz, Max F. 1964. "X-ray analysis of haemoglobin. Nobel lecture, December 11, 1962." In *Nobel Lectures Including Presentation Speeches and Laureates' Biographies: Chemistry 1942–1962*, 653–73. Amsterdam: Elsevier.

———. 1965. "Structure and function of haemoglobin: I. A tentative atomic model of horse oxyhaemoglobin." *Journal of Molecular Biology* 13: 646–68.

———. 1978. "Hemoglobin structure and respiratory transport." *Scientific American* 239: 92–125.

———. 1987. "The birth of molecular biology." *New Scientist* 114: 38–41.

Perutz, M. F., M. G. Rossmann, A. F. Cullis, H. Muirhead, G. Will, and A. C. T. North. 1960. "Structure of haemoglobin: A three-dimensional Fourier synthesis at 5.5-Å resolution, obtained by X-ray analysis." *Nature* 185: 416–22.

Rasmussen, Nicolas. 1977. "Midcentury biophysics: Hiroshima and the origins of molecular biology." *History of Science* 35: 244–93.

Richards, Frederic. 1968. "The matching of physical models to three-dimensional electron-density maps: A simple optical device." *Journal of Molecular Biology* 37: 225–30.

Schroeder-Gudehus, Brigitte, and David Cloutier. 1994. "Popularizing science and technology during the Cold War: Brussels 1958." In *Fair Representations: World's Fairs and the Modern World*, ed. Robert W. Rydell and Nancy Gwin, 157–80. Amsterdam: VU Press.

Science Lifts the Veil: A Series of Broadcast Talks on the Conquest of the Sub-Visible Universe. 1947. London: Longmans, Green and Co.

Seymour-Ure, Colin. 1991. *The British Press and Broadcasting since 1945*. Oxford: Blackwell.

Snow, C. P. 1981. *The Physicists*. London: Macmillan.

Watson, Herman. 1969. "The stereochemistry of the protein myoglobin." *Progress in Stereochemistry* 4: 299–333.

Watson, James D. 1968. *The Double Helix: A Personal Account of the Discovery of the Structure of DNA*. New York: Atheneum; London: Weidenfels and Nicolson.

Yoxen, Edward. 1978. "The social impact of molecular biology." Ph.D. dissertation, University of Cambridge.

———. 1982. "Giving life a new meaning: The rise of the molecular biology establishment." In *Scientific Establishments and Hierarchies*, ed. Norbert Elias, Herminio Martins, and Richard Whitley, 123–43. Dordrecht: Reidel.

CHAPTER THIRTEEN

Secrets Hidden by Two-Dimensionality: The Economy as a Hydraulic Machine

Mary S. Morgan and
Marcel Boumans

> It seems to me that the test of "Do we or do we not understand a particular subject in physics?" is "Can we make a mechanical model of it?" (Lord Kelvin, quoted in Duhem 1954, 71)

> Once upon a time there was a student at the London School of Economics studying for the B.Sc. (Econ), who got into difficulties with his Keynes and his Robertson and . . . with such questions as whether Savings are necessarily equal to Investment and whether the rate of interest is determined by the demand and supply of loanable funds or by the demand and supply for idle money; but he realised that monetary flows and stocks of money could be thought of as flows and tankfuls of water. (James Meade, writing about Bill Phillips in Meade 1951, 10)

A long-standing tradition presents economic activity in terms of the flow of fluids. This metaphor lies behind a small but influential practice of hydraulic modelling in economics. Yet turning the metaphor into a three-dimensional hydraulic model of the economic system entails making numerous and detailed commitments about the analogy between hydraulics and the economy.

The most famous 3-D model in economics is probably the Phillips machine, the central object in this chapter. Made in Britain in the late 1940s, this hydraulic model (shown with its creator Bill Phillips in Fig. 13.1) is 7 × 5 × 3 ft and represents the macroeconomy by flows and stocks of coloured water in a system of perspex tanks and channels. A small number of these machines was made (perhaps 14) and these found their way across the English Channel and the Atlantic, and even to the Antipodes. Though

Figure 13.1 Bill Phillips with the hydraulic machine, Mark II. Photograph, probably taken in the 1950s, from the Vrije Universiteit Amsterdam Archive. By permission of the Vrije Universiteit Amsterdam.

used for demonstrations and teaching in the 1950s and early 1960s, these machines fell into disuse by the end of that decade. Recently, several have been restored and have taken on the status of icons, given pride of place in their owning institutions. The Phillips machine remains one of the few 'objects' that the history of economic science can boast, and one of these restored machines has found a prestigious home opposite Charles Babbage's calculating engine in the London Science Museum.[1] Yet mention of the machine in economic circles still usually produces a grin on every face. For economists who saw and worked with the machine, that grin is accompanied by admiration and appreciation of its qualities and those of its creator.

Although economics was not a model-based science in the 1940s, when Phillips made his hydraulic model, it has since become one, with mathematics providing the dominant forms for modelling. Despite the fame of his machine, and recent evaluations of Phillips' work (Leeson 2000), the highly unusual physical form of his model, and its three-dimensionality, have hardly been discussed in the history of economics (but see Barr 1988). This chapter aims to understand the importance of the third dimension in modelling by exploring Phillips' machine and comparing it with metaphors and with some 2-D models in the hydraulic tradition. We portray the issue as one of 'secrets', meaning by this term the elements in modelling which remain unarticulated and so secret or hidden. We treat these secrets at three levels: the secrets which may remain hidden in 2-D models but have to be revealed in 3-D models; those things which may still remain hidden even in the 3-D case; and the kind of hidden knowledge which gets communicated *only by using* the 3-D model.[2]

PHILLIPS AND THE BIRTH OF HIS MACHINE

Alban William (Bill) Housego Phillips (1914–75) was born in Te Rehunga, New Zealand, into a farming family. At the age of 15, he left school and became an apprentice electrician with the Government Public Works Department. Till 1935 he worked at a hydro-electric station. Then a period started in which he took several different kinds of jobs and began studying electrical engineering in the evenings. In early 1937, he decided to go to Britain by way of China and Russia, but the Japanese invasion of China forced him to go via Japan and Siberia. While travelling to Britain he had been taking a correspondence course with the British Institute of Technology.

Shortly after his arrival in London in November 1937, he took examinations of the Institute of Electrical Engineers, becoming a Grad. I.E.E. in early 1938. He was appointed an assistant mains engineer with the County of London Electric Supply Company. Then he joined the Army, being taken prisoner of war by the Japanese (Blyth 1978; 1987; Barr 1988; Leeson 1994). It was only after World War II, at the age of thirty-two, that Phillips started a degree at the London School of Economics (LSE); he took sociology as his major with economics as subsidiary, and received a B.Sc., at a bare pass level, in 1949.

The second quote from the head of our chapter, describing Phillips' difficulties with economics, might be taken as applying to any student going to lectures on macroeconomics in 1949. In Britain, macroeconomics was based around mainly verbal elucidations and extensions of the ideas found in John Maynard Keynes' *General Theory* (Keynes 1936). Economics teaching was, in the 1940s, predominantly non-mathematical (Gilbert 1989, 110), and not just for pedagogical reasons; as a discipline economics was on the cusp between the use of verbal and mathematical language. In the 1930s Keynes' book had actually been regarded as highly mathematical. But it does not seem as if his limited algebra produced clarity, and the immediate response from a number of his contemporaries, both in Cambridge and further afield, had been to build small abstract graphic or algebraic models to clarify what they thought was the meaning of Keynes' system and show how it differed from the older classical system (Darity and Young 1995). Though these little mathematical models were useful, in several ways their media and forms could not represent fully the ideas and conceptions of Keynesian theory. For example, Phillips remembered one problem, the Keynesian identity between savings (S) and investment (I), in the following terms:

> [I]t was my dissatisfaction with [Lerner's] article using the S, I identity to 'prove' that the 'classical' theory was completely fallacious that started me off looking for a technique which would show the process more clearly than is possible with two-dimensional graphs (quoted in Barr 1988, 317–18).[3]

We focus on three particular problems. First, the Keynesian verbal approach presented the economy as a dynamic system, but these little mathematical models tended to be static, so that to understand the system, or show the implications of changes in it, required using 'the method of comparative statics'. This involved comparisons of 'before' and 'after' situations by shifting curves on diagrams or by making a change in one equation and following the causal impact through the series of equations in the system. Proposals to create dynamic mathematical models had already been made during the 1930s by the econometricians who introduced both lags and differential terms into their model equations. But the difficulty of solving these systems of equations and questions of interpretation remained.

Second, arguments of the day about how the macroeconomy worked often involved reference to both stocks and flows. Although such notions were well worked out in the context of monetary theory before the 1940s, the Keynesian theory dealt in terms of the aggregate national income, and here these notions were sources of confusion and not well represented in the mathematical models.

Third, the macroeconomic breakthrough of Keynesian economics involved the principle of effective demand operating in a continuous circular flow of macroeconomic activity, but this too was not easily represented in the little mathematical models. The circulation of income or goods, conceived as a flow of liquid, goes back at least to François Quesnay's mid-eighteenth-century *Tableau Economique* (Quesnay 1972). In this respect, Keynesian economics joined a long tradition in economics, but one that rarely operated beyond the level of metaphor.

To get a grip on this macroeconomic thinking, and to resolve his own difficulties in understanding, Phillips used his engineering skills to create the famous hydraulic machine. The machine-building began in early 1949 with the help of Walter Newlyn, who had been only a year ahead of Phillips at LSE but was already a lecturer in economics at Leeds University. Newlyn's head of department, Arthur Brown, provided £100 to fund the prototype machine (which is still on display in Leeds). The machine was made and assembled in the garage of friends of Phillips in Croydon, south of London. In engineering the machine, Newlyn was the apprentice, but often, it seems, they had to stop work while Newlyn explained the intricacies of economics to Phillips so that he could decide how to model the next part. By the end of November 1949, Phillips had demonstrated this Phillips/Newlyn prototype (Mark I) machine to the assembled faculty of economics at LSE.

In 1950 Phillips enlisted the help of an engineering firm to construct a more complicated Mark II machine (Fig. 13.1) which incorporated a number of additional features. The engineering innovations were created by Phillips, but now he relied on the economic advice of James Meade, a professor at LSE who had participated in developing small mathematical models of the

Keynesian system. Phillips was soon given a lectureship on the strength of his machine demonstrations and a 1950 article about the machine and its use in the LSE-owned journal, *Economica* (Phillips 1950). In 1950 also, Newlyn published an article in the *Yorkshire Bulletin* (Newlyn 1950), based on his experience using the Leeds prototype.

It seems fairly clear, from the descriptions of the machine-building process, that Phillips did not build his machine by starting from the extant mathematical models and simply translating them, but rather tried to resolve the problems we have indicated above by directly linking the economic ideas with hydraulic ones. This was in part an iterative process in which Phillips began to learn about and understand the macroeconomic system, with the help of Newlyn and then Meade, and embedded this understanding in the machine. The process also involved Phillips incorporating elements he found in various publications. One was Richard Goodwin's 1948 proposal to represent the dynamics of the economic process not by difference equations but by differential equations, in other words, not by discrete steps but as a continuous process (Goodwin 1948). Phillips referred to Goodwin's paper several times in his article and designed the machine's governing relations to incorporate Goodwin's formulations.

The stock-flow conceptualisation of the machine was probably inspired by Kenneth Boulding's *Economic Analysis* of 1948. To clarify how prices regulate production and consumption, he depicted a hydraulic-mechanical device (Boulding 1948, 117) showing economic stocks and flows in terms of a piece of domestic plumbing (Fig. 13.2). Phillips referred to this picture as the analogy on which his machine is based (Phillips 1950, 284), and the right-hand side of the Phillips machine (Fig. 13.3) is clearly an extension of Boulding's diagram.

These two ingredients—Goodwin's dynamics and Boulding's hydraulic design—plus the ideas about economics that Phillips had acquired from Newlyn and Meade, had to be integrated to create the machine.[4] It provided for a circular flow of national income, with the relationships between the elements in the economy to be represented by tanks of liquid with in- and out-flows, and by valves governing the stocks and flows. The machine represented the aggregate economy, and could be set up to model both Keynesian ideas and alternative theses about the economy. A fuller description of the 3-D machine and how it represents the macroeconomy is given below. First,

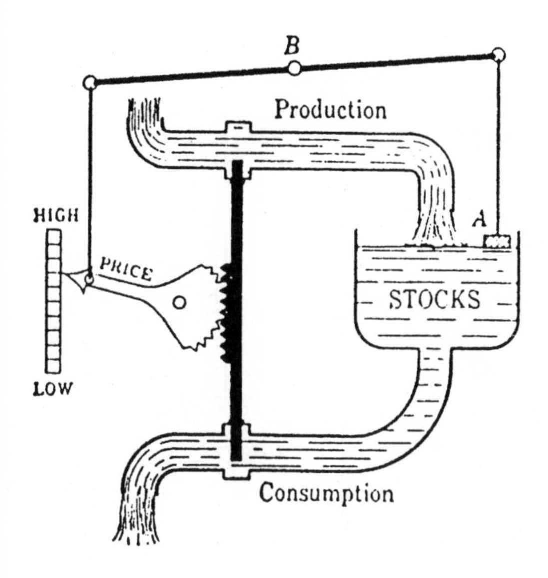

Figure 13.2 "How price regulates production and consumption". The figure shows how the price of a commodity may be compared to a valve, connected to a float *A* and a bar *B*, regulating the flows of production and consumption. Published in Kenneth E. Boulding, *Economic Analysis* (New York, 1948), 117, fig. 9. Source: Kenneth E. Boulding Collection, Box 3, Archives, University of Colorado at Boulder Libraries. By permission of the University of Colorado at Boulder.

though, we pause to discuss secrets, namely those things that are hidden but must be made explicit, even as we move from a metaphor to a 2-D model.

FROM METAPHOR TO 2-D MODEL

As we have noted, there is a long tradition in economics of understanding certain things as behaving like water and the economic system as acting according to the laws of hydraulics. For example, money is often thought of as liquid—it flows through one's fingers and leaks out of one's purse; in times of international financial crises, speculative flows of money or capital are reported daily in our financial newspapers. These figures of speech are usually and accurately referred to as metaphors.[5] Their interpretation is rather flexible and open; their usage restricts very little. The degrees of freedom are large and this encourages original insights. A metaphor might be called a one-dimensional model in the sense that it does not constrain

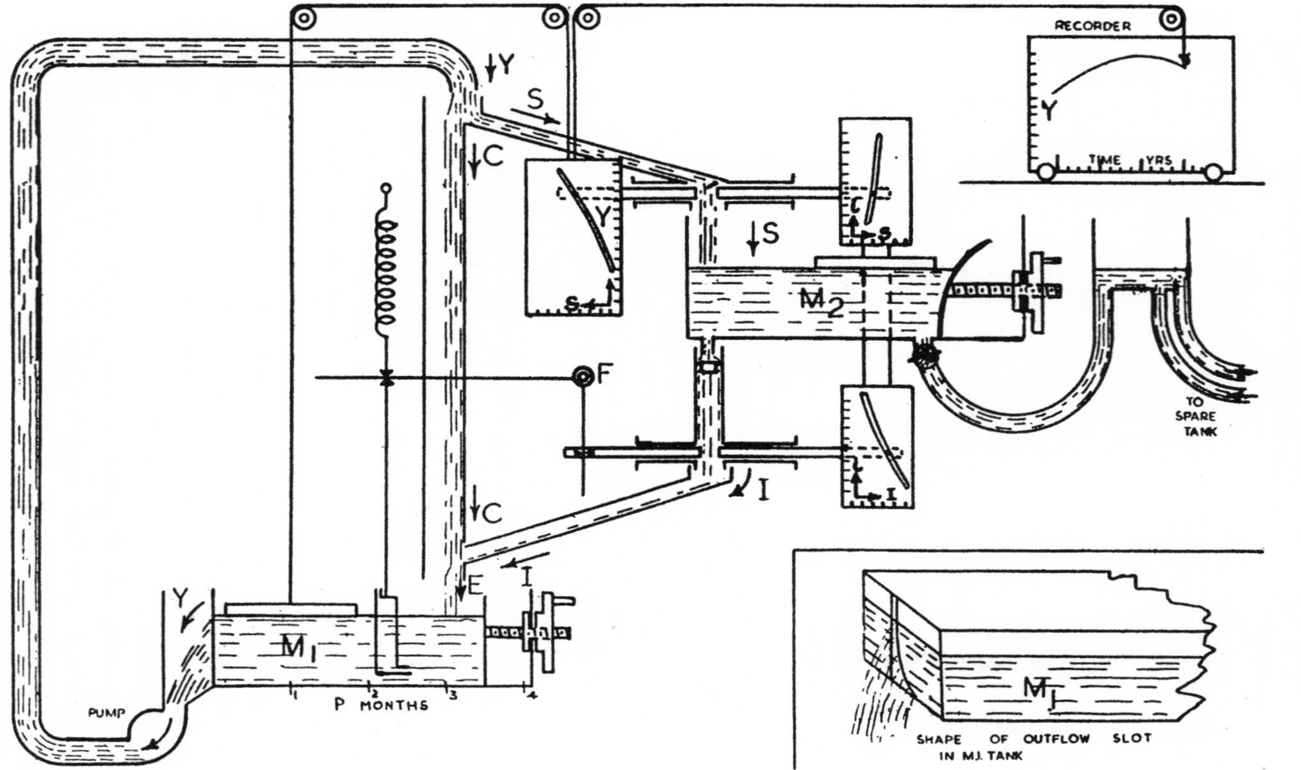

Figure 13.3 A simplified version of the hydraulic model showing an economy without foreign trade or government operations. The inset shows the shape of the outflow slot in the M_1 tank such that the rate of flow is proportional to the height of water in the tank. Source: A. W. Phillips, "Mechanical models in economic dynamics", *Economica* 17 (1950): 290, fig. 3. By permission of

our imagination, but rather gives it free rein even to the point of fantasy. When we come to build a model based on the metaphor, we have to make commitments about exactly what we mean.

First, moving from a metaphor to a model means you have to commit yourself to a particular hydraulic system, but there are several metaphors which do this, and choosing between them has different implications. What kind of an economy is this? And which analogical system would be most appropriate?[6] One possibility would be to think of the income in an economy as like a domestic plumbing system in which water comes out of the tap because of the head of water in the tank in the roof or in the town square. In economic terms this might be thought of as a supply-determined system. At first sight it does not seem quite right for the Keynesian economy, in which the great breakthrough was the idea that what made the economic world go around was the principle of effective demand. But of course, in such a plumbing system, gravity is also involved, and this might provide the analogy for the demand that pulls money or income around the system. Another possible metaphor might be the eighteenth-century view that likened the economic system to a human body with blood flowing around it. This contains the circulation idea Keynes wanted for economics, but we still have a problem with the motive power. The heart pumps blood around the body, but circulation may also depend upon demand elsewhere in the system.

A second aspect of this move from metaphor to model is that just as metaphors do not constrain very much, so they do not stretch very far. For example, in describing two markets as two connected ponds, we do not need to describe the ponds' shape and depth. To understand that the prices equalise, we can suggest several possibilities free of the constraints about how and what flows actually occur: buyers can move, goods can move, and so can money. But, as soon as we move beyond the level of metaphor to a model of such an hydraulic system, we are forced to be specific. We can show this clearly in the work by Irving Fisher, who, earlier in the twentieth century, designed several models on paper to represent the relationship between gold, the money stock, and the general level of prices (Fisher 1911).[7] His purpose was to understand the various different monetary standards that might be appropriate for the dollar, such as the gold standard and bimetallism (Fig. 13.4), which were the subject of considerable political and economic debate at that time.

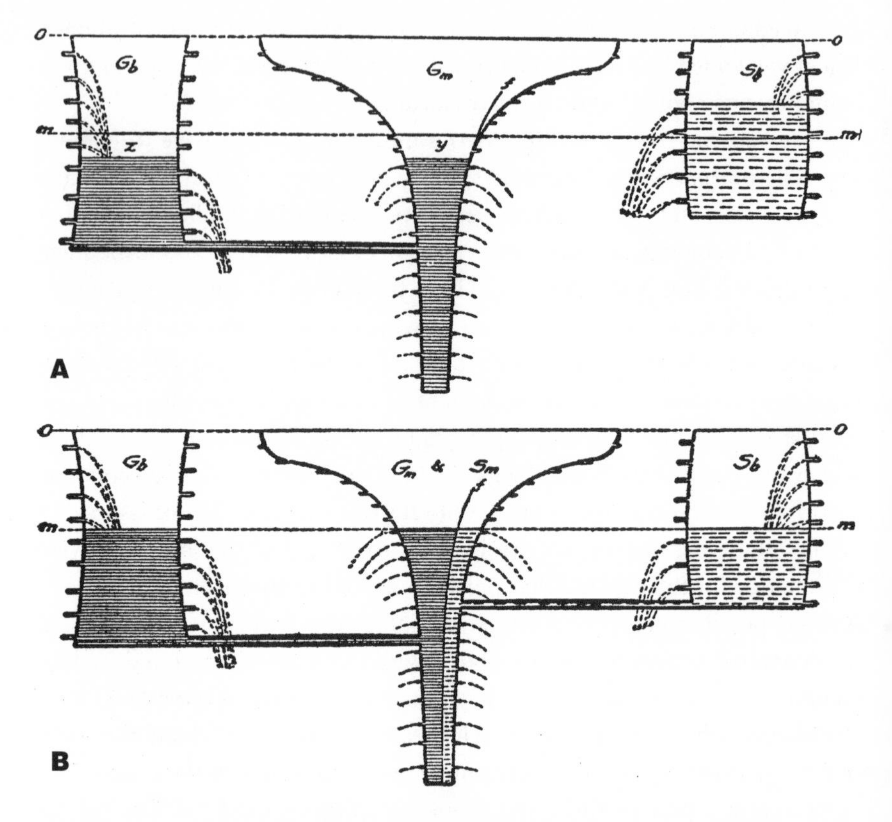

Figure 13.4 Bimetallism demonstrated by Irving Fisher's connecting reservoirs models. The contents of the reservoirs represent the stock of gold bullion (G_b), gold money (G_m), silver money (S_m), and silver bullion (S_b). The waters representing gold and silver money are separated by a movable film (f). Source: Irving Fisher, *The Purchasing Power of Money* (New York, 1911), 119, fig. 7.

In these fairly elaborate 2-D analogical models, Fisher designed a mock-laboratory system involving three vessels containing, from left to right, gold bullion, currency, and silver bullion, all in liquid form, and various inflow, outflow, and connecting tubes. In one respect, he was clever in filling in the space between the metaphor and his model—he was able to specify all the flows and stocks as well as the control valves in economic terms. By portraying various different arrangements of the pieces involved, he could use the model both to demonstrate various propositions in economic theory and to interpret the arrangements and results in terms of the observed monetary systems of the late nineteenth century and earlier times. But his ability to do all this successfully depended on the amounts of liquids in the vessels, and their shapes, positions, and relations. What were these vessels? Why should they be of any particular shape or arranged in any particular relation to each other? The answer is that Fisher designed their shapes and various arrangements 'just so' in order to make the demonstrations work—by thought experiments using the diagrams (Morgan 1999).

The point to notice is that in this 2-D paper modelling Fisher could fill in the metaphor in such a way that he could buy into the power of the hydraulics analogy when he wanted to, but could also finesse those parts of the system which did not have such a clear or ready-made analogy, or seemed strongly negative in their analogical connotations. He could, in part, choose where to have the hydraulic constraints bite and where he could or would ignore them.[8] This is not possible in a working 3-D model: all the elements in the metaphor have to be filled in for the model to work.

CONSTRAINTS AND COMMITMENTS: FILLING IN THE THIRD DIMENSION

Due to the complexity of Phillips' hydraulic machine, its description and an explanation of how it works are usually introduced with the aid of 2-D diagrams.[9] We shall follow this practice, but will argue later that the importance of the machine is not only its three-dimensionality, but also that in use it shows processes through time and so is, in effect, a 4-D model.

The first point to clarify is how the machine represents the aggregate economy as a circular flow of liquid which distributes itself through the system of pipes and tanks depicted in the rather simple 2-D flow chart shown in Figure 13.5. To imagine a single circulation of the liquid, we begin at the bottom tank, which contains a stock of coloured water representing 'transactions balances' (M_1). (We also give the symbolic names used in Figure 13.6 so that the same circulation can be followed in that more realistic drawing of the machine.) The outflow of water from this tank, that is, 'income' (Y), is pumped up to the top of the machine: this flow of income gushes down through the tanks and channels becoming expenditure (E) in the system and representing the stocks and flows of money in the economy. (Money is conceived as a real command over economic goods: there are no prices in the model.) Taking up the details of the story from the top of the machine, 'taxes' (T) are siphoned off leaving 'disposable income' (Y-T). A

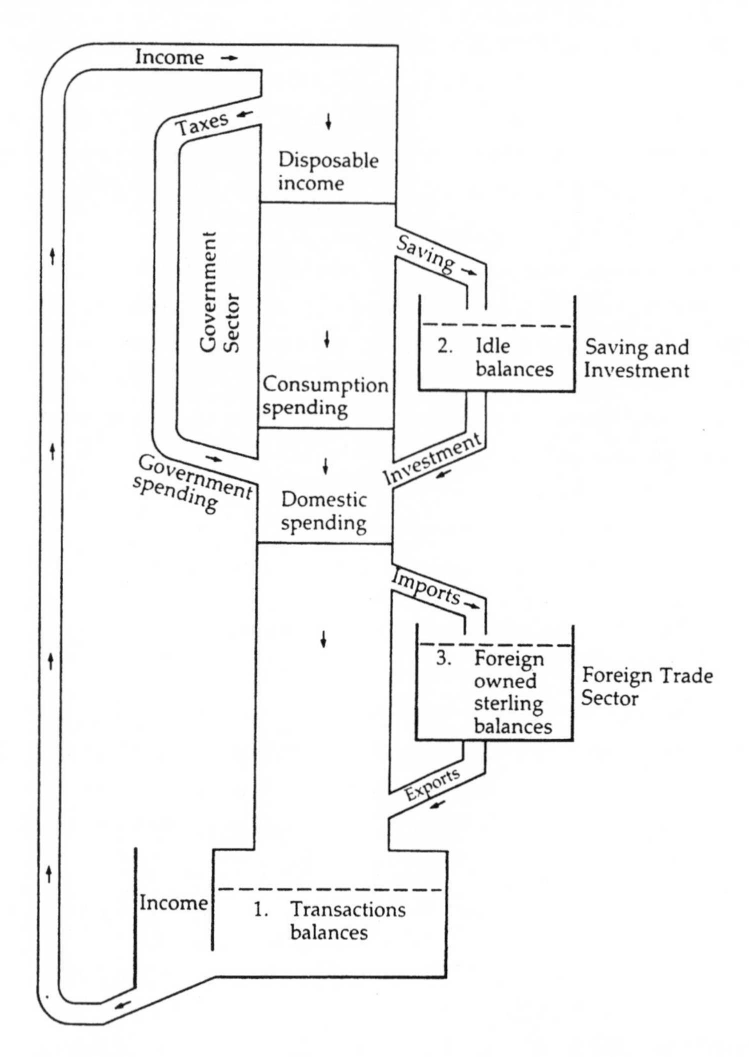

Figure 13.5 "A simplified view of the Phillips machine". Flow chart without control and servo mechanisms. Source: Nicholas Barr, "The Phillips machine", *LSE Quarterly* 2 (1988): 321, fig. 1. By permission of Nicholas Barr, the London School of Economics and Political Science, and Blackwell Publishing.

part of disposable income becomes 'consumption spending' (C) and the rest 'saving' (S). Savings are added to a stock of investments funds, called 'idle balances' (M_2), from which 'investment' (I) is withdrawn. (Figure 13.6 also shows a connecting side channel which allows the government to borrow from these balances, or repay debt to them.) Taxes can also be used by the

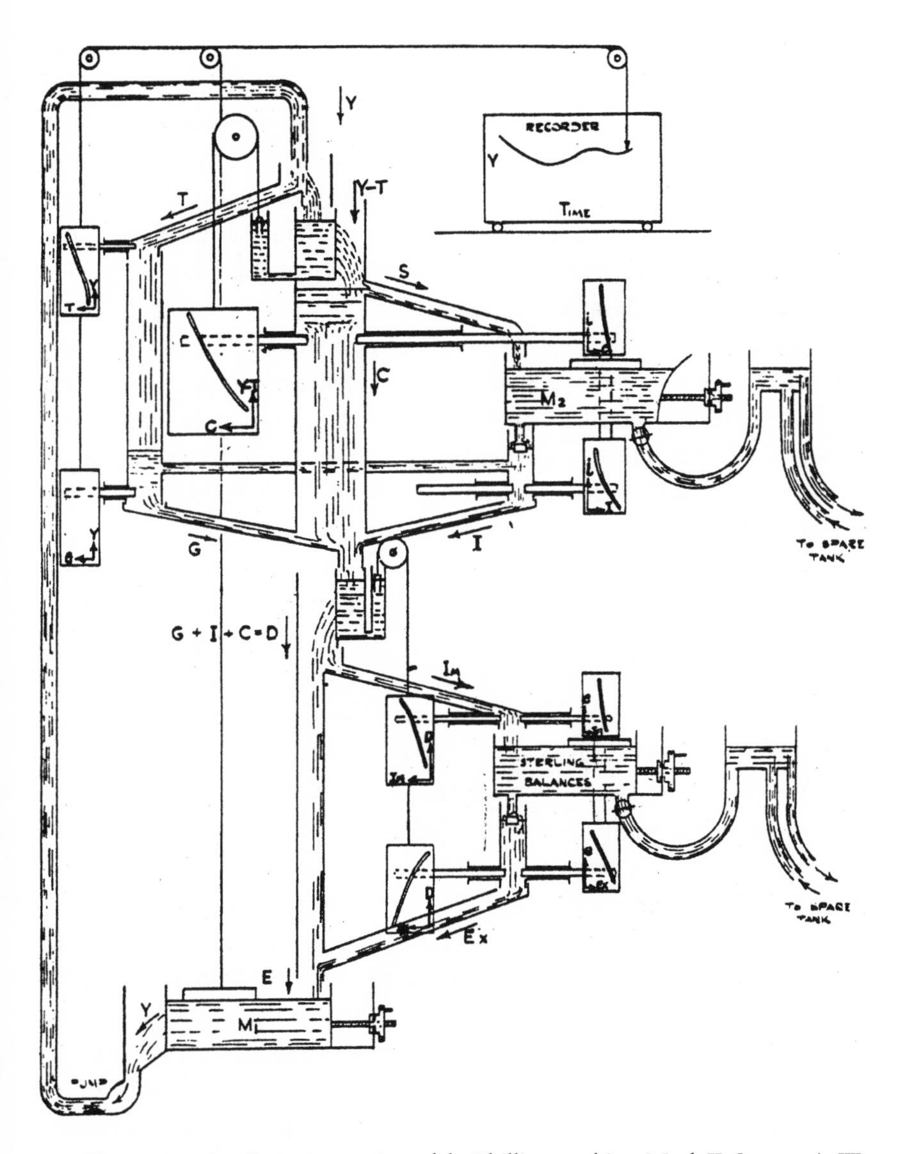

Figure 13.6 A 2-D representation of the Phillips machine, Mark II. Source: A. W. Phillips, "Mechanical models in economic dynamics", *Economica* 17 (1950): 302, fig. 4. By permission of *Economica*, the London School of Economics and Political Science, and Blackwell Publishing.

state in the form of 'government spending' (G). Investment expenditure and government spending are added to the flow of consumption spending to form 'domestic spending' ($D = G + I + C$). 'Imports' (Im) are removed from the stream of domestic expenditure and a flow of 'exports' (Ex) is added to form the flow of national expenditure (E) which enters the transaction balances tank (M_I) at the bottom of the flow chart. The change in the level of that tank, which amounts to the change in income, is registered in the recording device on the top right-hand side (Fig. 13.6).

The second characteristic of the machine that requires description is how it represents the governing economic relationships between the internal flows and stocks. These are shown in the more detailed diagrams in Figures 13.3 and 13.6. The machine has a series of valves that open and close depending on the level of water in the three tanks, and on the flow of water in the relevant part of the system. In some cases a float on the top of one of the tanks is connected to the relevant valve via a cord and a pulley (e.g., in Figure 13.6, the effect of idle balances on savings). In other cases (e.g., in Figure 13.6, the effect of domestic expenditure on imports) a similar effect is achieved by a small float connected to the relevant valve via a servo mechanism, which uses a small motor (at the back) to amplify any downward or upward movement of the float (and may also involve a lag in the adjustment). But the difference between these two float mechanisms says nothing of the economics. The forms of the economic relationships governing the system are found in the cut-out shapes inserted into the square or rectangular perspex slides connected to the other end of each float mechanism. It is these shapes that govern the extent of opening and closing of the valves in response to changes in stocks and flows. Thus, in economic terms, each of the inflows and outflows is governed by a specified economic function (e.g., in Figure 13.6, the consumption function, which relates consumption to income).[10] If you look carefully, first at Figure 13.3, you may get some sense of how these functions work. Figure 13.6 depicts all nine of these relations (look for the rectangular function holders), which together govern the stocks of money in the three tanks and the flows of income and expenditure through various parts of the system.[11] But understanding the diagram is not easy, and a detailed explanation of the economics of the whole machine using such a diagram requires an experienced machine-user and much time and space. It is even more difficult to imagine the ultimate behaviour of the flows of 'money', and changes in them, when the machine is at work.

This brief description of the economics and hydraulics of the 3-D model of Phillips shows how the gap between metaphor and model might be, and in another sense has to be, filled in. Phillips was required to specify the economic elements and relations of the macroeconomy in terms of the following physical elements:

(a) the flows and stocks in the system;
(b) the size, shape, and relative positions of the tanks (containing the stocks);
(c) how the flows go between different stocks including feedback loops;
(d) the nature of the connections between flows and stocks: valves, sluices, plugs, springs;
(e) the motive power(s) and their positions in the system;
(f) the viscosity of the fluid;
(g) the shape of the outflow slots in the tanks;
(h) the devices to maintain a constant head of water over the valves;
(i) measuring devices so that flows can be monitored (for which they are transformed into stocks) to regulate the valves via floats or servo mechanisms; and
(j) registration devices to see the effects of manipulations (which can sometimes be compared with non-machine calculations).

Items (a) to (c) in this long list involve commitments which already had to be made in the case of the 2-D models, as we can see from the detailed diagrams of the machine. Items (d) to (i) are in effect elements that might remain secret or hidden in the 2-D model, either because they can be more easily omitted or because they do not really have to 'work'. To put it another way, the list is a pile of boxes, each and all of which have to be opened and filled to make the 3-D machine work. All the hydraulic elements in the list, namely (a) to (i), are required to be specified fully, rather than partially as Fisher had done, for if the machine does not work successfully as a hydraulic system, it cannot function as a 3-D model.

But three points should be made about this process. First, there were various ways to make the model work. This points to implicit assumptions or decisions about the details of the economic equivalences, for the choices can be made to represent one or another interpretation of the relations thought to occur in the economy. Thus, the 3-D model that Phillips built was constrained not only by the laws of hydraulics, but also by the modeller's commitments to his account of the economic world being modelled.[12] The second point is that, although all the bits in the machine must be specified, not all the things in the economy must be specified in the machine. With the help of Newlyn and Meade, Phillips made his model to represent a set of elements and their economic relations, along with a set of controls and regulators, all built in line with economic theories of the day; he did not make a model to represent everything in the economy. Finally, as we shall discuss in the next section, not everything specified in the machine will necessarily have an economic meaning.

Choosing to model the economy as a 3-D hydraulic machine involves both a great many constraints imposed from the physical side and a whole lot of commitments about how the economics is physically represented. But these are not separate steps: each modelling decision involves both a physical constraint and an economic commitment at the same time. To make commitments about the analogue to the economy at the same time as working within the physical constraints requires tremendous creativity. Phillips demonstrated such creativity both in the craft skill he brought to bear in building the machine,[13] and also in the design choices he made so that the machine could be used to demonstrate many different (but by no means all) aspects of the macroeconomic thinking of his day.

WHAT STILL LIES HIDDEN IN THE MACHINE?

The constraints and commitments of 3-D modelling will reveal to the machine-builder elements which remain secret or hidden in 2-D models; such discoveries are inherent in the adventure of model-building. But to the onlooker, important elements are still hidden in the Phillips machine, and these we explore in this section. Even if we stand with the machine working in front of us, there remain invisible elements in the model as well as things hidden behind the model (Fig. 13.7); and, last but not least, there is tacit knowledge embedded in the modelling.

In Phillips' machine, two elements create the circular flow of liquid: an electric motor hidden at the back of the machine (Fig. 13.7), which has to be switched on to force the water to go upwards, and subsequently gravity which draws the water downwards through the system. One of the most important characteristics of 3-D physical objects is that they are subject to gravity. Gravity is the invisible hand that keeps our physical world together, and we become immediately aware of this characteristic if we want to use 3-D

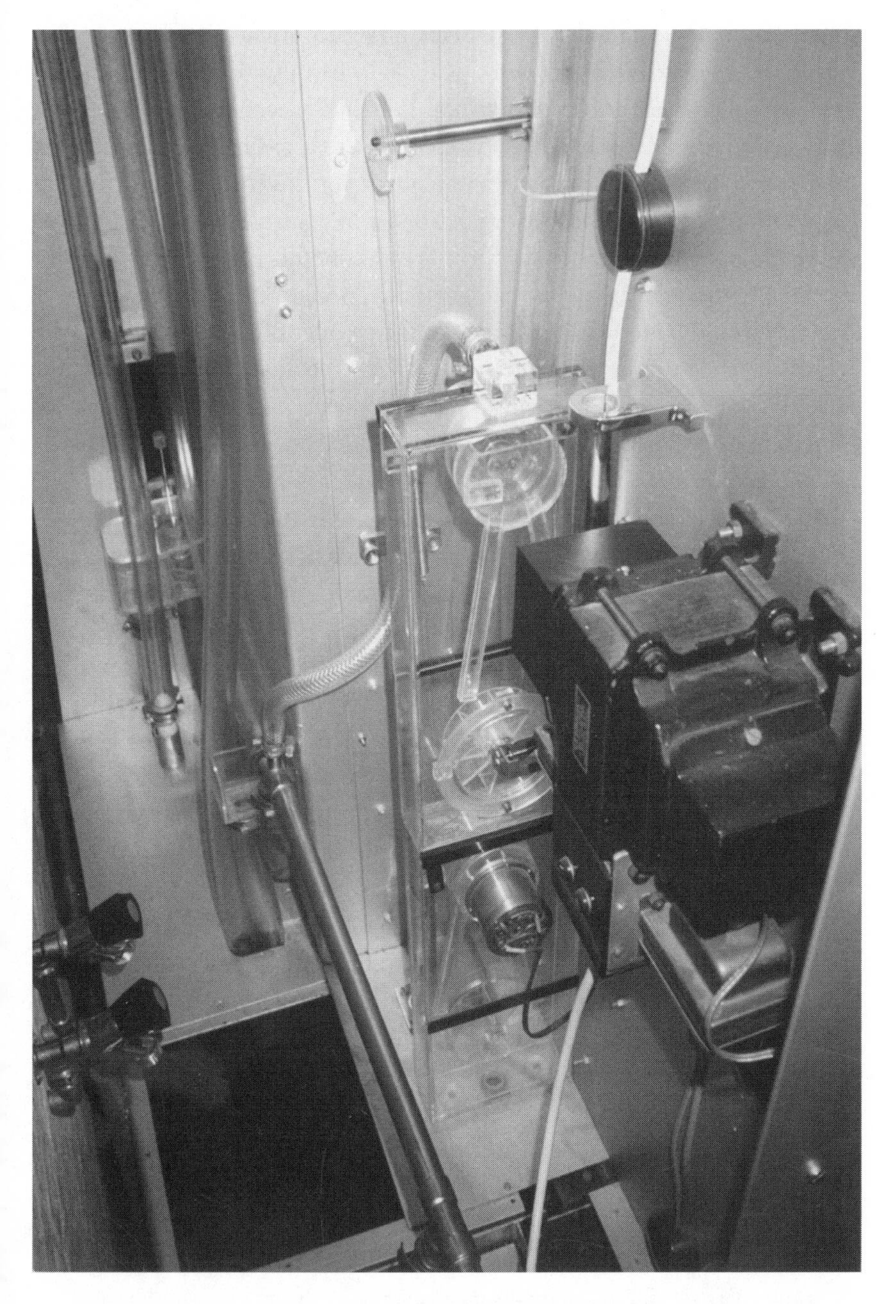

Figure 13.7 Behind the Phillips machine exhibited at the Erasmus University Rotterdam. Photo by Marcel Boumans.

objects as analogical models for understanding economics. Then we see that sometimes these 3-D objects work as they do because of this hidden force. One example where gravity works naturally in the economic model can be found in another of Fisher's 2-D models, where a mechanical balance and its dependence on gravity act as an analogue for the equation of exchange between money and goods (Morgan 1999). In contrast to the Fisher case, in the Phillips machine it seems there is no economic equivalent of gravity. It might be thought to be the analogue of the principle of effective demand, but that would be to negate the idea of the circular flow of the economy in which there should be no separation between the upstream (dependent on the electric motor) and the downstream (dependent on gravity). We therefore interpret gravity as an invisible element required for the machine to work, and the electric motor required to pump the water round as a hidden element.

Hidden elements may also be indicated by the bits of the machines the restorers did not understand. This was particularly true of the various connections between the elements and the system of sensors: "The facts behind these features were painstakingly deduced, though it would have been an impossible task without our comparison of the existing machines and the advice of those, particularly James Meade, who remembered operating the machine or helping when he first tackled these problems" (Moghadam and Carter 1989, 25). For Phillips, the back of the machine was just as important as the front, and we know that it posed the restorers an even greater challenge, perhaps because it was less well documented and its economic meaning remained unclear. This indicates further secret elements. The tacit knowledge hidden in, and behind, the machine was clearly immense, and had to be somehow reassembled by those rebuilding these machines.

In this list of 'secret' elements we should not forget that we also need a person to fill up the tanks, empty the overflow tanks (also hidden around the back), and turn the motor on. In fact, most times that the machines were used, they also needed an engineer in attendance, for they did not always work properly and often deposited red water on the floor.[14] If Phillips was there, all was well and good. But as the machines got older, getting them to work was increasingly difficult. Without Phillips' enormous tacit knowledge of engineering and of his machine, it was difficult to keep the machine alive.

Some of these secret elements need not have been hidden. There is no reason why the electric motor could not have been visible, for example. Nor is it the case that such elements are found only in 3-D models; they may equally appear in 2-D models. These elements, whether hidden or not, seem to have in common that we cannot easily give them an economic interpretation, yet they are critical to the working of the machine. But it would be wrong to see these elements as necessarily undermining the potential demonstrative power of the machine. Consider, as a comparison, the planetary-motion models of earlier times, for these bear the same kind of relation to the universe as Phillips' machine does to the economy. The planetary models were built with particular attention to the relationships between the parts that they were designed to understand and demonstrate. But neither the physical way the elements were joined up, nor the control systems which regulated their use, were necessarily particularly representative of the beliefs of the period about how the planets moved and why. Such models did not rely on the law of gravity, nor did their makers believe that the planets were made of wood or papier mâché and linked by metal rods. Similarly, economists did not believe that the economy consisted of perspex tanks and plastic tubes with red ink circulating around, but they did take seriously the notion that the main elements of the national income in the aggregate economy were rather liquid and that the stocks and flows thus conceived obeyed, in some ways or to some degree, the laws of liquids. For economists, this was more than a metaphor; it was an informative and substantive analogy.

Modelling is a process in which we try to represent some aspect of the world. We do not expect our model to represent the entire world, nor do we expect every part of the model to be represented. There are always some things, which are likely to be untranslatable or just plain wrong. But these elements do not necessarily cause difficulties in learning from the model—we willingly suspend disbelief in order to focus on the demonstrative power of those parts which do represent.

UNDERSTANDING THROUGH THE MIND'S EYE

Moving from one to two and finally to three dimensions involves an increase in commitments and constraints and ought to have some kind of pay-off in understanding. For what is the purpose of building such machines unless there are further insights to be gained? Does it alter the kind of knowledge we have? What are the secrets of understanding revealed by the 3-D machine that are not to be found in the 2-D diagram? We argue here that building and using such machines provides insights that are closely related to an engineer's way of understanding.

In *Engineering and the Mind's Eye*, Eugene Ferguson argues that good engineering is as much a matter of intuition and nonverbal thinking as of equations and computation. In nonverbal thinking, a special role is assigned to the "mind's eye", "the organ in which a lifetime of sensory information—visual, tactile, muscular, visceral, aural, olfactory, and gustatory—is stored, interconnected, and interrelated" (Ferguson 1992, 42). The mind's eye is that part of our memory built up by all the sensory information that we use in our efforts to understand the world. Because visualisations appeal to this memory, visual language is the *lingua franca* of the engineers: "It is the language that permits 'readers' of technologically explicit and detailed drawings to visualize the forms, the proportions, and the interrelationships of the elements that make up the object depicted" (Ferguson 1992, 41–42). As such, visualisations can function as an alternative for using the mathematical language, and they are particularly useful in helping understand complex systems. But, since a 2-D analogue is in principle a sufficient means of visualisation, it is still not clear why 3-D machines are built.

An engineer learns to understand the world through the "eyes and fingers", which are according to James Nasmyth "the two principal inlets to trustworthy knowledge in all the materials and operations which the engineer has to deal with" (quoted in Ferguson 1992, 50). There are many anecdotes of Phillips reminding us of his engineering approach to life:

> James Meade . . . recalled that Phillips stayed one summer with the Meade family in a cottage which contained a broken-down, out-of-tune piano: "We moaned what a pity it was so out of tune that we really couldn't use it. Bill went to the car, fetched his spanner and set to work tuning the piano . . . this anecdote shows how directly he was prepared to tackle any problem" (Leeson 1994, 606n).

Being an engineer, Phillips built the machine to learn and understand economics through his eyes and fingers rather than through mathematics and words. The quote from Kelvin which began our chapter gets at a similar kind of understanding: understanding the principles of something means being able to make a model of it, and, we would emphasise, 'a working model'.[15] But if Phillips built a model only for his own interest, an electronic analogue machine might have been a more obvious choice, because he was, after all, an electrical engineer. And in fact, in the same period as Phillips developed his machine, other 'electric analogues' were being built and used for economic investigations. The Aeracom analogue machine for studying oscillations was used to investigate inventory oscillations (Morehouse, Strotz, and Horwitz 1950). A simple electronic circuit for determining prices and exports in spatially distinct markets was used by Stephen Enke in 1951 (Enke 1951).[16]

We can think of the Phillips machine as an analogue computer but it is clear that Phillips' interest was not only (or primarily) in computer processing power and calculation, but also in how best to represent a macroeconomy so that its activities could be displayed and understood as a process (Swade 1995, 17).

> Fundamentally, the problem is to design and build a machine the operations of which can be described by a particular system of equations which it may be found useful to set up as the hypotheses of a mathematical model, in other words, a calculating machine for solving differential equations. Since, however, the machines are intended for exposition rather than accurate calculation, a second requirement is that the whole of the operations should be clearly visible and comprehensible to an onlooker. For this reason hydraulic methods have been used in preference to electronic ones which might have given greater accuracy and flexibility, the machines being made of transparent plastic ('perspex') tanks and tubes, through which is pumped coloured water (Phillips 1950, 283–84).

The mathematical model (like a computer) could be used to calculate solutions, but only for a restricted type of relations (those which could be easily solved). The mathematical models could, by the use of comparative statics, show certain steps in a dynamic process, but at the cost of missing much of the story. The analogue computer that Phillips built was both unrestrictive about the form of relations (any shapes could be cut into the governing slides) and it showed the full interactive and dynamic process, how things worked, directly.[17]

It is particularly important that Phillips aligned himself not with the mathematicians, but with gaining knowledge of the economy through the mind's eye, Ferguson's engineering mode of comprehension. Because of the inadequacies of words and mathematics to represent complex systems like the economy, models or machines are built to study both processes and outcomes. Because the machine involved a reasonably complex set of interconnected components and relationships, neither the processes nor the final outcomes were obvious. Using the hydraulic machine model makes these processes and outcomes evident and enables us to understand complexities of the economy through our eyes and fingers.

All the records show that the demonstrative power of the Phillips machine made a deep impression. Seeing the machine working is different from seeing pictures of it, as those who have seen the Philips machine working readily attest. Spectators could not only see the red water streaming through the pipes, but also hear the bubbling and splashing as it ran through the machine. They were able to see not a 2-D picture or system of equations, or even a static 3-D representation, but the kind of interrelated and dynamic cause–effect changes over time that economists suppose to happen in the circular flow of the aggregate economy. The working machine was a 4-D representation.

The Phillips machine was used to answer various questions about the dynamic economy it modelled.[18] As we have seen, the machine had valves that could be simply opened and shut manually. But they could also be controlled via the set of 'slides', each of which embodied a particular functional relationship between the relevant variables and could be changed. In principle, the relationships could be, and sometimes were, non-linear.[19] Once set going, the machine was used by changing the positions of one or more of the valves, and/or their governing relations. The user could then follow what happened and see the sequence of effects of such a change as they worked through the whole system, and thus understand how the immediate changes were transformed into longer-term effects as the circulations of income and expenditure continued. In this way, unlike the comparative statics of the mathematical demonstrations, the machine demonstrated a true dynamics. The effect of a change in policy or behaviour on national income could be 'measured' on a calibrated scale, and sceptics or those wishing to check the system could compare the result with arithmetical calculations using the mathematical relations expressed in the governing slides.[20] At first, it was also expected that the machine could be adjusted to represent the real timescale of changes, as we see in the bottom tank calibrations in Figure 13.3, but this seems to have been difficult to achieve. Finally, the flexibility and complexity of the machine offered many different options: by altering the governing slides, the machine could be set up and used as an experimental instrument to investigate various different theories about, and institutional arrangements of, the economy.

Phillips' faith in the ability of his machine to produce comprehension out of confusion proved correct. The machine as a large physical 'inscription' created 'optical consistency' in the sense Bruno Latour has defined: all the theoretical elements and institutional arrangements were made homogeneous in space in such a way "that allows you to change scale, to make them presentable, and to combine them at will" (Latour 1986, 7–8). The first time the prototype machine was demonstrated at LSE, in the weekly seminar for faculty and graduate students, has become the subject of folklore. The seminar was convened by Lionel Robbins, head of the economics department, who was sceptical about Phillips' hydraulic machine: "all sorts of people", he said, "had invented machines to demonstrate propositions which really didn't require machines to explain them" (Barr 1988, 310). From Robbins' and others' recollections, we learn that Phillips began by giving a lecture on the Keynesian system:

> They all sat round gazing in some wonder at this thing [the machine] in the middle of the room. Phillips, chain-smoking, paced back and forth and explained it in a heavy N.Z. drawl. Then he switched it on. And it worked! "There was income dividing itself into saving and consumption . . . ." He really had created a machine which simplified the problems and arguments economists had been having for years. "Keynes and Robertson need never have quarrelled if they had the Phillips machine before them".[21]

The demonstration convinced the participants that the on-going argument about the relative merits of the theories of Keynes and Dennis Robertson over the determination of the interest rate was settled. Previous arguments had claimed that these were alternative (rival) theories; or that they were essentially the same. According to the Phillips machine, they were to be understood as complementary claims about the stock of idle balances (the tank labelled M_2) and the liquidity preference function (represented in the side panel of the tank) and the flow of savings and investment funds (the I and S flows). Further, the continuous circular flow of the machine was able to demonstrate the initial changes, the chain of further changes, and their effects on the final equilibrium values.[22] The machine clarified, and sometimes 'settled', some of the endless arguments that arose from the verbal treatment of Keynesian economics and for which the available mathematical treatments seemed insufficient.

The machine showed things economists already thought they 'knew' via reasoning with words or with mathematics, but, to an extent (doubtless variable with different people) the understanding of that knowledge had remained both hidden and limited. From the demonstration of the machine, they really came to know and understand in their mind's eye the knowledge implicit in the macroeconomic thinking of the day. Thus, though neither mobile nor flat like Latour's 2-D inscriptions, the Phillips machine functioned in those same overlapping domains of visualisation and cognition which Latour ascribed to representation devices such as maps and diagrams (Latour 1986).

Richard Lipsey, who arrived as a student at the LSE soon after this demonstration, remembers how, from demonstrations with the Phillips machine, they immediately understood the notion of a model and how it could be used. The verbal reasoning and the comparative statics of mathematics and geometry were replaced by the real, not mathematical, dynamics. You could really see the economic process at work, you could see and understand the individual relations and the interactions between them in a way not available before.[23]

To understand why this machine use was so important in the 1950s, it helps to know something more about the design context. At that time, the Keynesian analysis was believed, at least in Britain, to offer governments a kind of instrumental control over the macroeconomic system via their ability to influence effective demand for goods and services. Keynes' success in managing the British war economy gave added credibility to this programme of macroeconomist as engineer. After the war, 'fine-tuning' government fiscal policy (by spending more or taxing more) was used to create multiplier effects in demand elsewhere in the economy. By such tuning, it was thought that the business cycles of the interwar economy could be mitigated, and the economy kept on a growth path. During the mid-1950s particularly, a stronger systems-control idea grew up in economic circles. Although such control-engineering notions of the use of Keynesian economics were relatively short-lived, they were particularly compatible with Phillips' background, and his work through the 1950s continued in this vein.[24]

The machine suggests the possibility of control, but at the same time it demonstrated the vulnerability of the relations and hence the dangers of such a project. James Meade regularly used the machine to demonstrate such policies in his teaching at LSE during the 1950s. One of his most enjoyable demonstrations was to appoint one student Chancellor of the Exchequer and another Governor of the Bank of England, and have them attempt to get the economy back to equilibrium, each using their own instrument of control to achieve a set target of national income. The usual outcome was confusion and leakages—from which students really understood the difficulties of fine-tuning and the importance of policy coordination.[25] The complicated dependencies in the machine (even if simpler than those of the real economy) showed just how difficult it was to co-ordinate or control the system.

WHEN VISUALISATIONS ARE ILLUSIONS

As we have seen, visualisations can lead to better understanding. But the opposite is also possible. If a visualisation contradicts the sensory information stored in our memory, it will lead to confusion, disorientation, or astonishment. Sometimes this is for aesthetic reasons, like the optical illusions in Escher etchings, or sometimes just for fun, like the Tom and Jerry cartoons. But water that streams upwards by itself never leads to a better understanding of either hydraulic or of economic principles. This was a little bit of a problem for the Phillips machine (remember the hidden pump). The problems are more clearly evident in FYSIOEN, a computer animation of a hydraulic system representing a macroeconomic model developed for PC usage, dating from the late 1980s.

FYSIOEN, a Dutch acronym for Physical–Visual Operational Economic Model for the Netherlands, represents a number of the most important mechanisms of the macroeconometric model (MORKMON) used in the Nederlandsche Bank (the Dutch Central Bank).[26] FYSIOEN was designed to help users gain understanding of the complex mathematical model by translating it into the visual domain: "The visual model affords more rapid and better insight—to more advanced model users as well—than a set of mathematical equations with explanatory notes" (Kramer et al. 1990, 159).

But there were problems in developing FYSIOEN. Although it required no motor to force the water round the system, what you saw in the computer animation were only coloured areas increasing or decreasing. You did not see a streaming of liquid, you heard no noise or splashing of running water. During the development of FYSIOEN several improvements to the animation, intended to make the 2-D visualisation behave more like a 3-D system such as the Phillips machine, were suggested (Reus 1987, 6–7). For example, the addition of noise and of arrows to show directions of flow, and the possibility of users manipulating the visual model. The PC computing technology of the time limited the possibilities. Some of these improvements were made, but even if all of the suggestions had been incorporated into FYSIOEN, the visualisation remained an illusion. For example, there were occasions when a cistern might fill up without the tap being opened. The equations of the economic model guiding the motion of the pictures were translated by a physics graphics package which interpreted them in terms of colours and sizes of predetermined shapes. The unfamiliar mathematics was translated into an animated representation that seemed to be in the more familiar language of hydraulics, yet it did not work according to the laws of hydraulics. Not every visualisation necessarily increases our understanding.[27]

Experience of such computer modelling suggests that more can be learnt from a 3-D model than a 2-D representation of a 3-D machine. Two points seem important. First, there appear to be certain aspects critical to the demonstrative success of 3-D models which remain hidden in various 2-D representations. Second, there are cognitive differences between using non-physical models (including computer-generated designs) and physical versions of the hydraulic model. At stake here are issues not just of dimensionality, but also of the familiarity of the representations.

ECONOMICS: THE RULES GOVERNING THE HOUSEHOLD

Hydraulic systems seem to be particularly important in analogical modelling just because we are all familiar with our domestic plumbing: this point is central to a *Punch* cartoon of 1953 by Rowland Emett in which the British public is invited to understand the Phillips machine through a vision of a domestic economy running on the circulation of cold tea around a Heath-Robinson kind of machine (Fig. 13.8).

Phillips' own childhood memories must have involved something like living inside a Phillips machine. His sister's description of their domestic economy tells how their mother first designed a system to provide running water in the house, and their father then adapted another water flow to provide electricity first in the milking shed, and then for lighting inside the house.[28] It was a kind of set-up rather like the *Punch* cartoon, in which

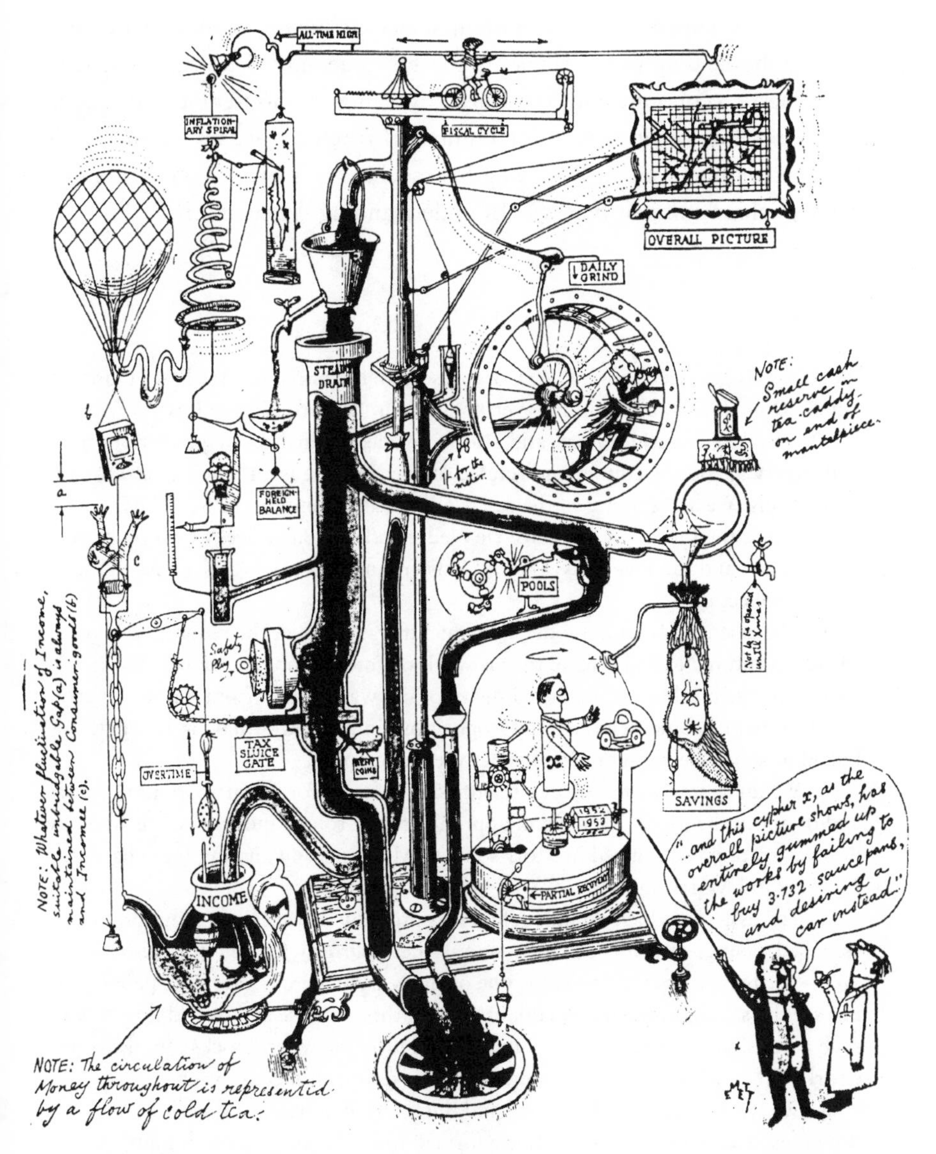

MACHINE DESIGNED TO SHOW THE WORKING OF THE ECONOMIC SYSTEM

Figure 13.8 "Machine designed to show the working of the economic system". Cartoon by Emett. Source: *Punch*, 15 April 1953, 456, from the Punch Library. By permission of Punch Ltd.

the luxuries denied to their neighbours, electric power, running water, etc., were all hooked up together via his parents' ingenuity. With this experience of domestic economy, and 'the rules governing the household', hydraulics was surely the obvious way for Phillips to reach an understanding of the macroeconomic system. By seeing his machine at work, other economists came to share some of that way of understanding the economy.

NOTES

We thank Roger Backhouse, P. A. Cornelisse, Peter van Els, Martin Fase, Francesco Guala, Dan Hausman, Kevin Hoover, Judy Klein, Richard Lipsey, Terry McDonough, Hans Martensen, and Doron Swade; seminar participants in Europe and America, and the editors and referees of this volume for all their questions and suggestions. We are grateful to the Science Museum, London; the STICERD Archive and the Library at the LSE; De Nederlandsche Bank; Delft University of Technology; and the Free University of Amsterdam for allowing us access to their archive collections.

1. The Science Museum model is unique. It was made at the request of James Meade (a future Nobel laureate for his work on international economics), as a 'mirror image machine' (configured the opposite way around from that in Figure 13.1) so that he could demonstrate propositions in his field by having two machines linked to show the interconnections between economies.

2. We understand a model as a scientific tool useful for learning about the world and/or about theories. This general approach to the role of models as a technology of investigation is discussed in Morrison and Morgan 1999 and in other chapters in Morgan and Morrison 1999.

3. The essay to which Phillips refers was probably Lerner's contribution to the so-called Savings–Investment Discussion in the *Quarterly Journal of Economics* in 1937–39 (Lerner 1938). Ironically, one of the reasons Lerner gave why people do not see the savings–investment equality is that they confuse stocks and flows. No such confusion arises in the hydraulic Phillips machine where stocks and flows are clearly differentiated and savings only equal investment in equilibrium.

4. We use the terms 'ingredients' and 'integration' advisedly, drawing on Boumans' account of how models are formed from the integration of many different kinds of elements (Boumans 1999).

5. Klamer and Leonard 1994 offers an insightful analysis of metaphors in economics.

6. The kind of analogies discussed in this chapter are substantive analogies, in contrast with formal analogies (Nagel 1961, 110). In the former, a system of elements possessing certain already familiar properties, assumed to be related in known ways, is taken as a recipe for the construction of a theory or model for some

second system. In the latter, the system that serves as the recipe is some familiar structure of mathematical relations.

7. On these 2-D models, see Morgan 1999. Fisher also had a working 3-D hydraulic model of the market system built for him, which is not discussed here.

8. The treatment here draws on Mary Hesse's classic work on models as analogies, but further suggests that negative analogies need to be taken seriously in analysing the way analogical models are used (Hesse 1966; see also Morgan 1997). Hesse actually refers to "hydraulic models of economic supply and demand" consisting of "pipes carrying colored fluids", clearly having the Phillips machine in mind (Hesse 1967, 354).

9. There are several different 2-D diagrams that we could have used for explanation, ranging from rather abstract illustrations (as in Figure 13.5) to rather realistic drawings. The most realistic one we found (in the STICERD Archive, LSE) is a detailed drawing labelled "Hydraulic Analogue of U.S. Money Flow by Phillips & Newlyn" (exact date and provenance unknown, but likely to be early 1950s given that Newlyn is still credited) in which you see the water whirl and splash; unfortunately the detail would not reproduce here. The best one we could reproduce is Figure 13.6.

10. These relations can also be represented in mathematical form, such as in the small mathematical model found in Phillips' 1950 essay. But that model does not provide a full mathematical description of the machine or of the flow of water around the machine, which would be no easy task for an hydraulic engineer.

11. Beside these functions there is also an ingenious mechanical connection between investment and the rate of change of income, the 'accelerator' seen in Figure 13.3 as the connection attached to the float with the curly spring.

12. Some of these differences in belief are reflected in Newlyn's discussion of the economic labelling of the three tanks and the flows in terms of Keynes' definitions, Robertson's definitions, and the definitions found in the National Income Accounts (see Newlyn 1950, 115–17).

13. The Phillips machine was beautifully engineered, a point on which technicians involved in restoration always insist.

14. This created amusement amongst students and faculty alike, but economists had a ready interpretation for this: everyone knows that money leaks from the regular economy into the black economy!

15. Francoeur and Segal's chapter in this volume discusses physical molecular models in a similar way. The importance of these physical models and the Phillips machine lies not just in their enhanced qualities of visualisation, but also in their manipulability compared to some of the other 3-D models discussed in this book.

16. Morris Copeland designed (but did not build) electric circuit models to represent the circulation of money in preference to hydraulic designs on the grounds that the latter involved a slow circulation with lags and delays (as appropriate for income circulation) (Copeland 1952).

17. The Eniac was already operational by the end of 1945, and the Edsac came into service at the University of Cambridge, England in June 1949 (Swade 1995), but both were digital computers and analogue computing still had such advantages as its ability to deal with non-linear equations (Gilbert 1989, 111–12).

18. A good description of the machine's working is given in Swade 1995.

19. Goodwin claims that he used the machine for non-linear formulae: "I spent years using it for teaching both linear and non-linear dynamics" (Goodwin 1992, 13).

20. It is probably this calculation to which the Phillips machine literature refers as giving an accuracy of ± 2% according to the manufacturers (Moghadam and Carter 1989, 25), but Phillips claimed only ± 4% (Phillips 1950, 284).

21. Robbins' recollections recorded by Shirley Chapman: "Some notes on Bill Phillips and his machine ... from a conversation with Lord Robbins, 1 Dec. 72", STICERD Archive, LSE, box 3.

22. This long-running debate is connected with the savings–investment debate referred to earlier. That late 1930s discussion was closed by F. A. Lutz with two sentences noteworthy in respect of Phillips' later demonstrations: "Those who think of things as happening in a certain order of time, and therefore try to link the future with the past (because they feel the desire to *visualize* the economic process step by step), will prefer Robertson's concepts. Those who think of things, not in process of happening but after the event, will favour Keynes' terminology" (Lutz 1939, 631).

23. Interview by Mary Morgan with Richard Lipsey on the topic of the Phillips machine, University of Bergamo, 15–17 October 1998.

24. See Phillips 1954 and 1957, in which electrical engineering diagrams replaced the hydraulic ones; and Allen 1954.

25. With the two linked machines (see note 1), the exercise involved the appointment of the two matching officials from the US system, and further confusion.

26. MORKMON is a 164-equation econometric model, meaning that the mathematical model has had its parameters estimated using statistical data and techniques. FYSIOEN was developed as a joint research project of the Bank's Econometric Research and Special Studies Department (P. Kramer, T. J. Mourik, and M. M. G. Fase) and the Control Engineering Laboratory of the Delft University of Technology (P. P. J. van den Bosch and H. R. van Nauta Lemke). Its dynamics however are based on MINIMORK, a simplified version of MORKMON.

27. Our discussion of FYSIOEN is necessarily incomplete and does not do justice to the ways in which the animation is helpful in understanding the mathematical model. Many of the problems we found might be overcome with the technical developments in computer animation of the last 15 years.

28. Carol Phillips, "A. W. H. Phillips, M.B.E.; 1914–1975, A.M.I.E.E.; A.I.L.; Ph.D. Econ., Professor Emeritus; Sibling Memories, Press Cuttings, Selected Biographical Notes" (undated), STICERD Archives, LSE, box 7, file 6.

REFERENCES

Allen, R. G. D. 1954. "The engineer's approach to economic models." *Economica*, new ser., 22: 158–68.

Barr, Nicholas. 1988. "The Phillips machine." *LSE Quarterly* 2: 305–37.

Blyth, C. A. 1978. "A. W. H. Phillips, M.B.E.; 1914–1975." In *Stability and Inflation*, ed. A. R. Bergstrom, A. J. L. Catt, M. Peston, and B. D. J. Silverstone, xiii–xvii. Chichester: Wiley.

———. 1987. "Phillips, Alban William Housego." In *The New Palgrave*, ed. J. Eatwell, M. Milgate, and P. Newman, vol. 3: 857. London: Macmillan.

Boulding, Kenneth E. 1948. *Economic Analysis*, rev. edn, New York: Harper.

Boumans, Marcel. 1999. "Built-in justification." In *Models as Mediators: Perspectives on Natural and Social Science*, ed. Mary S. Morgan and Margaret Morrison, 66–96. Cambridge: Cambridge University Press.

Copeland, Morris A. 1952. *A Study of Money Flows*. New York: National Bureau of Economic Research.

Darity, W., and W. Young. 1995. "IS-LM: An inquest." *History of Political Economy* 27: 1–41.

Duhem, Pierre. 1954. *The Aim and Structure of Physical Theory*. Princeton: Princeton University Press.

Enke, Stephen. 1951. "Equilibrium among spatially separated markets: Solution by electronic analogue." *Econometrica* 19: 40–47.

Ferguson, Eugene S. 1992. *Engineering and the Mind's Eye*. Cambridge, Mass.: MIT Press.

Fisher, Irving. 1911. *The Purchasing Power of Money*. New York: Macmillan.

Gilbert, Christopher L. 1989. "LSE and the British approach to time series econometrics." In *History and Methodology of Econometrics*, ed. N. De Marchi and C. Gilbert, 108–28. Oxford: Clarendon Press.

Goodwin, Richard. 1948. "Secular and cyclical aspects of the multiplier and the accelerator." In *Income, Employment, and Public Policy: Essays in Honor of Alvin H. Hansen*, ed. L. A. Metzler, 108–32. New York: Norton.

———. 1992. "Foreseeing chaos." *Royal Economic Society Newsletter* no. 77 (April): 13.

Hesse, Mary. 1966. *Models and Analogies in Science*. Notre Dame: University of Notre Dame Press.

———. 1967. "Models and analogy in science." In *The Encyclopedia of Philosophy*, ed. Paul Edwards, vol. 5: 354–59. New York: Macmillan and Free Press.

Keynes, John Maynard. 1936. *The General Theory of Employment, Interest and Money*. London: Macmillan.

Klamer, Arjo, and Thomas C. Leonard. 1994. "So what's an economic metaphor?" In *Natural Images in Economic Thought*, ed. Philip Mirowski, 20–51. New York: Cambridge University Press.

Kramer, P., P. P. J. van den Bosch, T. J. Mourik, M. M. G. Fase, and H. R. van Nauta Lemke. 1990. "FYSIOEN, macroeconomics in computer graphics." *Economic Modelling* 7: 148–60.

Latour, Bruno. 1986. "Visualization and cognition: Thinking with eyes and hands." *Knowledge and Society: Studies in the Sociology of Culture Past and Present* 6: 1–40.

Leeson, Robert. 1994. "A. W. H. Phillips M.B.E. (Military Division)." *The Economic Journal* 104: 605–18.

———, ed. 2000. *A. W. H. Phillips: Collected Works in Contemporary Perspective*. Cambridge: Cambridge University Press.

Lerner, A. P. 1938. "Saving equals investment." *Quarterly Journal of Economics* 52: 297–309.

Lutz, F. A. 1939. "Final comment." *Quarterly Journal of Economics* 53: 627–31.

Meade, James. 1951. "That's the way the money goes." *LSE Society Magazine*, January: 10–11.

Moghadam, Reza, and Colin Carter. 1989. "The restoration of the Phillips machine: Pumping up the economy." *Economic Affairs*, October/November: 21–27.

Morehouse, N. F., R. H. Strotz, and S. J. Horwitz. 1950. "An electro-analog method for investigating problems in economic dynamics: Inventory oscillations." *Econometrica* 18: 313–28.

Morgan, Mary S. 1997. "The technology of analogical models: Irving Fisher's monetary worlds." *Philosophy of Science* 64: S304–14.

———. 1999. "Learning from models." In *Models as Mediators: Perspectives on Natural and Social Science*, ed. Mary S. Morgan and Margaret Morrison, 347–88. Cambridge: Cambridge University Press.

Morgan, Mary S., and Margaret Morrison, eds. 1999. *Models as Mediators: Perspectives on Natural and Social Science*. Cambridge: Cambridge University Press.

Morrison, Margaret, and Mary S. Morgan. 1999. "Models as mediating instruments." In *Models as Mediators: Perspectives on Natural and Social Science*, ed. Mary S. Morgan and Margaret Morrison, 10–37. Cambridge: Cambridge University Press.

Nagel, Ernest. 1961. *The Structure of Science*. London: Routledge and Kegan Paul.

Newlyn, Walter T. 1950. "The Phillips/Newlyn hydraulic model." *Yorkshire Bulletin of Economic and Social Research* 2: 111–27.

Phillips, A. W. 1950. "Mechanical models in economic dynamics." *Economica* 17: 283–305.

———. 1954. "Stabilisation policy in a closed economy." *Economic Journal* 64: 290–323.

———. 1957. "Stabilisation policy and the time-form of lagged responses." *Economic Journal* 67: 265–77.

Quesnay, François. 1972 [1758]. *Quesnay's Tableau économique*, ed. Marguerite Kuczynski and Ronald L. Meek. London: Macmillan.

Reus, J. A. 1987. *Suggesties ter verbetering van de MORKMON-visualisatie* [Suggestions to improve the MORKMON-visualisation] T87.042, Vakgroep voor Regeltechniek, Faculteit der Elektrotechniek, Technical University of Delft.

Swade, Doron. 1995. "The Phillips economics computer." *Resurrection*, no. 12: 11–18.

CHAPTER FOURTEEN

From Model Kits to Interactive Computer Graphics

Eric Francoeur and
Jérôme Segal

The past four decades have seen an exponential increase in the use of computers for both the analysis of experimental data and the simulation of natural systems in all branches of scientific activity. The power of computers has also been harnessed to organise the enormous flow of data resulting from modern simulation and experimentation into a graphic form, so as to make this information easier to analyse and assimilate (Pickover and Tewksbury 1994). The study of the complex three-dimensional structure of proteins and other macromolecules has not escaped this trend, and any scientist wanting to visualise such a structure is likely to turn to a computer. Anyone with a minimum of equipment (a desktop computer, internet connection, and the appropriate software)[1] can access sites such as the Protein Data Bank and call up on a monitor the structure of any protein structure found therein (Fig. 14.1). Nor does the visualised structure need to be one of the thousands resolved experimentally by protein crystallographers. It could also be a theoretical structure generated through a variety of increasingly sophisticated molecular simulation techniques.

This approach to the visualisation of molecular structures is known as interactive molecular graphics—the term 'interactive' referring here to the user's capacity to 'manipulate' and transform in real time the structure presented on the computer monitor. Interactive molecular graphics shares characteristics with both 2-D and 3-D modes of representation. The screen of a computer monitor is undoubtedly a 2-D surface, yet molecular graphics, like many of the modes of representation discussed in this book, facilitates

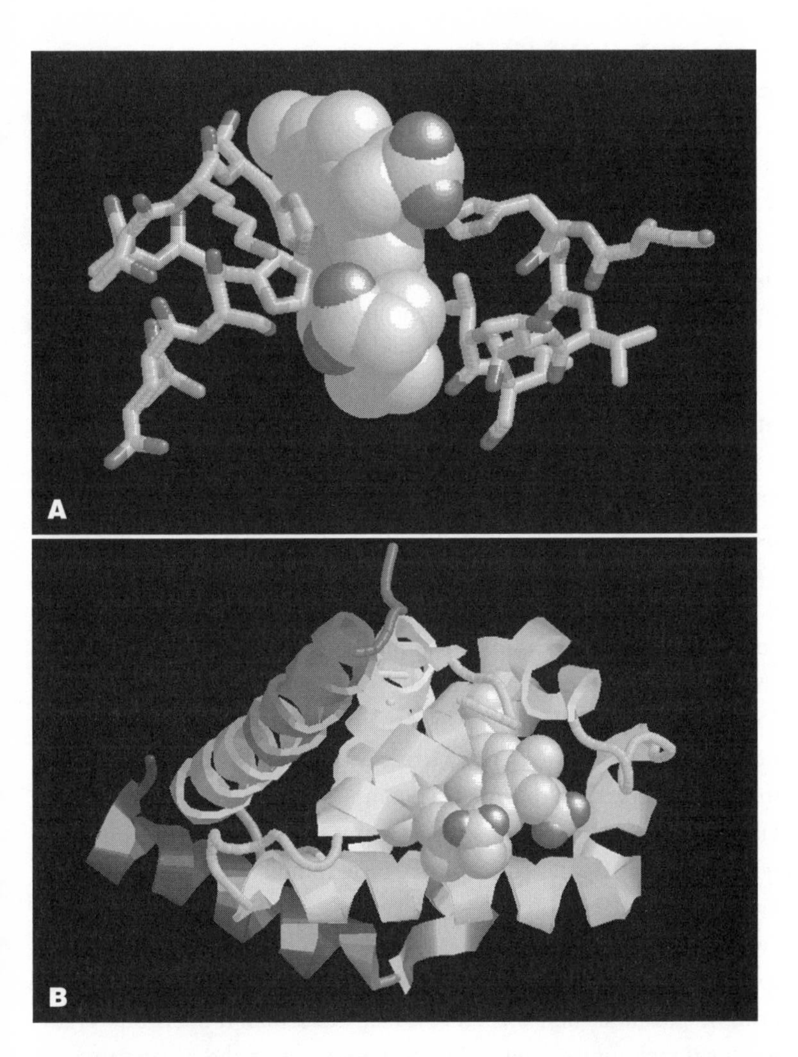

Figure 14.1 Two images of the structure of the protein myoglobin produced by the authors using currently available molecular graphics software, MDL's Chemscape Chime (originals in colour). (A) The haem oxygen-carrying group, as space-filling atoms, with portions of the surrounding polypeptide chain represented as 'stick' models. (B) The same haem group, surrounded by a diagrammatic rendering of the polypeptide chain. Compare Figure 14.2B.

the study of spatial relationships by freeing the user from the constraints of perspective, by not imposing a single point of view. It calls upon various techniques, such as rotation, shading, stereoscopy, and occlusion, to ensure a 3-D perception of the structure. Interactive molecular graphics is also a dynamic method of displaying structural information: a variety of viewing options, based on colour coding and graphical conventions, allow the user to highlight (or hide) some of the features of the structure displayed. One can zoom in to observe specific details or zoom out to take in the overall shape of the molecule. Most software will also allow specific elements to be labelled, so as to facilitate their recognition. Through this technology, the visualisation of a protein structure becomes a performance determined by the data defining the structure, the computer software and hardware, and the parameters provided by the user (Mitchell 1998).

In this chapter, we explore the first attempt at coupling the computer modelling of molecular structures with an interactive graphic display, examining the work of a small group of researchers headed by the molecular biologist Cyrus Levinthal (1922–90), who in the mid-1960s sought to take advantage of the avant-garde computer technology available at the Massachusetts Institute of Technology (MIT) to further their study of protein and nucleic acid structures.[2] At the centre of their work was one of the first time-shared mainframe computers, the Project MAC computer system (MAC stands for both 'Multiple-Access Computer' and 'Machine-Aided Cognition'), as well as a newly developed system to display and manipulate virtual objects on a computer-controlled cathode-ray tube screen, the forerunner of today's graphic workstation. At the time, the development of an interactive molecular graphic display was a clear departure from the conventional means of representing macromolecular structures, i.e., molecular models and traditional graphical techniques used to lay down images of these structures on paper.

By the 1960s, molecular models already had a long history in the study of macromolecular structures (Francoeur 1998). Their production was, in the early days of protein crystallography, an intrinsic part of interpreting the electron-density maps provided by the X-ray analysis of protein crystals (de Chadarevian, this volume). These models provided a local solution to the problem of visualising and studying macromolecular structures; they were both an embodiment of the crystallographic results and a key component of the process by which structures and their internal relationships were made analysable. Distant scientists interested in a particular structure could also, given the proper information, assemble a model of it from commercially available components. Such model-building activities became a favoured means of studying and teaching protein structures. Producing illustrations of these complex structures, especially for publication, raised its own problems. Techniques and conventions borrowed and adapted from chemistry (sometimes associated with stereoscopy) provided a means to illustrate in detail small portions of a particular structure (Johnson 1964), and schematic diagrams proved useful in providing illustrations of their general features (Richardson 1985). As early as the 1950s, computers had been used to generate static images of molecules (Langridge 1988)—illustrative, as opposed to interactive, molecular graphics—and the technique was applied early to proteins.[3] It is against the background of these available techniques that the early proponents and developers of interactive molecular graphics characterised and situated their own practices.

In the following pages we describe the development and use of the interactive molecular graphics system devised by Levinthal and his collaborators at MIT in 1964, in the context of Project MAC and of the early development of computer graphics technology. We focus primarily on the importance accorded to the interactive graphic display as an interface between the user and the numerical simulation of these structures, showing how this approach was conceived as both an extension of the physical modelling of molecular structures and as a means of circumventing the major problems associated with this practice. Finally, we look at the subsequent development of interactive molecular graphics, emphasising the role of Levinthal and his colleagues as 'pioneers' in the field, the importance of funding institutions in sustaining this development, and the links with the nascent computer graphics industry.

PROJECT MAC AND THE ORIGINS OF INTERACTIVE MOLECULAR GRAPHICS

In late 1963, MIT became the host of Project MAC (Fano and Corbató 1966; Norberg and O'Neill 1996). Funded by the Department of Defense through the Advanced Research Projects Agency (ARPA), Project MAC offered the first interactive time-shared computing system in an academic environment. It could be accessed by up to 24 users through a variety of terminals. The key feature of this system was that each of the users had in effect the impression of having at their disposal a powerful well-equipped personal computer on which to write, compile, and run programs in real time. A key goal of Project MAC was not simply to provide the MIT research community with this new form of computing, but also actively to promote its use, in the hope of improving research performance.

Cyrus Levinthal was Professor of Biophysics in the Department of Biology at the time Project MAC was getting under way. Originally trained in nuclear physics, he had turned to molecular biology, like many of his colleagues in the 1950s, and took up his position at MIT in 1957. Levinthal's laboratory was engaged at the time in fairly standard 'wet' molecular biology work, mainly complementation studies of hybrid alkaline phosphatase (Levinthal 1966c, 46). In late 1963, Robert Fano, director of Project MAC, introduced Levinthal to a graphic-display terminal, the Electronic Systems Laboratory (ESL) Display Console, otherwise affectionately known as the 'Kluge' (or sometimes 'Kludge').[4] While the vast majority of Project MAC terminals were teletypewriters (often modified electric typewriters), the Kluge could display 3-D 'objects' on a cathode-ray tube—the ancestor of today's computer monitors. The user could interact in real time with the displayed objects through a variety of interfaces, notably a keyboard and a light-pen, performing operations such as rotation, translation, and magnification. A vector-based display, the Kluge could represent only white lines on a black background. In other words, although limited by today's standards, it had the familiar trappings of a graphics workstation (Stotz 1963). The Kluge was a prototype: interactive computer graphics was in its infancy, having emerged in the early 1960s "as part of the search for new ways to improve communication between humans and computers" (Norberg and O'Neill 1996, 199), or in the language of the time, to improve "man–machine communication" (Licklider 1960). MIT was then a major player in the field (Norberg and O'Neill 1996, chap. 3). Then, as today, Computer-Assisted Design (CAD) was considered the main area of potential application for computer graphics (Lewin 1967; Machover 1978). This was reflected in some of the work performed with the Kluge at MIT during the 1960s, such as that of Steve Coons on engineering design (Coons 1966) and of Michael Dertouzos on interactive circuit design (Dertouzos 1967). The 'open' environment of Project MAC proved particularly conducive to the emergence of new areas of application for interactive graphics, which included, for example, a vocal tract display for speech analysis developed by Kenneth Stevens and his group (Wildes and Lindgren 1985, 351), and the case that concerns us here, interactive molecular graphics. MIT was also the home of *Spacewar*, the first 'video' game, designed to showcase the capacities of the Kluge's ancestor, a DEC 30 graphic console.[5]

In his own words, Levinthal became rapidly enthused at the idea of employing this display system to make models of protein structures, which could then be used in their complementation work and in "a variety of other macromolecular modelling problems".[6] Assisted by Project MAC personnel, he set out to learn programming, and within a few months he and his collaborators were able to develop a set of programs "to construct, display, and analyse macromolecules on the MAC system's ESL display" (Levinthal 1965, 15). The key features of their system were the 3-D visualisation of macromolecular structures through the computer-controlled display combined with real-time control and alteration of these structures. The result was described as a new, exciting way of looking at macromolecular structures; Levinthal and his collaborators "went on to study proteins, protein crystals, and whatever structural data we could get our hands on".[7] Levinthal's research agenda changed radically. Although his molecular biology research program was maintained, much effort and energy was now also dedicated to the computer modelling of protein structures. These two research programmes remained fairly autonomous, i.e., each maintained its own particular set of research problems and the results of one were not intended, nor suitable, to further the aims of the other.

The molecular structures were displayed on the Kluge in a 'skeletal' form through vectors representing the covalent bonds between atoms (Fig. 14.2); since the Kluge could only draw lines, this was the only possible representation. 3-D perception of the displayed structure was achieved by having it 'rotate' on the CRT screen. The user could control the direction and rate of rotation of the image with a device known as the 'globe', a plastic dome free to rotate about three orthogonal axes on which the user placed his whole hand, the ancestor of the trackball (Fig. 14.3). A light-pen was also used to interact with the displayed structure, to select, for example, specific atoms on which operations were to be performed. Special equipment was also available to photograph or film the images produced on the Kluge display. This system was to remain relatively stable in its general features over the three years that the molecular model-building project lasted at MIT, although improvements and refinements were constantly introduced in both

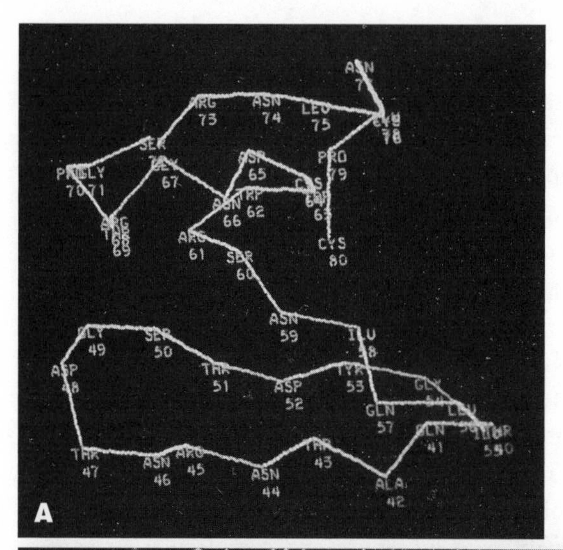

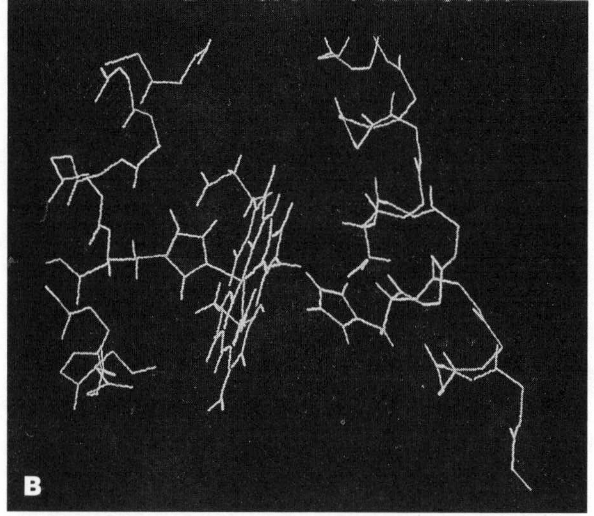

Figure 14.2 Details of two polaroid snapshots taken from the Kluge interactive display at MIT, circa 1966. (A) Segment of the lysozyme structure with vectors connecting each labelled and numbered amino acid residue. (B) The haem group of myoglobin surrounded by portions of the polypeptide chain, as in Fig. 1A. The absence of volume and shading makes it very difficult to perceive the three-dimensionality of the structure. Films of structures rotating on the Kluge display can be seen at http://www.purl.org/efranc65/movie. The structures of myoglobin and lysozyme were the first to be resolved at the atomic level and served as models for Levinthal's work on protein structure and folding. Photographs courtesy of Martin Zwick.

Figure 14.3 (A) Details of the Kluge interactive display system at MIT, circa 1966, with the 'globe', which allowed the user to control the direction and speed of rotation of the displayed structure. (B) An overview of the 'hardware platform', with the PDP-7 computer in the background. Notice the space-filling model in the foreground. Photographs courtesy of Martin Zwick.

the 'software package' and the hardware platform (Levinthal 1965; 1966b; 1967). The most important change came in early 1966, when the original Kluge, which was hardwired to the Project MAC mainframe, was replaced by a 'satellite' display console controlled by a PDP-7 mini-computer.

In the years following the creation of this interactive molecular graphics system, Levinthal and his collaborators studied a variety of questions related to macromolecular structures. Levinthal himself was initially concerned with the investigation of protein structures. His key interest rapidly became the question of protein folding, of how a protein achieved its native 3-D conformation (Levinthal 1966a). He was soon joined by Robert Langridge (then at Harvard Medical School and the Children's Cancer Research Foundation), whose concern was the refinement of nucleic acid structures, i.e., the alteration of postulated structures through minimisation algorithms so as to make the structures agree with X-ray crystallographic evidence and stereochemical constraints (Langridge and MacEwan 1966).[8] How both Levinthal and Langridge sought to take advantage of the interactive graphics system in their research will be discussed in the next section. Edgar Meyer, a postdoctoral researcher with a background in X-ray crystallography and an experienced computer programmer, developed software to display and study the structure of small molecules, using vitamin B_{12} as a test structure (Levinthal 1967). Under Levinthal's supervision, two graduate students from the Electrical Engineering Department, William Brody and David Avrin, developed software to display on the CRT the 3-D electron-density functions resulting from X-ray crystallographic studies of proteins, from which protein structures are derived (Brody 1966; Avrin 1967).

VISUALISATION AND MANIPULATION: INTERACTIVE MOLECULAR GRAPHICS AS A RESEARCH TOOL

As we mentioned in the introduction, molecular models were by the 1960s a privileged tool for the visualisation and investigation of macromolecular structures. But in one of his early public presentations of the interactive molecular graphics system, Levinthal highlighted the limitations he perceived in molecular models:

> Many of the recent developments in molecular biology have depended on understanding of the 3-D structure of large molecules. Such understanding is usually reflected in the ability to construct a 3-D model of the molecule under

> consideration. For really large molecules, however, such construction can be extremely difficult and time consuming, and in addition, it is virtually impossible to enumerate all of the many small interactions which contribute to the molecular stability. Thus, if a structure is known, physical models can be built, but the complexity of the process prevents one from using such models for examining a large number of different configurations (Levinthal 1966a, 315).

In this passage, Levinthal introduced an argument that became a leitmotif of the molecular graphics literature. Over the next two decades it would become common for authors to introduce the topic of interactive molecular graphics by pointing out both the importance and limitations of molecular (physical) models in macromolecular structural studies (Bond 1972; Barry and North 1971a; Langridge et al. 1981; Tollenaere and Janssen 1988).

It is likely that very few researchers involved with macromolecular structures at the time would have disagreed with Levinthal's assessment of molecular models—but they might not have agreed that interactive molecular graphics was an obvious solution to these problems. Two types of models were commonly used by researchers at the time: skeletal and space-filling (Francoeur 2001). In skeletal models, metal rods representing the covalent bonds between atoms are joined at appropriate angles to form atoms or atomic groups and then assembled using special connectors. The atoms themselves are simply where these rods intersect. Such models proved particularly useful for X-ray crystallography and to anyone interested in the internal features of molecular structures. Space-filling models represent atoms by hard spheres of which the diameter is proportional to the 'volume' of the atoms represented, a volume that cannot be occupied by another atom.[9] Sections are cut off from these spheres at appropriate angles and depths so as to allow bonded atoms to approach each other at a proper distance. Space-filling models have traditionally been used to study steric hindrance and intermolecular interaction, i.e., the ways in which the volume of the atoms in a structure constrains its possible conformations and interactions with other molecules.

Working with models of short polypeptide chains was already a challenge. Building a physical model of a known or postulated protein structure, let alone exploring its various possible conformations, was a much more difficult, and more frustrating, proposition. Skeletal models had to be supported by metal rods or suspended wires and even then could easily be displaced (or sag out of place), reducing greatly the precision of the resulting structure (Barry and North 1971b). The same was true of space-filling models, the problem being compounded by the sheer bulk of their components. Gravity was the model-builder's enemy number one. This was vividly illustrated to us by C. David Barry, who joined Levinthal's group in 1966 as a postdoctoral fellow. In the early 1960s, as a graduate student in physics at the University of Manchester, Barry was involved in an ad hoc project studying the possible conformation of the enzyme ribonuclease. An amino acid chain of the protein was built out of Courtauld space-filling models (Hartley and Robinson 1952), which, given some basic knowledge of the molecule obtained through physical measurements, they then attempted to fold into a plausible conformation:

> [W]e were doing that in a dry and very dusty basement, and one day all the elastic bands [holding the polypeptide chain into conformation] suddenly unzipped and the thing just fell on the floor. The thought was "Well, this is a difficult thing to do". We'd even thought of having to build it in the bottom of a swimming pool so that gravity doesn't have much to do with the structure, because most people found that putting up sort of a framework and hanging springs on to things, you basically chose a conformation that you could actually build . . . not the one that nature might be going for. So Herzenberg had us investigate building this at the bottom of a swimming pool, and found out that the plastic was neutral density, and so on. It might have worked, but we decided that it was a pretty horrible thing to have to work underwater to cancel the effect of gravity.[10]

By the mid-1960s, the computer modelling of molecular structures provided a possible solution to the problems of physical modelling. The group of Harold Scheraga at Cornell University was among the pioneers of this approach to the study of protein structure (Nemethy and Scheraga 1965; Scheraga et al. 1967). Not only were computer models free from the physical constraints and limitations inherent to physical models, but computer simulations of molecular structures also allowed a more complex representation of the forces at play within molecules. In other words, the simulated structures could have a 'behaviour' that was judged to conform more closely to what was known of molecules. Computer modelling involved a passage to an abstracted, numerical representation of molecular structures amenable to treatment by the computer. Protein and nucleic acid structures could thus be represented as a set of orthogonal coordinates indicating the position of each atom in a given structure. Additionally, a protein structure could be entered as a series of torsional angles around the main carbon atoms of the polypeptide chain, which along with the rotation angles of the side chain are its only degrees of freedom. In the absence of any other form of interface, the interaction between user and computer was in the form of these numerical representations. Again, such representations were not very useful if one wanted to visualise the relative spatial position and relations between the various elements of a structure. But one could always have the computer print an illustration of the structure, or even build a physical model from the computed coordinates.

It was in the context of these difficulties associated with both physical and computer modelling of molecular structure that interactive molecular graphics was judged to be particularly useful by its first users. When they started making use of the display system at MIT, Langridge and his assistant Andy MacEwan were already using computers for the refinement of nucleic acid (DNA) structures, i.e., the alteration of postulated structures through energy minimisation algorithms so as to make the structures agree with both X-ray crystallographic evidence and known stereochemical constraints. The following passage indicates that the interactive display made their task easier and allowed them to visualise the structure in a way similar to that of physical models:

> The unique Project MAC display system . . . is of great value, especially in debugging programs. Even though a bug may cause gross errors in the molecular configuration, such errors are not easily detectable by inspection of lists of numbers, and it usually is necessary to draw a diagram. But a display of the resultant structure quickly reveals the problem . . . . The 'three-dimensional' property of the display enables us to view the structure in various directions, particularly along important bonds, whose configuration is of interest. It has almost completely eliminated our use of wire models (Langridge and MacEwan 1966, 311).

Levinthal reported a similar experience with debugging in his work on protein structure.[11] This did not mean that physical models were completely thrown aside. Levinthal and his collaborators had devised a clever procedure for the computer to rapidly detect atoms lying too close to each other (the "cubing" routine), but overall the display was very poor at showing the "space-filling properties" of molecular structures (i.e., the way atoms are packed together). In these circumstances, it is not surprising that space-filling models were kept at hand to supplement the display (Fig. 14.3B).[12]

The application of interactive graphics to computer modelling struck a chord with C. David Barry, who after narrowly escaping a swim with the ribonuclease model in Manchester had turned to computer modelling of small polypeptides as part of his Ph.D. (Barry 1965). He was unpleasantly surprised when he started to organise the data for his thesis:

> I needed to put [in] pictures of these conformations, so I produced them on a line printer. Basically I made the machine print where the atoms were and I connected them by hand. To my horror, I found out that occasionally there were 21 residues, not 20, in the simulation. That had slipped through completely. It did not invalidate the work, but it was very unnerving that even the debug program wasn't debugged.[13]

In December 1965 Barry travelled to London to attend the Annual Meeting of the Biophysics Society, where his path crossed that of Levinthal, there to introduce MIT's newly developed interactive molecular graphics system. Barry recollects being immediately seduced by the technology: "At that point it just leapt out at me that if for no other reason than the need to be able to see what conformations the computer was generating on line—rather than three months later after you'd run the simulation and you had to make drawings of the structure and look at them in stereo, and found that you had 21 residues instead of 20 occasionally—it needed to be there and then."[14] Barry approached Levinthal and a few months later joined his group as a postdoctoral fellow.

We can see how the visual display became a central means of assessing a program's behaviour, exploiting the user's ability to recognise and assess specific patterns, here molecular structures. As the epiphenomenon of the 'mathematical model' in the computer, it served to assess the correct implementation of this model, through the generation of structures recognisable as correct or incorrect to the trained eye. But the interactive display could also be a means of intervening in the course of molecular simulation. Levinthal believed in particular that the interactive graphic display could be useful in addressing the problem of protein folding. In the early 1960s it had become widely believed that no information other than the linear sequence of amino acids was required for the polypeptide chain of a specific protein to fold into its native (functional) 3-D conformation (Epstein, Goldberger, and Anfinsen

1963). In what became known as the thermodynamic hypothesis of protein folding, it was held that the native conformation of a protein corresponds to its state of lowest free energy. This hypothesis opened up the possibility of predicting the 3-D conformation of a protein from its amino acid sequence alone—information relatively easy to obtain by the 1960s. However, checking systematically for all the possible configurations of a specific polypeptide chain to find the one with the lowest energy was not an option, as the number of such possible configurations, even for a short chain, was astronomical (Levinthal 1969). By the mid-1960s, some researchers, in particular the group of Harold Scheraga, mentioned above, had started to devise computer simulations in which very short virtual polypeptide chains were submitted to energy minimisation algorithms (Scheraga et al. 1967). These produced conformations with lower total energy, but they could only be expected to alter the structure to the bottom of the local minimum (Levinthal 1968, 44), rather than reach the global energy minimum corresponding to the conformation of the polypeptide chain in its 'native' state—just as a skier going down a slope can come to a halt in a 'dip' on the hillside rather than reach the base of the mountain. Levinthal thought that combining the skills and knowledge of a user with the analytical capacity of the computer, through the graphic display, might provide a solution to this specific problem. The key was to have the user identify the cause of the local minimum and make alterations to the structure so as to return the algorithm to a 'downhill' path. Levinthal characterised this interactive work, performed in real time, as a combination of "manual manipulation and energy minimisation" (Levinthal 1966a, 320). In essence, the system capitalised on two different modes of structural representation. The first was numerical representation, in the form of orthogonal coordinates or torsion angles that make structures amenable to mathematical operations, namely energy minimisation routines, or routines to perform structural changes as specified by the user. The second mode was graphical, or pictorial, reproducing to a certain extent the experience of looking at and interacting with 3-D physical models, while forming the basis for the user's evaluation and understanding of the conformation of postulated structure. The two were tightly bound in the real-time, interactive set-up of the system, and it was in the visual cues provided by the display that the user's ability "to make changes in the commands that control the sequential flow of the program itself" resided (Levinthal 1966a, 320). It was through this approach that Levinthal attempted to predict the structure of cytochrome C, a protein of which Richard Dickerson was in the process of resolving the structure crystallographically at Caltech. Levinthal's ultimate goal, it seems, was not to replace X-ray crystallography, but to make possible the determination of protein structures with less expensive, less time-consuming low-resolution data (Levinthal 1966b). By 1967 only a handful of protein structures had been determined and although each new structure had taken less time to determine than the previous ones, the length of the task was still measured in years.

The attempt to predict the structure of cytochrome C eventually failed. A prediction, which Levinthal later described as "plausible" (Levinthal 1976), was made (Levinthal 1966b), but it did not prove to be unique—i.e., several equally plausible structures could be postulated.[15] This prediction was never published and when Dickerson solved the structure of cytochrome C, it proved to have "absolutely no resemblance to reality" (Levinthal 1976, 3).

In retrospect, it is easy to comment that Levinthal's project was far too ambitious. This work did not distinguish itself by the physics or the chemistry that stood behind it, but rather by its sheer scale. It certainly stood in sharp contrast to the work of the Scheraga group, which had adopted a far more cautious 'bottom-up' approach by working on short polypeptide chains whose level of complexity was well below that of full-fledged protein polypeptide chains. As none of the work on cytochrome C was ever published, it is difficult to see how the effort was judged at the time. But one thing is clear: nobody had foreseen how complex and time-consuming the protein-folding problem would prove to be. More than three decades later, the ultimate goal of prediction remains elusive. As a recent National Science Foundation report pointed out: "The protein folding problem attracts some of the best minds in biological physics and statistical mechanics, but it will require much more efforts to reach satisfactory results" (Commission on Physical Sciences 1999, 215).

LIFE AFTER THE KLUGE

Cyrus Levinthal left MIT in 1967 to head the Biology Department at Columbia University. With this departure (and that of most of his group) work on the development and use of molecular graphics effectively ceased at MIT.

Three-and-a-half years of interactive molecular graphics work had yielded, in strictly bibliometric terms, very little. Very few of the results from the work on protein and nucleic acid structures performed with the system were ever published or presented in public—an exception being Martin Zwick's study of the contribution of electrostatic interactions to the stability of protein configuration, presented at the 1966 Biophysical Society Meeting (Zwick and Levinthal 1966). According to Zwick "[m]any of the kinds of analyses published by other workers in that period, such as Ramachandran plots, helical wheels (hydrophobic clustering), secondary structure prediction schemes, etc., were done independently then by me or other members of our group, but we never published them separately because they were all done in the context of the larger (and unattainable) goal of predicting folding".[16] Levinthal realised more or less rapidly that he had been "over-optimistic" in his expectations that the interactive graphic display would lead to tangible results concerning the problem of protein folding. His interest in protein structures had somewhat waned by the time he moved to Columbia, "because it seemed to me that most of the activities which were directed at predicting structure from sequence data were, at the time, more game playing than serious science".[17] This became a recurrent theme in Levinthal's writing. Discussing years later another computer graphics application he had devised with his colleagues at Columbia (the 3-D reconstruction of nerve cell images from microscopical serial sections), he wrote that for an interactive system to be valuable, "the investigators must avoid the trap of playing games with the nice pictures" (Levinthal et al. 1974), by which he seems to have meant that more attention had to be paid to the collection and analysis of quantitative data. As for Langridge's work on the refinement of DNA structures, he had rapidly become much more interested in the "computational ends of things" than in publishing research results.[18]

Levinthal did publish two short and very famous notes on the concept of 'folding pathways', the idea that specific local interactions guide protein folding (Levinthal 1968; 1969).[19] These were general theoretical statements that included discussions of the methodology behind his computer study of protein folding, rather than the presentation of any specific results obtained from this modelling. Finally, a series of descriptions of the interactive graphics system and of its workings was published in computer science and general science forums (Langridge and MacEwan 1966; Levinthal 1966a; 1966c; Levinthal et al. 1968).

Computer scientists and the computer industry showed some interest in what Levinthal and his colleagues had achieved.[20] Joseph C. R. Licklider, the foremost authority in man–machine interaction—and the main sponsor of Project MAC through his position as programme director at ARPA—later hailed it as an exemplar of early achievement in interactive computing.[21] At a time when computer graphics was considered an expensive cure "for no known disease" (Machover 1999, 45), any potential field of application was bound to attract the attention of the computing community. On the other hand, it seems that interactive molecular graphics was judged rather equivocally by biochemical and biophysical researchers both at and beyond MIT. That the system was not taken up by anyone at MIT after Levinthal's departure shows that its advantages were either not obvious or not suited to the research styles and problems of the biophysicists, molecular biologists, and crystallographers at this institution. Langridge remembers that after presenting the system at a Gordon Conference in 1965, he was roundly criticised by colleagues. A big issue was the cost: the interactive molecular graphics system depended on a unique computer system costing well in excess of two million dollars and at the time not likely to be replicated in many institutions. A crystallographer also objected that a graphics display would simply not do as a substitute for physical models, since "he had to have his hands on something, something physical, so that he could understand it".[22] A major problem, according to Langridge, was that the photographs and even films of molecular structures produced to present the system in public forums (such as journals or conferences) did not do justice to its capabilities:

> [N]ow, it's so ubiquitous that you don't have to explain it, but, standing up at a conference and showing 16mm movies, in the early days, was really not a good substitute for sitting in front of the computer and actually using it. When you first got your hands on that crystal ball at Project MAC and moved the thing around in three dimensions it was thrilling. There was no question.[23]

Among those who saw the Kluge in action was Bruce Waxman, then at the Division of Research Facilities and Resources (DRFR) of the National Institutes of Health (NIH). He was apparently impressed and praised Levinthal's work as a prime example of the use of computer graphics in data analysis (Waxman 1966). Waxman introduced Levinthal's work to one of his associates within the DRFR, William Raub. At the time, Raub was involved in the development of what became the PROPHET System—a time-shared, interactive, graphics-oriented database management system used by laboratory and clinical scientists studying the relationships between molecular structures and biological function (Raub 1975). According to Raub, it was after seeing what Levinthal had achieved at MIT that the idea arose of adding a graphic display to the system. At Raub's initiative, NIH entered into a contract with Levinthal "to extend the capabilities for three-dimensional representation and display of macromolecular models".[24] The contract financed the acquisition of an Adage interactive graphics terminal (Hagan 1969) by Columbia University. Raub also arranged for the NIH Division of Computer Research and Training (DCRT) to acquire a similar terminal, in the hope of fostering collaboration between the two groups. DCRT personnel, however, at first showed little enthusiasm for molecular graphics, and it was not until 1971 that molecular graphics software was developed for the DCRT Adage terminal, by the computer scientist Richard Feldmann (Feldmann, Heller, and Bacone 1972), in response to the explicit request from an NIH scientist who wanted to visualise specific protein structures.[25] Once the contract between the NIH and Levinthal came to term, the Adage facility at Columbia was maintained as a National Research Resource facility through NIH (peer-reviewed) grants. At Columbia, Levinthal continued his work on molecular graphics (Katz and Levinthal 1972), and led efforts to apply computer graphics to other fields, especially the above-mentioned reconstruction of nerve cell images from microscopical serial sections.

In 1969, Robert Langridge established the Computer Graphics Laboratory at Princeton as another NIH National Research Resource at a cost of around one million dollars (Bond 1972; Langridge 1988). The facility was based on a state-of-the-art Evans and Sutherland LDS-1 graphic terminal. Langridge moved this facility to the University of California in San Francisco in 1976 and headed it until his retirement in 1994. Barry brought the experience acquired in his two years at MIT to Washington University and Oxford, where he helped in the development of molecular graphics facilities (Barry et al. 1969; Barry and North 1971b). In the early 1980s he joined the computer firm Bolt, Beranek, and Newman, working on the "application of state-of-the-art computer technology to the biomedical sciences",[26] and spent the final years of his career at AstraZeneca, working on biomedical imaging (Barry et al. 1997). Edgar Meyer took on a faculty position at Texas A&M University and became a research collaborator at the Brookhaven National Laboratory, where he used the Brookhaven Raster Display to produce 3-D pictures of molecular structures, and later contributed to the development of the Protein Data Bank (Meyer 1970; 1971). He established, and still heads, the Biographics Laboratory at Texas A&M University. Martin Zwick was not involved in molecular graphics after leaving MIT, but rather continued the work in computational crystallography which had formed the bulk of his Ph.D. thesis.

The interactive graphics system at MIT did not lead to great scientific breakthroughs. Indeed, some of the work performed with the system might have looked, on Levinthal's own admission, more like game-playing than serious research. Yet, out of this project emerged not only the possibility of using interactive graphics in the study of molecular structures, but also a group of scientists dedicated to using it: scientists who in various institutions played a key role in the establishment and development of interactive molecular graphics as an approach to the study of molecular structures. The subsequent development of molecular graphics was by no means spectacular, but nevertheless steady. By 1974, 18 computer modelling systems with molecular graphics capabilities had been described in the literature (Marshall et al. 1974), a fair achievement if one considers the cost of each of these facilities, in terms not only of hardware, but also of software development. There was no off-the-shelf molecular graphics software, and basic hardware incompatibilities made it difficult to 'export' software from one platform to another. Funding agencies, in particular the NIH, were a key factor in this development. Through its research grant program, the NIH actually provided most (although not all) of the money for the development and maintenance of early molecular graphics facilities. These grants were not reviewed by peers from the biophysical or biochemical sciences, but rather by the Computer Research Study Section, composed of computer scientists and computer-experienced biomedical scientists. There is little doubt that pushing the envelope of computer application in scientific research was one of the key concerns of this study section and molecular graphics appeared as a promising direction. The development of a computer graphics industry, which provided the necessary hardware, also proved crucial. No fewer than sixteen computer graphics companies were founded or became involved in the field between the mid-1960s and mid-1970s (Machover 1978). The nascent molecular graphics community was not involved directly in the technological and commercial development of computer graphics, but constituted a group of software developers and 'power users' which sought the state-of-the-art technology and pushed what was available (or affordable) to its limit. Molecular graphics constituted a relatively small market, and it is unlikely that it ever provided much of the "impetus for 3D computer displays" as an author of the time claimed (Nelson 1987, 91), at least in comparison to clients such as the "military–entertainment complex" (Lenoir 2000). The interpretation and display of X-ray crystallographic data (in particular protein structure) remained the principal area of research and application for interactive molecular graphics throughout the 1970s (Collins et al. 1975; Jones 1978). By the early 1980s, 'rational drug design' became another prominent field of application (Cramer 1983). The Molecular Graphics Society was created in 1982, and its *Journal of Molecular Graphics* first appeared in 1983 (Morfew 1983), marking a turning point in the institutionalisation of the field.

SEE IT, TOUCH IT, FEEL IT

We have seen how interactive molecular graphics has been mainly presented as a solution to the problems associated with the physical modelling of macromolecular structures, while on the other hand there was an early insistence on preserving the type of 'interactions' with structures afforded by physical models. This testifies to the entrenchment and importance of physical modelling practices in chemistry and biochemistry, and there is little doubt that these inspired and shaped, at the conceptual level, the early development of interactive molecular graphics.

One can also argue that interactive graphics eventually came to replace physical models in a number of specific roles, an important one being the interpretation and refinement of crystallographic electron-density data—a task which was again considered at best difficult with molecular models (Richards 1997). Yet this substitution did not take place overnight. One can find, for example in 1985, published side-by-side in the same volume, two articles discussing respectively improvements in the physical-modelling and computer-graphics approaches to the refinement of crystallographic data (Jones 1985; Salemme 1985). The high price and limited capabilities of computer graphics constituted for a long time a major hindrance to the diffusion of molecular graphics.

Yet it is not sufficient to consider molecular graphics simply as an improvement on physical models. Molecular graphics did not just insert itself into existing practices, it also transformed them, opening up opportunities for old questions to be answered in a different way, raising the possibility of asking new questions by combining the analytical power of the computer with the 'insights' provided by the visual display, for example in rational drug design. In short it allowed scientists to do things differently. Furthermore, we are forced to consider that physical models, although far less used today than they were twenty years ago, have not been completely replaced by computers. Molecular graphics are deemed to have specific limitations relative to physical models, namely the absence of haptic feedback, as the biophysicist Jane Richardson recently made clear:

> Computer versions of CPK [space-filling] models have successfully imitated their appearance and most of their disadvantages (the fact that the inside is completely hidden and the difficulty of identifying an atom or group), without, so far, imitating the real virtue of CPKs, which is the physical feel for the bumps, constraints, and degrees of freedom one obtains by manipulating them (Richardson et al. 1992).

The lack of a haptic dimension may not be considered of much consequence by most users of molecular graphics, but computer modelling is not necessarily incompatible with haptic feedback. In the late 1980s a group from the University of North Carolina first described an interactive graphics system equipped with a force feedback system through which a user could feel the attractive and repulsive forces between two simulated molecules—a drug molecule and its receptor, for example (Brooks et al. 1990; Ouh-young et al. 1988)—forces which in fact could not be simulated by physical models. More recently, the San Diego Supercomputer Center has started using a rapid-prototyping machine to produce solid models of protein structures (Svitil 1998; Bailey, Schulten, and Johnson 1998). Whether these last two cases are examples of specialised solutions to very narrow, specific problems, or approaches that will be reappropriated by various researchers as opportunities are grasped and possibilities explored, is too early to say, just as the future of molecular interactive graphics was not settled in the late 1960s.

CONCLUSION

Interactive molecular graphics emerged in the mid-1960s from the chance encounter of two seemingly unrelated technological and scientific trends. On the one hand, there was growing interest, in computer engineering, in issues such as interactive computing, computer graphics, and the general improvement of 'man–machine interaction'—issues that formed the core of

Project MAC at MIT. On the other hand, one finds among biochemists, biophysicists, and molecular biologists a growing interest in the study of the structure of proteins and nucleic acids and in areas such as protein folding. Levinthal and his collaborators brought these two trends together.

Interactive molecular graphics presented itself as an alternative to two other modes of macromolecular representation: molecular models, which had traditionally been used to study molecular structures but proved difficult to adapt to the representation of giant molecules; and the symbolic (or numerical) representations used in computer simulations of molecular structures, which were abstract and difficult to interpret. Molecular graphics preserved the advantage of the latter (the possibility of complex simulation and modelling) while making the visual representation of the 3-D structure a site of intervention into, and evaluation of, the simulation and modelling activities of the computer. It remained for many years an exceptional solution, available only to those who could afford or had access to relatively expensive computer equipment. Its subsequent development beyond MIT was linked, not only to a small group of researchers committed to this new approach to molecular representation, but also to technological and commercial development in the field of computer graphics and the financial support of funding organisations such as the National Institutes of Health.

NOTES

We wish to thank Soraya de Chadarevian, Nick Hopwood, and our colleagues at the Max Planck Institute for the History of Science for their many useful comments. Our gratitude also goes to the scientists who provided us with comments, information, and documents, in particular David Avrin, C. David Barry, William Brody, Joël Janin, Robert Langridge, Françoise Levinthal, Edgar Meyer, Hans Neurath, Anthony C. T. North, Joseph Weizenbaum, and Martin Zwick. This research was made possible by a postdoctoral fellowship from the Centre National de Recherche Scientifique (JS), a postdoctoral fellowship from the Max Planck Institute for the History of Science (EF) and a John J. Pisano travel grant from the National Institutes of Health (EF).

1. A variety of structure visualisation software is available at no cost on the World Wide Web. MDL's Chemscape Chime (http://www.mdli.com/chemscape/chime/chime.html) and the Swiss-PDB viewer (http://www.expasy.ch/spdbv/mainpage.html) are two particularly popular programs.

2. We here follow Cyrus Levinthal's own claim to have created interactive molecular graphics. He produced a short manuscript, circa 1989, entitled "The

Beginnings of Interactive Molecular Graphics", apparently for that purpose. This manuscript is available online at http://www.purl.org/efranc65/levinthal. Although never published, it was widely distributed (Françoise Levinthal, personal communication). His claim is supported implicitly or explicitly by the literature in the domain. See, for example, Langridge 1995; Lattman 1995; and Marshall et al. 1974.

3. A short description of a system devised by Bill Busing to produce such illustrations can be found in Pepinsky 1961, 305. Created in 1964 by Carroll Johnson of the Oak Ridge National Laboratory (Johnson 1964), ORTEP has become over the years a very popular program for producing illustrations of molecular structures.

4. According to the Hacker Jargon Lexicon (http://www.purl.org/efranc65/jargon), the term 'kluge' refers to an inelegant, yet effective, solution to a software or hardware problem. The Oxford English Dictionary Online (2nd edn) gives a similar definition. It is not clear why this epithet came to be attached to the ESL display console.

5. J. M. Graetz, *The Origin of Spacewar*, available at http://www.purl.org/efranc65/spacewar.

6. Levinthal, "Beginnings" (see note 2).

7. Ibid.

8. At that time Langridge was already an experienced computer user. As a member of Maurice Wilkins' research team at King's College, London in the second half of the 1950s, he pioneered the use of computers to calculate the Fourier transforms of nucleic acid structure in the course of refining the structure of DNA proposed in 1953 by Watson and Crick (Langridge et al. 1960).

9. This value is often referred to as an atom's 'van der Waals radius'. On the difficulty of defining this value for modelling purposes, see Francoeur 1997, 29.

10. C. David Barry, telephone interview with the authors, 2 October 2000.

11. Levinthal, "Beginnings" (see note 2).

12. Martin Zwick, email to the authors, 9 August 2001.

13. Barry interview (see note 10).

14. Ibid.

15. Ibid.

16. Martin Zwick, email to the authors, 15 July 1999.

17. Levinthal, "Beginnings" (see note 2).

18. Robert Langridge, telephone interview with the authors, 5 August 1999.

19. Levinthal postulated these pathways in view of the fact that proteins fold relatively rapidly into their native state, despite the astronomical number of possible configurations that their polypeptide chains could theoretically take. This discrepancy became rapidly and widely known in the protein-folding literature as the 'Levinthal paradox', a formulation for which Levinthal himself did not much care (Lattman 1995).

20. Langridge interview (see note 18).

21. Oral History 150, conducted by William Aspray and Arthur Norberg on 28 October 1988, Cambridge, Mass., Charles Babbage Institute Center for the History of Information Processing, University of Minnesota, Minn., tape 3 side 1.

22. Langridge interview (see note 18).

23. Ibid.

24. William Raub, email to the authors, 11 November 1999.

25. Richard Feldmann et al., "Modelling protein structure: Research program description", NIH Historical Office and Museum, Richard J. Feldmann File.

26. C. David Barry, unpublished *curriculum vitae,* circa 1985, made available to the authors.

REFERENCES

Avrin, David E. 1967. "Computer display of protein electron density." M.Sc. thesis, MIT.

Bailey, Michael J., K. Schulten, and J. E. Johnson. 1998. "The use of solid physical models for the study of macromolecular assembly." *Current Opinion in Structural Biology* 8, no. 2: 202–8.

Barry, C. David. 1965. "Many body aspects in molecular theory." Ph.D. thesis, University of Manchester.

Barry, C. D., C. P. Allott, N. W. John, P. M. Mellor, P. A. Arundel, D. S. Thomson, and J. C. Waterton. 1997. "Three-dimensional freehand ultrasound: Image reconstruction and volume analysis." *Ultrasound in Medicine and Biology* 23, no. 8: 1209–24.

Barry, C. D., R. A. Ellis, S. M. Graesser, and G. R. Marshall. 1969. "Display and manipulation in three dimensions." In *Pertinent Concepts in Computer Graphics: Proceedings of the Second University of Illinois Conference on Computer Graphics*, ed. M. Faiman and J. Nievergelt, 104–53. Urbana: University of Illinois Press.

Barry, C. David, and A. C. T. North. 1971a. "Molecular modelling and computer graphics." *Biochemical Journal* 121: 3.

———. 1971b. "The use of a computer-controlled display in the study of molecular conformation." *Cold Spring Harbor Symposium in Quantitative Biology* 36: 577–84.

Bond, Peter J. 1972. "Interactive computer graphics and macromolecular structures." *Computer Graphics* 6, no. 4: 13–26.

Brody, William R. 1966. "Computer display of three-dimensional scalar functions." M.Sc. thesis, MIT.

Brooks, F. P., Jr., M. Ouh-young, J. J. Batter, and P. J. Kilpatrick. 1990. "Project Grope: Haptic displays for scientific visualisation." *Computer Graphics* 24, no. 4: 165–70.

Collins, D. M., F. A. Cotton, E. E. Hazen, E. F. Meyer, and C. N. Morimoto. 1975. "Protein crystal structures: Quicker, cheaper approaches." *Science* 190: 1047.

Commission on Physical Sciences, Mathematics and Applications. 1999. *Condensed-Matter and Materials Physics: Basic Research for Tomorrow's Technology*. Washington, DC: National Academy of Science.

Coons, Steven A. 1966. "Computer graphics and innovative engineering design." *Datamation* 12: 32–34.

Cramer, R. D. 1983. "Computer-graphics in drug design." *Pharmacy International* 4, no. 5: 106–7.

Dertouzos, Michael L. 1967. "Circal: On-line circuit design." *Proceedings of the IEEE* 55, no. 5: 637–54.

Epstein, Charles J., Robert F. Goldberger, and Christian B. Anfinsen. 1963. "The genetic control of tertiary protein structure: Studies with model systems." *Cold Spring Harbor Symposium in Quantitative Biology* 28: 439–49.

Fano, Robert M., and Fernando J. Corbató. 1966. "Time-sharing on computers." *Scientific American* 215, no. 3: 128–40.

Feldmann, Richard J., Stephen R. Heller, and C. R. T. Bacon. 1972. "An interactive, versatile, three-dimensional display, manipulation and plotting system for biomedical research." *Journal of Chemical Documentation* 12: 234–36.

Francoeur, Eric. 1997. "The forgotten tool: The design and use of molecular models." *Social Studies of Science* 27: 7–40.

———. 1998. "The forgotten tool: A socio-historical analysis of the development and use of mechanical molecular models in chemistry and allied disciplines." Ph.D. thesis, Sociology, McGill University, Montreal.

———. 2001. "Molecular models and the articulation of structural constraints in chemistry." In *Tools and Modes of Representation in the Laboratory Sciences*, ed. Ursula Klein, 95–115. Dordrecht: Kluwer.

Hagan, Thomas G. 1969. "Adage Inc." In *Interactive Graphics: Where Is the Market?*, ed. L. K. Grodman, 75–80. Watertown, Mass.: Keydata Corporation.

Hartley, G. S., and Conmar Robinson. 1952. "Atomic models: Part 1. A new type of space filling models." *Transactions of the Faraday Society* 48: 847–51.

Johnson, Carroll K. 1964. *ORTEP: A Fortran Thermal Ellipsoid Plot Program for Crystal Structure Illustrations. Oak Ridge National Laboratory Report 3794*. Oak Ridge, Tenn.: Oak Ridge National Laboratory.

Jones, Alwyn T. 1978. "A graphic model building and refinement system for macromolecules." *Journal of Applied Crystallography* 11: 268–72.

———. 1985. "Interactive computer graphics: Frodo." In *Diffraction Methods in Biological Macromolecules*, ed. H. W. Wyckoff, C. H. W. Hirs, and S. N. Timasheff, 157–71. Orlando, Fla.: Academic Press.

Katz, Lou, and Cyrus Levinthal. 1972. "Computer representation of complex structures." *Annual Review of Biophysics and Bioengineering* 1: 465–504.

Langridge, Robert A. 1988. "The early days of molecular graphics." In *Molecular Structure: Chemical Reactivity and Biological Activity*, ed. J. J. Stezowski, J. L. Huang, and M. C. Shao, 583–87. Oxford: Oxford University Press.

———. 1995. "Molecular modeling (Letter to the Editor)." *The Scientist* 9, no. 15: 11.

Langridge, R. A., T. E. Ferrin, I. D. Kuntz, and M. L. Connolly. 1981. "Real-time color graphics in studies of molecular-interactions." *Science* 211, no. 4483: 661–66.

Langridge, Robert A., and Andrew W. MacEwan. 1966. "The refinement of nucleic acid structure." In *IBM Scientific Computing Symposium on Computer Aided Experimentation*, 53–59. Yorktown Heights, N. Y.: IBM.

Langridge, R. A., D. A. Marvin, W. E. Seeds, H. R. Wilson, C. W. Hooper, M. H. F. Wilkins, and L. D. Hamilton. 1960. "The molecular configuration of deoxyribonucleic acid: II. Molecular models and their Fourier transforms." *Journal of Molecular Biology* 2: 38–64.

Lattman, Eaton E. 1995. "Remembering Cy Levinthal." *Proteins: Structure, Function, and Genetics* 23: 1.

Lenoir, Timothy. 2000. "All but war is simulation: The military–entertainment complex." *Configurations* 8: 289–335.

Levinthal, Cyrus. 1965. "Molecular model building." In *Project MAC: Progress Report II, July 1964 to July 1965*, 15–21. Cambridge, Mass.: MIT.

———. 1966a. "Computer construction and display of molecular models." In *IBM Scientific Computing Symposium on Computer Aided Experimentation*, 315–21. Yorktown Heights, N. Y.: IBM.

———. 1966b. "Molecular model building." In *Project MAC: Progress Report III, July 1965 to July 1966*, 41–45. Cambridge, Mass.: MIT.

———. 1966c. "Molecular model-building by computer." *Scientific American* 214 (June): 42–52.

———. 1967. "Molecular model building." In *Project MAC: Progress Report IV, July 1966 to July 1967*, 39–45. Cambridge, Mass.: MIT.

———. 1968. "Are there pathways for protein folding?" *Journal de Chimie Physique* 65: 44–45.

———. 1969. "How to fold graciously." In *Mossbauer Spectroscopy in Biological Systems: Proceedings of a Meeting Held at Allerton House, Monticello, Illinois*, ed. J. T. P. DeBruneer and E. Munck, 22–24. Champaign: University of Illinois Press.

———. 1976. "Introduction: A brief history of computer graphics." Paper read at *Workshop on Computer Graphics in Biology*, at Columbia University, New York.

Levinthal, Cyrus, C. David Barry, Steven A. Ward, and Martin Zwick. 1968. "Computer graphics in macromolecular chemistry." In *Emerging Concepts in Computer Graphics—1967 University of Illinois Conference*, ed. D. Secrest and J. Nievergelt, 231–53. New York: Benjamin.

Levinthal, Cyrus, E. Macagno, and C. Tountas. 1974. "Computer-aided reconstruction from serial sections." *Federation Proceedings* 33: 2336–40.

Lewin, Morton H. 1967. "An introduction to computer graphic terminals." *Proceedings of the IEEE* 55: 1544–52.

Licklider, Joseph Carl Robnett. 1960. "Man–computer symbiosis." *IRE Transactions on Human Factors in Electronics* 1: 4–11.

Machover, Carl. 1978. "A brief, personal history of computer graphics." *Computer* 11, no. 11: 38–45.

———. 1999. "What does it mean to be a CG pioneer?" *Computer Graphics* 33, no. 2: 45–48.

Marshall, Garland R., Heinz E. Bosshard, and Robert A. Ellis. 1974. "Computer modeling of chemical structures: Applications in crystallography, conformational analysis, and drug design." In *Computer Representaton and Manipulation of Chemical Information*, ed. W. T. Wipke, S. R. Heller, R. J. Feldmann, and E. Hyde, 203–37. New York: Wiley.

Meyer, Edgar F. 1970. "Three-dimensional graphical models of molecules and a time-slicing computer." *Journal of Applied Crystallography* 3: 392–95.

———. 1971. "Interactive computer display for the three-dimensional study of macromolecular structures." *Nature* 232: 255–57.

Mitchell, William J. 1998. "Picture this: Build that." *Harvard Design Magazine*, Fall: 8–11.

Morfew, Andrew J. 1983. Editorial. *Journal of Molecular Graphics* 1, no. 1: 3–4.

Nelson, Theodor H. 1987. *Computer Lib/Dream Machine*. Redmond, Wash.: Tempus Books of Microsoft Press.

Nemethy, George, and Harold A. Scheraga. 1965. "Theoretical determination of sterically allowed conformations of a polypeptide chain." *Biopolymers* 3: 155–84.

Norberg, Arthur L., and Judy E. O'Neill. 1996. *Transforming Computer Technology: Information Processing for the Pentagon, 1962–1986*. Baltimore: Johns Hopkins University Press.

Ouh-young, M., M. Pique, J. Hughes, N. Srinivasan, and F. P. Brooks, Jr. 1988. "Using a manipulator for force display in molecular docking." Paper read at *IEEE International Conference on Robotics and Automation*, Philadelphia, Pa.

Pepinsky, Ray, J. M. Robertson, and J. C. Speakman. 1961. *Computing Methods and the Phase Problem in X-Ray Crystal Analysis*. Oxford: Pergamon.

Pickover, Clifford A., and Stuart K. Tewksbury, eds. 1994. *Frontiers of Scientific Visualization*. New York: Wiley.

Raub, William F. 1975. "The Prophet system." In *Computers in Life Science Research*, ed. W. Siler and W. A. B. Lindberg. Bethesda: Faseb; New York: Plenum.

Richards, Frederic. M. 1997. "Whatever happened to the fun? An autobiographical investigation." *Annual Review of Biophysics and Biomolecular Structure* 26: 1–25.

Richardson, Jane S. 1985. "Schematic drawings of protein structures." *Methods in Enzymology* 115: 359–80.

Richardson, Jane S., David C. Richardson, Neil B. Tweedy, Kimberly M. Gernert, Thomas P. Quinn, Michael H. Hecht, Bruce W. Erickson, Yibing Yan, Robert

D. McClain, Mary E. Donlan, and Mark C. Surles. 1992. "Looking at proteins: Representations, folding, packing, and design." *Biophysical Journal* 63: 1186–209.

Salemme, F. R. 1985. "Some minor refinements on the Richards optical comparator and methods for model coordinate measurement." In *Diffraction Methods in Biological Macromolecules*, ed. H. W. Wyckoff, C. H. W. Hirs, and S. N. Timasheff, 154–56. Orlando, Fla.: Academic Press.

Scheraga, Harold. A., R. A. Scott, G. Vanderkooi, S. J. Leach, K. D. Gibson, and T. Ooi. 1967. "Calculations of polypeptide structures from amino acid sequence." In *Conformation of Biopolymers*, ed. G. N. Ramachandran, 43–60. London: Academic Press.

Stotz, Robert. 1963. "Man–machine console facilities for computer-aided design." In *AFIPS (American Federation of Information Processing Systems) Conference Proceedings* 23: 323–28.

Svitil, Kathy A. 1998. "A touch of science." *Discover* 19, no. 6 (June): 80–85.

Tollenaere, J. P., and P. A. J. Janssen. 1988. "Conformational analysis and computer-graphics in drug research." *Medicinal Research Reviews* 8, no. 1: 1–25.

Waxman, Bruce D. 1966. "Foreword." Paper read at *Image Processing in Biological Science*, UCLA School of Medicine, Los Angeles.

Wildes, K.L., and N.A. Lindgren. 1985. *A Century of Electrical Engineering and Computer Science at M.I.T., 1882–1982*. Cambridge, Mass.: MIT Press.

Zwick, Martin, and Cyrus Levinthal. 1966. "Charge interactions in a computer-generated protein." Paper read at *1966 Biophysical Society Annual Meeting*, New York.

Commentaries

CHAPTER FIFTEEN

Three-Dimensional Models in Philosophical Perspective

James Griesemer

The essays in this book on three-dimensional models in wax, wood, wire, plaster, and plastic challenge the approaches that philosophers have taken towards scientific knowledge. Model-based approaches to science have become a popular way around philosophical difficulties with a purely linguistic account of the theories and explanations that codify knowledge. On the model-based view, words are means of comparing models—abstract or concrete, symbolic or physical—with worldly phenomena, but knowledge is produced in the contact between models and world. Connections between representations and the concrete world cannot themselves be exclusively abstract if science is to yield empirical knowledge. Yet, while the approach is open in principle to a wide variety of models, philosophers of science have tended to cleave to familiar shores—mainly mathematical models and 2-D visual representations—with only occasional nods to 3-D physical models. Study of 3-D models may thus provide a fresh philosophical purchase on the relations between representations and knowledge.

A history of the philosophy of scientific representations since the 1940s would reveal movement towards higher dimensionality: from 1-D linguistic or symbolic expressions in the work of most logical empiricists, towards 2-D, non-linguistic, pictorial, diagrammatic, and graphical displays post-Kuhn. Instead of reconstructing theories, the new work aims to interpret a variety of representational practices in parallel with increased attention in cognitive psychology to mental maps and 'visual thinking', and in sociology to scientific practice.

Consideration of 2-D representations served well to criticise the logical empiricist philosophy of scientific theories, but replacing linguistic analysis with a semantic view of theories helped only a little in understanding the history and variety of representational practices in the sciences. The movement showed how non-linguistic tools such as set theory could be useful, yet switching formal tools was only one step; wood, wire, plaster, and wax remain to be accommodated.

The studies of 3-D models in this book cast doubt on the utility of critiques from two-dimensionality for illuminating representations in more dimensions. Rather, the philosophy of scientific representations needs rethinking, perhaps from a (4-D?) perspective on science as a process, or from cognitive theories that distinguish among representational contents, targets, and attitudes (Cummins 1995). 3-D models have more to contribute than evidence against linguistic accounts of science-as-theory and knowledge-as-fully-inscribable. Their making and uses challenge the possibility of comprehending scientific knowledge through articulation of word–world representation relations alone.

Knowledge has a tacit (Polanyi 1962), gestural (Sibum 1995), even muscular side (Cat 2001), which is as important as word–world relations for understanding science. Scientists have often been clearer about this than philosophers. James Clerk Maxwell, for example, wrote, "There are . . . some minds which can go on contemplating with satisfaction pure quantities presented to the eye by symbols, and to the mind in a form which none but mathematicians can conceive. There are others who feel more enjoyment in following geometrical forms, which they draw on paper, or build in the empty space before them" (Maxwell 1890, 220, quoted in Cat 2001, 407).

Since *models*—not just words—must be made and compared to produce knowledge (Teller, 2001; Giere 1988; 1997), knowing how models are made and deployed is as important to the philosophy of scientific representation as knowing that models are linguistically connected with phenomena by hypotheses in specified respects and degrees. Moreover, knowledge acquired through performance (Sibum 1995, 28) can by its nature be put into words only imperfectly and even then requires conventional notations that themselves can be understood only in terms of their associated performances (Goodman 1976). Often in science, for reasons of commerce, priority, protocol, or tact, it has proved undesirable to express the connections in words anyway.

If ideas cannot be comprehended without a history of the gestural knowledge and the objects through which they came to be expressed, and to which the terms of their expression most directly refer, then history of scientific ideas is a poor history indeed. A philosophy of representation that takes account only of texts is similarly inauspicious. But how is one to understand scientific practice if our tools for understanding are built solely out of logics of representation? The problem remains even if logic could be extended beyond words and mathematical symbols to handle pictures and diagrams. These essays show that there is a world of models and representations beyond descriptions and inscriptions. If philosophy of science is to contribute seriously to multi-disciplinary investigation of scientific representations, it must consider gestural as well as symbolic knowledge and the variety of means and modes of making, experiencing, and using models.

Evelyn Fox Keller characterises models as tools *for* various kinds of scientific activities, such as material intervention or concept and theory development, in addition to their role in representing objects (or phenomena) already in existence—models *of* things (Keller 2000; see also Sismondo 1999). Malcolm Baker, in his chapter on models in eighteenth-century England, draws a similar distinction between models *for* the construction of objects such as vehicles, and models of—taking *after*—objects already constructed. Models do more than simply 'stand for' something else.

The histories in this book show how 3-D models can be both models *of*—objects that stand for others that are worthy of inclusion in epistemologies of science—and models *for*—objects that facilitate various scientific activities. Although recent philosophical literature on models as mediators of word–world or theory–phenomenon relationships has usefully complicated naïve correspondence views of scientific knowledge (van Fraassen 1980; Cartwright 1983; Lloyd 1988; Giere 1988; 1997; Morrison and Morgan 1999), it has not really come to grips with the dual origins of philosophical talk about models, arising on the one hand from philosophies of language, truth, and logic (particularly model theory), and on the other from scientists' shop talk and use of models as guides for action (Wimsatt 1987).

One move took theories of theories away from formal, syntactic analysis towards semantics and pragmatics, opening up for philosophers terrain explored in parallel ways by sociologists who rejected functionalism. The syntactic view focused on theory structure and logic of explanation at the cost of attention to meanings, modes, and purposes of representation. The new emphasis on models brought to the fore problems with formalist accounts, but without seriously questioning the representational and other aims of modelling.

The second move rode a Kuhnian wave of historical contextualisation of scientific knowledge, raising history to the status of constraint on philosophising: science ought to be accurately described rather than whiggishly normed. Model-based views of mediation between words and world seemed closer to ways scientists talked and worked than did the 'received view' of scientific theories (van Fraassen 1980; Wimsatt 1980; Lloyd 1988; Morrison and Morgan 1999).

Philosophers tried to patch up or replace received views of scientific theories and explanation while tending, ironically, to perpetuate the legacy of logical empiricism by expanding the search among representations for analytical routes to word–world relations; on this view models are clearly in the middle. However, a model-based view of science must be incomplete if it offers only a model-based view of scientific theories without a model-based view of scientific practice.

Nelson Goodman's *Languages of Art* outlined a broader philosophy of representations. His nearly forgotten but still useful observations on diagrams, maps, and models appeared as part of a general exploration of a theory of notation applicable to both science and art. He wrote,

> While scientists and philosophers have on the whole taken diagrams for granted, they have been forced to fret at some length about the nature and function of models. Few terms are used in popular and scientific discourse more promiscuously than 'model'. A model is something to be admired or emulated, a pattern, a case in point, a type, a prototype, a specimen, a mock-up, a mathematical description—almost anything from a naked blonde to a quadratic equation—and may bear to what it models almost any relation of symbolization. In many cases, a model is an exemplar or instance of what it models: the model citizen is a fine example of citizenship, the sculptor's model a sample of the human body, the fashion model a wearer, the model house a sample of the developer's offerings, and the model of a set of axioms is a compliant universe. In other cases, the roles are reversed: the model denotes, or has as an instance, what it models: the car of a certain model belongs to a certain class. And a mathematical model is a formula that applies to the process or state or object modelled. What is modelled is the particular case that fits the description (Goodman 1976, 171–72).

Attention to 3-D models should help philosophers confront the full scope of this promiscuity. Models demand as much attention from philosophers as do the phenomena from the scientists who wish to describe, represent, intervene in, manipulate, construct, or control them. The essays in this book begin to give 3-D models the sustained, varied, and nuanced attention that pictures have received (Jones and Galison 1998).

In the remainder of this commentary, I shall address two ways in which taking seriously the implications of the present book might challenge philosophical talk about models, and hence the role of models in philosophies of scientific knowledge: (1) the concept of a model is historical, and (2) we can appreciate the significance of 3-D models only in making and use.

'MODEL' IS A HISTORICAL CONCEPT

Like objectivity (Daston and Galison 1992) and experiment (Hacking 1983), the concept of a model and its representational powers should be historicised. This would challenge the application of univocal, ahistorical standards to models in science and help philosophers who aim to understand science as a process rather than a 'body' of knowledge. Reading the chapters in this book, I detect two major shifts in meaning of the word 'model', the first in the late eighteenth and early nineteenth centuries, and the second in the late twentieth century.

In the eighteenth century, 'model' meant primarily an original, such as an ancient church, that served as a pattern for a modern copy (Baker, this volume). It played dual roles as a representation *of* an antiquity and *for* the construction of a new work. 3-D figures *of* agricultural machines as well as cathedrals paved the way for a shift of emphasis to objects that represent ideas to be realised and therefore serve as representational tools in discussions with patrons, fabricators, users, and eventually wider publics. In the transition between ideas and realisations, models took on a political role also, in establishing rights over works of art and nature (Schaffer). A key innovation may thus have been the fruitful running together of political and scientific notions of one thing 'standing for' another.

In the nineteenth century, 'model' came to mean primarily, not a subject worthy of representation, but the representation itself. Along the way, models variously stood for the scientific work of representation (Evans), the relation of representation in shows of force (Schaffer) and, finally, those bits of reality worth understanding, explaining, or controlling. Meinel's chapter on chemical models shows that, rather than representations of an external reality, up to mid-century chemical moulds were used as tools of construction and aids to thought that drew on formal notations of chemical composition that were emphatically not to be taken as corresponding to reality.

'Model', as the representation itself, replaced neither the ancient meaning (model as subject of imitation) nor the transitional eighteenth-century one (model as tool for presentation of projects). The chapters on nineteenth- and twentieth-century anatomical waxes (Hopwood), moulages (Schnalke), and economic machines (Morgan and Boumans) point to ways in which the new meaning enhanced rather than replaced old conceptions of mediation. In the twentieth century, a 3-D molecular model functioned *both* as a representation of worldly molecules and as a subject worthy of contemplation, a tool of engineering and patronage, and an instrument of pedagogy and entertainment (de Chadarevian). These models not only mediate between words and worlds (Morrison and Morgan 1999) but also between patrons, producers, and publics (Secord, Nyhart). How, though, did mediation among audiences and allies come to depend on the representational functions of models? Many of these essays hint that this was linked to the nineteenth-century emergence of cultures of construction. If so, the philosophical claim that models mediate must be historicised too.

A second shift of meaning has occurred recently as computer simulation began to overtake material 3-D practices. In this shift, '3-D' changed reference from physical, space-filling objects to 2-D visual displays of 3-D information, especially in interactive graphic displays on computer screens (Francoeur and Segal). The use of 3-D language to describe literally 2-D images and the varieties of associated visual and tactile experience raise additional philosophical challenges. Interactive graphic displays fall conceptually between the static 2-D displays of traditional graphics and the dynamic interactivity of 3-D physical models. Interactive computer graphics free users from the contraints of a single, imposed visual perspective by allowing user-controlled rotation and manipulation of images. They couple tactile and visual experience through control of the screen image via user-input devices such as the 'globe' in a 1960s molecular graphics project at MIT (Francoeur and Segal) and the now ubiquitous 'mouse'. Although interactive graphics extended the tradition of physical modelling, they also constituted a new mode of interaction with numerical data, allowing users to intervene kinaesthetically in the simulation process. This is terra incognita for conventional philosophies of scientific knowledge.

3-D MODELS IN MAKING AND USE

Understanding scientific models of any dimensionality requires study of how they are made and used in addition to analysis of their formal representational properties. We want to know, for example, how 2-D drawings at the laboratory bench guide theory and channel experiment and hypothesis testing (Griesemer 2000a).

A less well-known tradition in philosophy of science has focused on heuristic processes of 1-D and 2-D model-building as a way of classifying and understanding processes and products of this activity (Wimsatt 1980; 1981a; 1981b; 1987; Griesemer 2000b). This approach aims to ensure that represented phenomena are robust to the idealising and falsifying assumptions of models by requiring a variety of models making different idealising assumptions. "[O]ur truth is the intersection of independent lies" (Levins 1966) because, as Nancy Cartwright has argued, to be explanatory, our representations of phenomena must falsify (simplify, idealise), but the truth "doesn't explain much" (Cartwright 1983, 44; see also Wimsatt 1987).

3-D models present challenges even to the heuristic perspective because knowledge depends not only on the robustness of representations to falsifying assumptions but also on the robustness of our workings of models in varying and generally uncontrolled conditions of use. Gestural heuristics that govern swinging hammers, gluing plastics, cutting waxes, and writing software are needed, not only to build 3-D models but also to use them. These have not been targets of heuristic accounts in philosophy of science. No doubt gestural heuristics are required for making and using models of any dimensionality, but the demand for tactility and multiple perspectives in making and using 3-D models pushes this need to prominence.

The importance of 3-D models to a philosophy of gestural heuristics is particularly evident in the work of Morgan and Boumans on the Phillips machine. To understand the Phillips machine as a hydraulic model of the dynamic macroeconomy is to use it, to set the machine into activity, to see and hear the liquid flow around the machine, and to witness its leaks and failures. To describe the machine, Morgan and Boumans had to set it into figurative motion by narrating the flow through the articulated parts of 2-D cartoons and photographs. Merely describing or looking at the machine (or a photograph of it) is hardly heuristic or fruitful at all.

The argument from gestural knowledge is explicit also in the chapters on anatomical models. Mazzolini writes of Felice Fontana's 'artificial anatomy' in wax and wood. Fontana wrote of the artists he employed as instruments of his anatomical work and later in his career of the virtues of wood over wax models: because the former could be disassembled and reassembled, the body of the model could be physically entered. Fontana's contemporary, Condillac, offered a theory of sensation that justified this conception of anatomy as learning by doing: the origin of ideas lies in the coordination of the senses, particularly sight and touch. Hopwood's chapter on embryological waxes appeals to similar forms of knowledge in explaining how anatomists coped with the introduction of serial sectioning in the last third of the nineteenth century. This technique revealed structural details never before seen, but threatened to deprive microscopists of the physical engagement with bodily form that dissection under a magnifying glass had provided. Eventually, research in embryological anatomy produced composites of wax-plate models constructed from sectioned specimens plus books that included pictures of the models. By the end of the century, leading anatomists argued that the models, which could be seen from all sides and even 'dissected' with hot wires, were more important publications than printed works.

But why would scientists find gestural knowledge valuable and how would a philosophy of this kind of knowledge differ from a purely representational one? What does working in wax, or wire, or wood add? One view is that working with models is concrete, so the problem of representation cannot be solved from the side of abstraction alone because it is in the nature of representations to connect the abstract and the concrete (Cat 2001). The tactile, muscular, kinaesthetic experience of working with a model brings experience to mind in more and different ways than the merely visual. In other words, the gestural heuristics for using 3-D models may differ from those for representations in 2-D, and philosophy should take account of these differences.

A philosophy of scientific knowledge that includes an account of gestural heuristics requires comprehension of the variety of scientific experiences of models of different dimensionalities as well as analysis of models of any particular dimensionality. If tacit gestural knowledge brings experiences to mind, the question remains how those experiences can be translated into marks on paper, images, publishable plastic models, or reports by historians of science. Who or what mediates and is mediated in the production of scientific models? Because representation is itself an aim of science, these problems of representation have no philosophical end points. There are no knowledge products that can be fully described and analysed in abstraction from the processes of idealisation and abstraction that produced them. Thus, it would be wrong to take 3-D models out of their historical and social contexts of making and use to attempt philosophical closure. So rather than sum up, I invite the reader to engage with these essays and, indeed, with the models themselves.

REFERENCES

Cartwright, Nancy. 1983. *How the Laws of Physics Lie.* Oxford: Oxford University Press.

Cat, Jordi. 2001. "On understanding: Maxwell on the methods of illustration and scientific metaphor." *Studies in History and Philosophy of Modern Physics* 32B: 395–441.

Cummins, Robert. 1995. *Representations, Targets, and Attitudes.* Cambridge, Mass.: MIT Press.

Daston, Lorraine, and Peter Galison. 1992. "The image of objectivity." *Representations* 40: 81–128.

Giere, Ronald. 1988. *Explaining Science: A Cognitive Approach.* Chicago: University of Chicago Press.

———. 1997. *Understanding Scientific Reasoning*, 4th edn. Forth Worth: Harcourt Brace College Publishers.

Goodman, Nelson. 1976. *Languages of Art: An Approach to a Theory of Symbols.* Indianapolis: Hackett.

Griesemer, James. 2000a. "Development, culture, and the units of inheritance." *Philosophy of Science* 67 (Proceedings): S348–68.

———. 2000b. "Reproduction and the reduction of genetics." In *The Concept of the Gene in Development and Evolution: Historical and Epistemological Perspectives*, ed. Peter Beurton, Raphael Falk, and Hans-Jörg Rheinberger, 240–85. Cambridge: Cambridge University Press.

Hacking, Ian. 1983. *Representing and Intervening: Introductory Topics in the Philosophy of Natural Science.* New York: Cambridge University Press.

Jones, Caroline, and Peter Galison, eds, with Amy Slaton. 1998. *Picturing Science, Producing Art.* New York: Routledge.

Keller, Evelyn Fox. 2000. "Models of and models for: Theory and practice in contemporary biology." *Philosophy of Science* 67 (Proceedings): S72–82.
Levins, Richard. 1966. "The strategy of model building in population biology." *American Scientist* 54: 421–31.
Lloyd, Elisabeth. 1988. *The Structure and Confirmation of Evolutionary Theory.* New York: Greenwood.
Maxwell, James Clerk. 1890. *The Scientific Papers of James Clerk Maxwell*, vol. 2, ed. W. D. Niven, Cambridge: Cambridge University Press.
Morrison, Margaret, and Mary Morgan. 1999. "Models as mediating instruments." In *Models as Mediators: Perspectives on Natural and Social Science,* ed. Mary S. Morgan and Margaret Morrison, 10–37. Cambridge: Cambridge University Press.
Polanyi, Michael. 1962. *Personal Knowledge.* Chicago: University of Chicago Press.
Sibum, Heinz Otto. 1995. "Working experiments: A history of gestural knowledge." *The Cambridge Review* 116: 25–37.
Sismondo, Sergio. 1999. "Models, simulations, and their objects." *Science in Context* 12: 247–60.
Teller, Paul. 2001. "Twilight of the perfect model model." *Erkenntnis* 55: 393–415.
van Fraassen, Bas. 1980. *The Scientific Image.* Oxford: Clarendon.
Wimsatt, William. 1980. "Reductionistic research strategies and their biases in the units of selection controversy." In *Scientific Discovery*, vol. 2: *Case Studies*, ed. T. Nickles, 213–59. Dordrecht: Reidel.
———. 1981a. "Units of selection and the structure of the multi-level genome." In *PSA-1980*, vol. 2, ed. D. Asquith and R. N. Giere, 122–83. Lansing, Mich.: Philosophy of Science Association.
———. 1981b. "Robustness, reliability, and overdetermination." In *Scientific Inquiry and the Social Sciences*, ed. Marilynn Brewer and Barry Collins, 124–62. San Francisco: Jossey-Bass.
———. 1987. "False models as means to truer theories." In *Neutral Models in Biology*, ed. Matthew Nitecki and Arie Hoffman, 23–55. London: Oxford University Press.

CHAPTER SIXTEEN

Material Models as Visual Culture

Ludmilla Jordanova

Models have long been an important issue in the history of science, medicine, and technology, thanks particularly to the concern of philosophers and sociologists with models as heuristic devices for scientific thinking. Nevertheless, for historians to problematise models, as this volume does, is to do something original. Despite the long-standing interest in scientific and medical *thinking*, strikingly little attention has been paid to the *physicality* of models as distinct from their role as bearers of concepts. *Models: The Third Dimension of Science* both brings into prominence a distinctive type of cognition in science, medicine, and technology through a focus on material models, and provides a wealth of historical examples to suggest how they worked in practice. As a result it offers fresh insights to those in other fields who are concerned with visual and material culture and whose assumptions about aesthetic values and productive interpretations are likely to be somewhat different.

We might want to consider the extent to which the distinction between conceptual and material models reflects academic practice in the humanities and social sciences. Professional scholars place great, if generally unconscious, emphasis on two-dimensional items. Words, pieces of paper, computer screens, and, to a lesser extent, images are our bread and butter. I would suggest that virtually every person who has received a humanistic education feels more comfortable working in two dimensions than in three. Aspects of some disciplines, archaeology and anthropology for example, proceed differently, but they are exceptions. So working with three dimensions, which is commonplace in most sciences, could be seen as the province of some specialists, while most other academics operate with two.

There is a similar and persistent schism closer to home: the professional groups who study scientific instruments and those who deal with the rest of science and medicine remain distinct. To say 'scientific instruments' still suggests objects that require a particular kind of expertise; it can imply the need for less 'theory' and a more 'nuts and bolts' approach. In other words, the conceptual and the material are often treated as different phenomena and as requiring different kinds of input. And in the history and philosophy of science, the conceptual has generally been accorded considerably higher status than the material. So a book that insists on the importance of studying material models precisely because they give access to a wide range of scientific and medical practices, including those that would previously have been assigned exclusively conceptual status, is historiographically significant.

There are disciplines where we might expect the relationships between ideas and objects to be especially carefully conceptualised. One of these is art history, another is the relatively new domain of material culture studies. That the two cases are rather different becomes apparent when we compare leading journals in the two fields, *Art History* and the *Journal of Material Culture*, for example. Scholars working under the rubric 'material culture' certainly place objects centre stage. The key players are archaeologists and anthropologists, and they speak to particular constituencies such as the growing band working on museums, heritage, and display. Yet since material culture studies can include so many different kinds of objects, it is hard to identify a conceptual core. Thus while serious attempts are being made here to get away from the dominance of words and texts, perhaps because so much is being covered the precise problem of understanding 3-D thinking in general and models in particular is swamped. The issue has been marginalised further by the dominance of literary critical methods in humanities and social-science disciplines, which has reinforced their propensity for 2-D thinking.

Art history does have a core—the concept of 'art'—despite the difficulty of reaching any consensus on how that term is to be defined. This difficulty has kept conceptual matters primary for many art historians, who have been concerned to give prominence to theoretical and critical debates about the nature of art and the value of specific works (Nelson and Shiff 1996; Harrison and Wood 1992; 1998; Harrison, Wood, and Gaiger 2000). As a result, the materiality of art is often neglected, and so is 3-D art that demands distinctive frameworks (but see Arscott and Scott 2000, 69–89; and Potts 1994). Regrettably, the history of sculpture is a rather small and specialised field. The discipline of art history is still often divided up in terms both of the type of object studied, such as design, decorative arts, crafts, and prints, and of the status of that type, in an implied hierarchy of artistic value. To a degree, this reproduces historic hierarchies within art itself, although these have certainly changed markedly with time. Painting and sculpture have generally been assigned to the top positions; within the former, history painting was given pre-eminence, within the latter, large-scale work, especially in marble. By contrast, porcelain items, for example, are not given much prominence in most art history degrees. So in practice 3-D items we could either call models or see as akin to models are considered marginal, coming up, if they do at all, principally in connection with processes of making. The valuable approach that Malcolm Baker, a distinguished historian of sculpture, takes in his contribution to the volume, i.e., integrating models into a broad cultural history, is in my experience unusual.

It is significant that Baker's work, inspired in part by Michael Baxandall's influential writings (Baxandall 1980), has consistently looked at the processes whereby sculpture is made, avoiding a connoisseurial and anecdotal approach, and seeking precisely to present them as rich in social, economic, political, and cultural history (Bindman and Baker 1995; *Art History* 1998; *Return to Life* 2000; Baker 2000). In this respect some rather special art-historical work resonates with an important historiographical style in the history of science, seen, for example, in Simon Schaffer's piece in this volume. In both cases 'practice' is foregrounded because it holds a commitment to unravelling social processes. It is there, in 'practice', that relations of power and authority, both political and intellectual, make themselves manifest. With this in mind, there are three ways in which art-historical approaches could be useful for our purposes here, and they can be summed up in the following concerns: to understand aesthetic phenomena; to evaluate the usefulness of 'visual culture' as an alternative to 'art'; and to unravel the complexities of viewing.

All fields have habits of mind; they take certain things for granted that other domains of study find unfamiliar and astonishing. Interdisciplinary work is valuable in challenging those habits precisely because, in working across disciplines, they become more visible and hence more available for critical inspection. Engaging with the modes of thought of other fields is one way of divesting oneself of settled approaches and of cultivating an 'innocent eye'. Historians of science, medicine, and technology tend not to respond to or dwell on the sensual qualities of objects. When pushed, we frequently fall back on common-sense explanations. Presenting wax anatomical models in terms of the need to see accurate body parts when cadavers are in short supply is an excellent example. Yet the physical properties of the models are far in excess of any 'need' for information. So we might ask knowingly naïve questions, such as why do they look like *that*? Why do they have real hair, naturalistic eyes, necklaces, alluring flesh, and sexually inviting postures? Art history can help to answer such questions by offering a range of ways in which visual experiences are generated and spoken about, how they can be set in context, and interpreted. These can be called aesthetic issues; while there are a number of ways of defining aesthetics, we can take it as a term that conveys the value of carefully analysing the nature of works of art and the responses they elicit. Aesthetics tends to proceed philosophically rather than historically and contextually (Feagin and Maynard 1997). But the attention to a meticulous understanding of art that the term implies is nevertheless important. So, art history's engagement with aesthetics is a potentially rich source for other disciplines.

Art historians' exploration of 'visual culture' is the second way in which the discipline can be useful to others. Among other things, this term signals a commitment to looking at horizontal slices rather than vertical lineages. 'Visual culture' carries diverse connotations; one is certainly a commitment to looking at the range of items and images produced at a given time and place, without placing them within an artistic hierarchy and then giving more attention to those that come out at the top. It is difficult to avoid the exclusive implications of the term 'art'. 'Visual culture' is less overtly judgemental. Its value as a term has much to do with the possibility of seeing relationships between apparently disparate forms of culture. A ready example would be Michael Baxandall's suggestion that dancing, preaching, and gauging help us to understand fifteenth-century Italian paintings (Baxandall 1988). Visual habits and motifs are historically specific, and finding them in a range of contemporary cultural items helps in defining just what is specific and in delineating those habits and motifs. Since no *a priori* judgements are being made about cultural or aesthetic importance, historians are free to find and study the 'conversations' between visual forms. Several chapters in this book document and discuss just such conversations between scientific and medical models and other artefacts (see especially Secord, Meinel, Mehrtens, this volume).

To turn to the third way in which art historical approaches can be useful: we might mention the recent insistence that viewing is a complex process and that art should not be thought to work at only one level, a position which owes much to feminism. Feminist art historians insisted that a gendered reading of works of art reveals precisely their capacity to be viewed from a range of standpoints, including class and race. Some scholars have gone so far as to suggest that artists knowingly work with distinct viewpoints in mind. Thus Ronald Paulson has proposed that Hogarth's print series comparing the lives of the industrious and the idle apprentices invites two distinct readings (Paulson 1979, chap. 2). One takes the settled position of masters and upright citizens and treats the series as a straightforward moral tale. The other subversively identifies with the marginalised, 'idle' boy and concludes that privilege attracts further privilege while others become progressively disadvantaged. Models offer possibilities for diverse types of viewing, and to be alert to this is to take them seriously as visual and material objects (e.g., Schnalke, Nyhart, this volume).

As many contributions to this book note, 'model' is a complex concept. It is most obviously complex by virtue of referring to diverse objects and ideas; the range of examples in this book is truly dazzling. But 'model' possesses other complexities. 'Model' is what can be called an incomplete concept in implying the existence of something else, by virtue of which the model makes sense. This 'something else' might already be in existence or yet to come. It might be larger or smaller, more or less complete, sophisticated, or accessible. Models then, however verisimilitudinous, beautiful, or satisfying, always refer onwards. As a result there are interpretative gaps for viewers to fill in, the 'beholder's share' in Gombrich's words (e.g., Gombrich 1977). In the case of the models used in scientific, medical, and technological practice, precisely how this interpretation is accomplished becomes absolutely vital. Viewers are not on the whole totally free to make what they will of such models, which were designed to demonstrate something to which the maker had a set of commitments. It is essential to stress, as many contributors do, that these commitments were an intricate blend of intellectual, political, ideological, commercial, and personal factors. Model-making is an intentional activity, hence rhetorical strategies repay scholarly attention. Yet, however attentive makers were to the intended uses of models, they can in fact be viewed in a variety of ways. The simple fact is that wax anatomical models *can* be viewed pornographically; I suspect they often were, and still are

(Kemp and Wallace 2000). Thus in thinking about models we have to take on board the most sophisticated perspectives on diverse viewing positions. Such positions were available because (most) models are inherently complex, visually and materially. They are so because they are part of broader visual and material cultures, which are in perpetual exchange and conversation with objects in other social spaces. Models should indeed be seen in terms of visual and material culture, which is in effect to think laterally, to compare them, in relation to production and reception, with other 3-D forms such as porcelain figures, theatrical props, the lay figures used by artists, and toys. Comparison with 2-D items is equally important (Hopwood and others, this volume).

Clearly models are representations; in a sense they exemplify something, make it material, or give it a more accessible and tangible form. Since they exemplify, 'model' as verb and adjective refers to standards, even to perfection. A phrase such as 'the model husband' implies unusually high attainment, as does 'x is modelling themselves on y'. The point is important because although it is always productive to recognise (in general) just how complex taken-for-granted ideas are, the idealisation present within 'model' (in particular) indicates clearly a kind of longing that is implicit too, I think, in models as material objects (Stewart 1984).

Models give diverse pleasures, and those pleasures contain many clues to their nature. People use models to think with, hence one pleasure is the intellectual mastery that models permit. Models in art are part of production processes during which problems are solved; this is thinking, as it were, with the hands. Making models, especially small ones, poses technical difficulties, hence the more elaborate and detailed, the more their makers and viewers can take pleasure in the achievement, in the obstacles overcome, in the very scale itself. What is small is appealing for that very reason, as Edmund Burke made clear when defining the beautiful (Burke 1958). To be more concrete, we might say that children are alluring as smaller models of ourselves, of adults yet to come. It is not surprising that models of embryos exercise special fascination. These pleasures are to do on the one hand with techniques of making and on the other with those of scale.

One problem with an emphasis on visual pleasure is that it apparently neglects the other, very different emotions inspired by visual experience. For example, the moulages discussed by Schnalke could provoke disgust in viewers. Indeed some items of visual culture were specifically designed to remind their viewers of intense pain and suffering. Scenes of martyrdom and depictions of the crucifixion are cases in point. And it is always salutary to remember that the wide range of beholders' reactions include revulsion. Yet visual pleasure may still be involved. People sometimes seek out that which horrifies and terrifies them, and it is perfectly possible to appreciate the skills involved in making objects from which we recoil.

Models hold promises—the delight, for example, of seeing inside, and of taking things to bits, either in actuality or in the head (e.g., Mazzolini, this volume). The pleasure of intellectual penetration is most evident in those anatomical models that have removable parts, and I would suggest that Christopher Wren's architectural plans, showing not the appearance of a dome but its structure, function in a similar way. The ability to contemplate normally-hidden structures, such as skeletons and molecular architecture, which at least some models permit, has its own frisson. There is a distinctive interchange here between revelation and concealment. Similarly, models purport to reveal a reality without being the real thing themselves. Many of these pleasures have a strong visual component. Hence models need to be analysed in just the same ways as we analyse other works of art.

The argument about visual pleasure can be taken in yet another direction. To consider a model next to a book illustration reminds us of the running conversations between 2-D and 3-D items. For many centuries books have come in rather standard formats, and for most of that time most illustrations have been in black and white. They are more or less devoid of texture and can be viewed from only one angle. The contrast with a model could not be greater. The vast majority of models can be viewed from a wide variety of angles. Many can be touched, indeed were designed to be touched. They come in many shapes and sizes, numerous colours, textures, and materials. In other words they invite distinctive *bodily* reactions in their audiences.

Flat art elicits bodily reactions as book illustrations do—we respond somatically, to scale, for example. The illustrations to William Hunter's 1774 obstetric atlas are striking not just by virtue of their high level of verisimilitude; their sheer size impresses the viewer. Certainly big images hung in a large room, like small-scale intimate Dutch art in a confined space, work at a somatic level. So do the postures adopted by human figures in 2-D representations. But models invite a wider range of bodily reactions, including the endlessly alluring comparisons between the model, ourselves, and the 'original'.

For all these reasons, historians of scientific and medical models do well to compare them as widely as possible. In their own ways, dolls' houses and the accoutrements of theatres, as well as plays themselves, may be grist to their mill. This way of putting it runs the risk of sounding empiricist, of advocating a search for cases in the absence of a guiding framework. While there can be no single framework for such work, it is worth keeping in mind some of the concepts that are most useful here. The notion of 'display' is as an apt example—it evokes the inseparability of demonstration and audience, and hints at the theatricality implicit in modelling. Several chapters are attentive to the uses of and audiences for models. They also document the forms of display involved (e.g., Secord, Nyhart, de Chadarevian, this volume). If these are understood in historically specific terms, there is much that can be done with them. However, display is an exceptionally wide-ranging concept, and that in part accounts for its appeal, but it can be weakened if it is assumed to apply too widely—it loses its bite. There are two ways in which 'display' can be useful in relation to scientific and medical models. First, it does indeed draw attention to the uses and audiences that are implied in both the initial set-ups and the subsequent deployments of models. It invites scholars to specify how displayed items actually work. Second, displays generally have a rhetoric built into them. Items should be understood as parts of ensembles, not as isolated fragments. 'Display' can be a shorthand for such concerns: it indicates how very rich are the *visual* properties of *material* models in science, medicine, and technology. The term signals a domain of growing interdisciplinary interest: its affinities with the project this book explores reveal the wide significance of *Models*.

REFERENCES

Arscott, Caroline, and Katie Scott, eds. 2000. *Manifestations of Venus: Art and Sexuality*. Manchester: Manchester University Press.

Art History. 1998. Special issue dedicated to Michael Baxandall. 21, no. 4.

Baker, Malcolm. 2000. *Figured in Marble: The Making and Viewing of Eighteenth-Century Sculpture*. London: V&A Publications; Los Angeles: J. Paul Getty Museum.

Baxandall, Michael. 1980. *The Limewood Sculptors of Renaissance Germany*. New Haven: Yale University Press.

———. 1988. *Painting and Experience in Fifteenth-Century Italy*, rev. edn. Oxford: Oxford University Press.

Bindman, David, and Malcolm Baker. 1995. *Roubiliac and the Eighteenth-Century Monument: Sculpture as Theatre*. New Haven: Yale University Press.
Burke, Edmund. 1958 [1757]. *A Philosophical Enquiry into the Origin of Our Ideas of the Sublime and the Beautiful*, ed. J. Boulton. London: Routledge.
Feagin, Susan L., and Patrick Maynard, eds. 1997. *Aesthetics*. Oxford: Oxford University Press.
Gombrich, Ernst H. 1977. *Art and Illusion: A Study in the Psychology of Pictorial Representation*, 5th edn. London: Phaidon.
Harrison, Charles, and Paul Wood, eds. 1992. *Art in Theory, 1900–1990: An Anthology of Changing Ideas*. Oxford: Blackwell.
Harrison, Charles, and Paul Wood, with Jason Gaiger, eds. 1998. *Art in Theory, 1815–1900: An Anthology of Changing Ideas*. Oxford: Blackwell.
Harrison, Charles, Paul Wood, and Jason Gaiger, eds. 2000. *Art in Theory 1648–1815: An Anthology of Changing Ideas*. Oxford: Blackwell.
Kemp, Martin, and Marina Wallace. 2000. *Spectacular Bodies: The Art and Science of the Human Body from Leonardo to Now*. London: Hayward Gallery.
Nelson, Robert S., and Richard Shiff, eds. 1996. *Critical Terms for Art History*. Chicago: University of Chicago Press.
Paulson, Ronald. 1979. *Popular and Polite Art in the Age of Hogarth and Fielding*. Notre Dame: Indiana University Press.
Potts, Alex. 1994. "Dolls and things: The reification and disintegration of sculpture in Rodin and Rilke." In *Sight and Insight: Essays on Art and Culture in Honour of E. H. Gombrich at 85*, ed. John Onians, 354–78. London: Phaidon.
Return to Life: A New Look at the Portrait Bust. 2000. Leeds: Henry Moore Institute.
Stewart, Susan. 1984. *On Longing: Narratives of the Miniature, the Gigantic, the Souvenir, the Collection*. Baltimore: Johns Hopkins University Press.

INDEX

Page numbers in bold refer to figures.